TK 2181 N37
3 0250 003
D0338551

PRENTICE-HALL ELECTRICAL ENGINEERING SERIES

WILLIAM L. EVERITT, editor

Electromagnetic
Energy Conversion
Devices and Systems

PRENTICE-HALL INTERNATIONAL, INC., *London*
PRENTICE-HALL OF AUSTRALIA, PTY. LTD., *Sydney*
PRENTICE-HALL OF CANADA, LTD., *Toronto*
PRENTICE-HALL OF INDIA PRIVATE LTD., *New Delhi*
PRENTICE-HALL OF JAPAN, INC., *Tokyo*

S. A. NASAR

Associate Professor of Electrical Engineering
University of Kentucky

Electromagnetic Energy Conversion Devices and Systems

PRENTICE-HALL, INC.

Englewood Cliffs, New Jersey

Current printing (last digit):
10 9 8 7 6 5 4 3 2 1

13–249003–X
Library of Congress Catalog Card No. 79–95757
Printed in the United States of America

Dedicated
to the memory
of my father

Preface

The purpose of this book is to introduce the principles and techniques of study of electromagnetic energy conversion devices and systems. It has been developed keeping in mind the current trends in electromechanical energy conversion, and the recently revised curricula in this area at various schools. Use of computers—analog and digital—is, therefore, encouraged and emphasized. Computer programs, in FORTRAN, are given for a number of examples leading to numerical end results. Likewise, the application of electromagnetic field theory to energy conversion is stressed.

This book is the outcome of courses in energy conversion taught by the author at the University of California at Berkeley (1959–63); at E. P. University of Engineering and Technology, Dacca, Pakistan (1963–66); and at Gonzaga University, Spokane, Washington (1966–68). As such, it should be suitable for a one semester course in electromagnetic energy conversion at the junior/senior level. There are, however, topics of advanced nature. Sections containing these are marked with †, and may be skipped without loss of continuity. Likewise, for completeness, review material has been included in sections marked with *.

As prerequisites, the student is expected to have had a first course in circuits (or linear systems, preferably), an introduction to electromagnetic fields, and elementary vector operations. Some familiarity with matrix algebra and partial differential equations is desirable, but not absolutely necessary. Because of the range of topics covered in the book, there is an inherent and, perhaps unavoidable, unevenness in the level of presentations.

For guidance, the pertinent sections contributing to the variation in the difficulty of the various topics are marked with an asterisk or dagger.

To briefly review the contents of various chapters, in Chapter 1 some examples of electromagnetic energy conversion devices and systems are given and the nature of associated problems is qualitatively discussed.

Most of the review material is given in Chapter 2. This includes a resumé of electromagnetic field theory and linear systems. Various methods of formulating the equations of motion of electromagnetic devices are discussed, linearization techniques, and methods of "solving" linearized equations are included in this chapter. A brief discussion on state variables is also presented.

In Chapter 3, a number of electromagnetic and electrostatic transducers are studied. Current-excited as well as voltage-excited systems are considered; and the significance of instantaneous, average, and RMS values of time-varying quantities, as related to mechanical forces of electrical origin are discussed.

Chapter 4 deals with devices having continuous media such as conducting discs, liquid metals, etc. The traveling-field theory is introduced and three different methods of determination of forces in eddy-current devices are given. A number of devices, such as liquid-metal pumps, acyclic generators, and MHD generators are discussed to illustrate the applications of electromagnetic field theory.

In Chapter 5, rotary energy converters are viewed as coupled circuits in relative motion. Brief qualitative descriptions of conventional dc, induction, and synchronous machines are given. Next, an idealized two-phase machine is introduced; and, using the concept of current sheets and the field equations, the inductances of the machine are derived. The student is introduced to the various methods of calculating machine inductances, from which the equations of motion are then derived and solved using certain linear transformations. Here linear transformations are introduced from a mathematical viewpoint.

Dc machines, induction machines, and synchronous machines are discussed in Chapters 6, 7, and 8, respectively. Here the emphasis is on obtaining the dynamic characteristics of these machines and machine-systems, using the techniques developed in Chapters 2, 4, and 5. Physical phenomena in, and certain conventional aspects of, electrical machines are also discussed in Chapters 6 through 8 and physical interpretations of linear transformations are given wherever possible.

Realistic considerations are taken up in Chapter 9 and applications of analog computers, digital computers, graphical field mapping, numerical methods, etc., are illustrated by means of a number of examples. While it is not practically possible to include diverse kinds of realistic considerations, three important topics—space harmonics, saturation, and leakages—are treated briefly in this chapter. The references at the end of this chapter provide material for research on the topics discussed.

In each chapter, a number of worked-out examples are given either to illustrate the application of the theory or to develop the theory from the example. A list of references is given at the end of each chapter and the student should be encouraged to consult these.

At numerous places many steps have been skipped in various derivations. These are posed as problem exercises. In order to get the maximum benefit from the book, the student should work out the problems at the end of various chapters.

Many publications have contributed to this book. Among others, the author was much influenced by the pioneering work of White and Woodson: *Electromechanical Energy Conversion* (Wiley, 1959), and this is reflected in a number of places throughout the book.

Deep felt gratitude is expressed to Professor Robert M. Saunders, of the University of California at Irvine, who not only introduced to the author many of the topics treated in this book, but also gave permission for the use of some of his unpublished work. Acknowledgement is made to Professor Robert W. Newcomb, of Stanford University, and to Dr. Wolf H. Koch, of Ford Motor Company, for many helpful discussions. The permission granted by the Institution of Electrical Engineers, London; The Institute of Electrical and Electronics Engineers, Inc., New York; and the General Electric Company, Schenectady, to use some of their published material is gratefully acknowledged. The author wishes to thank Dr. Harold A. Foecke, former Dean of Engineering, and the Jesuit Research Council of Gonzaga University for the help and support in the preparation of the manuscript. Finally, thanks are due to John Nelsen, a former student at Gonzaga University, who proof-read the entire manuscript.

A Note to the Instructor

The material presented in this book has been used at the junior/senior level at four different schools. The students involved had widely different levels of preparation. From past experience, for a one-semester course in electromechanics (at the junior level) for students having had a first review course in electrical engineering [Typical texts: R. J. Smith, *Circuits, Devices, and Systems* (Wiley); H. A. Foecke, *Introduction to Electrical Engineering Science* (Prentice-Hall)], the following sections are recommended:

Chapter 1	
Chapter 2	Sec. 2.1, 2.2, 2.4–2.6
Chapter 3	
Chapter 4	Sec. 4.1–4.3, 4.6, 4.7, 4.9
Chapter 5	Sec. 5.1, 5.2, Examples 5E–3 and 5E–4, Sec. 5.5, 5.6, Examples 5E–5, 5E–6, Sec. 5.7.5, 5.7.6
Chapter 6	
Chapter 7	Sec. 7.1–7.5
Chapter 8	Sec. 8.1–8.5
Chapter 9	Sec. 9.1–9.3

At the senior level, for students having had a course in electromagnetic fields and a course in networks [Typical texts: W. H. Hayt, Jr., *Engineering Electromagnetics* (McGraw-Hill); M. E. Van Valkenburg, *Network Analysis* (Prentice-Hall)], the review sections, such as Sec. 2.1, 2.5.1–2.5.4, may be skipped. Most of the remaining sections could be covered with the exceptions of specialized topics, such as Sec. 7.6, 7.7, 8.6, 8.7, and 9.4.

Many derivations in the text have been posed as problems. It is strongly recommended that the students work out these problems.

Contents

Electromagnetic
Energy Conversion
Devices and Systems

1

Introduction

1.1 General Remarks

Of all means and devices for converting energy from one form to another, those components converting electrical to mechanical energy (and vice versa) are some of the most important in engineering systems. We encounter this energy-conversion process in almost every walk of life. The basic principles governing the operation of the largest machines—such as hydroelectric generators—as well as of the smallest energy-processing devices—such as the microphones and loudspeakers—are one and the same. Thus, we see that the theoretical background to such a wide variety of applications is necessary to the engineer concerned with the modulation and control of the flow of energy while interchanging the energy from electrical to mechanical forms.

Electromagnetic energy-conversion processes involve macroscopic electrodynamic interactions between current-carrying conductors and electromagnetic fields. Energy conversion occurs when coupling electromagnetic fields are disturbed. A system which is provided with electromagnetic coupling fields such that the energy stored in the fields changes with the position of some moving member of the system can, therefore, be considered an *electromagnetic energy converter*. Some examples of energy converters based on the above principle are solenoids, relays, motors, generators, etc.

The last paragraph indicates that the presence of electromagnetic fields, current-carrying conductors, and associated stored energies are necessary for energy conversion. If the stored energy is expressed in terms of circuit constants, such as inductances and capacitances, currents and voltages, the energy-conversion device can be analyzed in terms of coupled circuits in relative motion. On the other hand, a direct application of field equations such as Maxwell's equations also leads to a complete analysis of a large class of electromagnetic energy-conversion devices. Thus, we have two useful methods for studying electromagnetic energy-conversion devices: lumped-parameter circuit methods and the distributed-parameter field methods. Each has its merits and neither can be substituted for the other. Rather, they complement each other, and a combination of the two methods is the most powerful tool for the quantitative analysis of all sorts of electromagnetic energy converters under various steady-state and dynamic operating conditions.

In the following chapters we shall attempt to identify the fundamental features common to all types of electromagnetic (and electrostatic) energy converters. We shall also try to demonstrate how the appropriate physical laws, when formulated in terms of mathematical concepts, provide a sound basis for the study and analysis of various kinds of devices.

Electromechanical energy conversion is in most cases a reversible process. A device that converts energy from mechanical to electrical form and modulates in response to an electrical signal is a *generator*. When the conversion

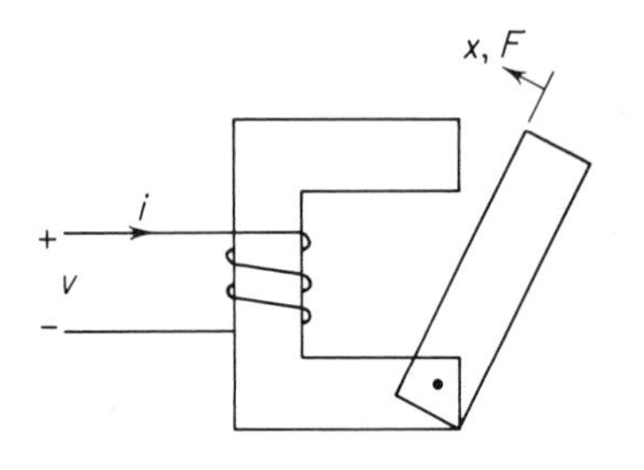

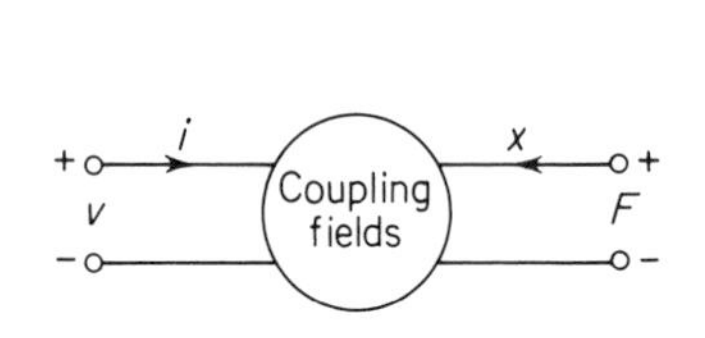

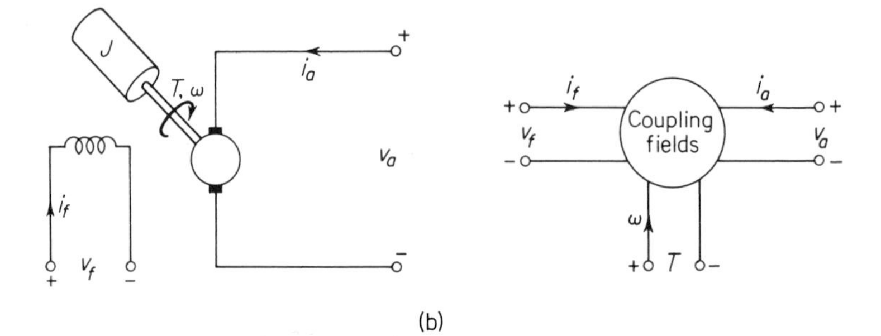

Fig. 1-1 Two- and three-port devices: (a) an electromagnetic relay as a two-port device; (b) a dc motor as a three-port device.

involved is from electrical to mechanical energy and the modulating signal is electrical in nature, the component accomplishing such conversion is a *motor*. Incremental-motion electromechanical energy converters, whose main function is to process energy, are called *transducers*. For example, a microphone can be considered a transducer. Generators and motors are usually 3-port (terminal-pair) devices, whereas a transducer is a 2-port device, as illustrated in Fig. 1-1. In the following discussions, the terms "generator," "motor," and "transducer" are used in accordance with the above definitions; and an "electromagnetic energy-conversion device" is meant to imply either a rotating machine, a linear-motion machine, or an incremental-motion transducer.

1.2 Some Examples

Perhaps the most commonly encountered energy converter is the electric motor. However, we wish to consider here a less common example, an elementary magnetohydrodynamic (MHD) generator, which illustrates the applications of a number of devices to be discussed in subsequent chapters. An *MHD generator* is based on *Faraday's law* of electromagnetic induction; namely, that an electromotive force (EMF) is induced in a conductor when it moves in a magnetic field and thereby "cuts" the magnetic lines of force. In an MHD generator the conductor is an ionized gas at an elevated temperature. Although it might not be the most practical example, let us consider the combined nuclear-reactor–steam–MHD power plant[1] shown schematically in Fig. 1-2, where the MHD generator is used as a topping unit. Even in this simplified representation we can clearly identify the components involving electromagnetic energy-conversion devices: the ac generator, the MHD generator, the homopolar generator, and the liquid-metal pump. The liquid-metal pump is an electromagnetic pump required to pump sodium-potassium coolant, which has been found useful in high-temperature applications because of its high rate of heat transfer (of the order of several kilowatts per square inch). In subsequent chapters we shall study and analyze the behavior of these and other devices.

To illustrate another aspect of our study let us consider the automatic voltage-regulating system shown in Fig. 1-3. Without going into the details of the system, we notice that it is a *closed-loop system*. The changes in the output voltage at the generator terminals are monitored by the transformer. This information is fed back to the exciting circuit through the rectifier and control field No. 1. The current in the generator field finally adjusts according to the output-voltage fluctuations such that the output voltage remains within specified limits. Since sudden load changes result in sudden voltage changes, a system of this nature interests us because of its *transient response*. We also

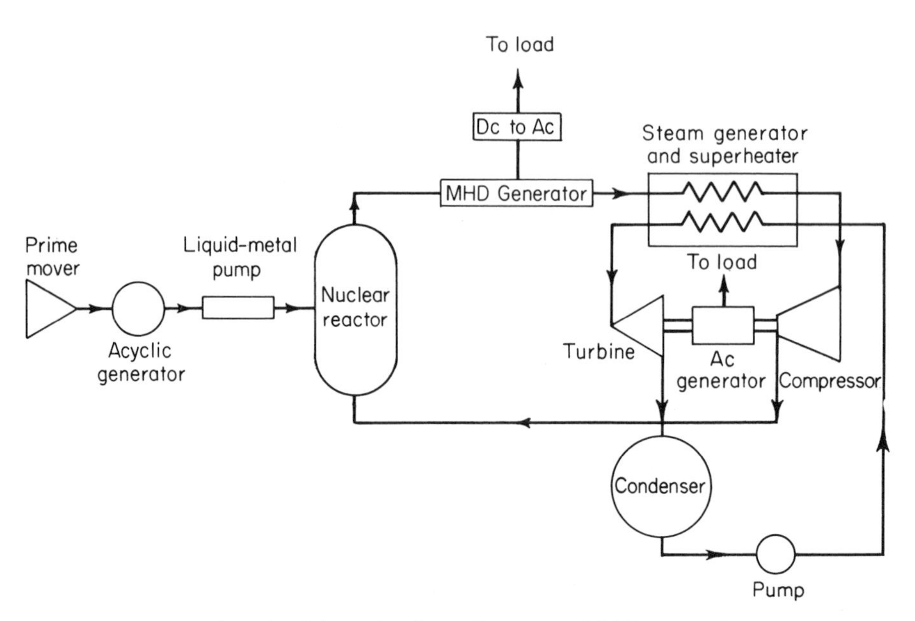

Fig. 1-2 Schematic of a nuclear–steam–MHD power plant.

wish to know the *steady-state error*, that is, the deviation in the output voltage from the ideal (when there is no deviation, regardless of the load). In this example, we see that electromagnetic devices are used as elements of a closed-loop (control) system. In order to learn about the behavior of the system it is important that we understand the dynamics of each component.

A slightly more complicated example of the applications of energy-conversion devices, in conjunction with electronic control circuitry, is the

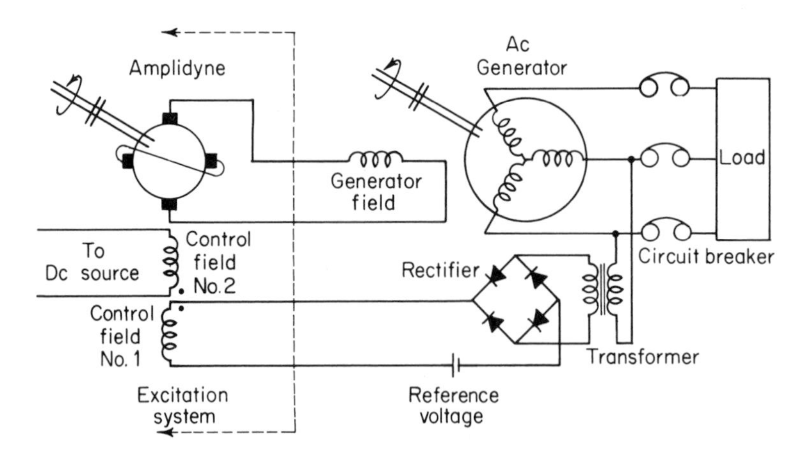

Fig. 1-3 An automatic voltage-regulating system.

electric car,[2] illustrated in Fig. 1-4. The major energy-converting component here is the *induction motor*. Various considerations have led to the conclusion that the induction motor is not the ideal drive for an electric car. Perhaps a better drive might be a (new kind of) *dc motor* without any moving contacts and a large power-to-weight ratio. In an electric car, the control of the drive is as important as the drive itself. The operating characteristics of motors, whether ac or dc, vary considerably with the waveform of the supply voltage. Because the voltage waveforms are far from sinusoidal when electronic switching circuits are incorporated for the control of the drive motor in an electric car, it is most important that we thoroughly understand the dynamics of the drive as well as the dynamics of entire system.

The three examples considered in this section, then, lead us to investigate the nature of the problems and the general purpose of the study of energy-conversion devices. These are discussed in the following section.

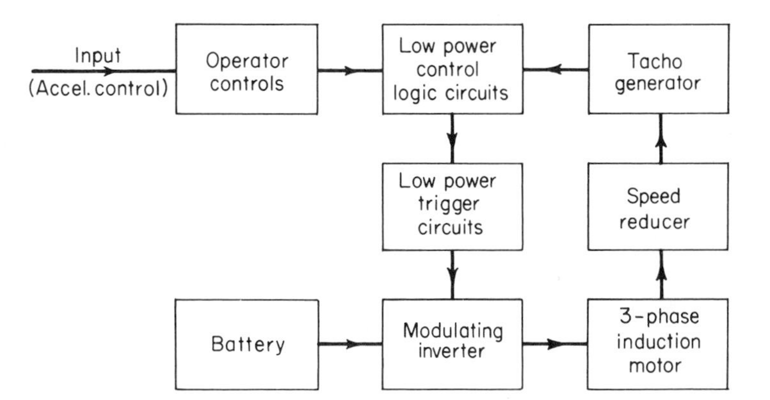

Fig. 1-4 Control circuitry of an electric car (schematic).

1.3 Nature of the Problem and Scope of the Study

In general terms, the problems relating to electromagnetic energy-conversion devices involve the physical description and characteristics of the device, as well as the formulation and solutions of the dynamical equations of motion. The main objective of the study of the energy-conversion properties of an electromagnetic device is to predict quantitatively the behavior of the device under specified operating conditions. Sometimes, the response of the device to frequency-variable inputs may be desired; and, for *linear* or linearized *systems*, the input–output relationships are expressed in terms of *transfer functions*, from which the transient (or step) and steady-state (or frequency) response are directly obtainable. An alternative, more general approach toward obtaining the dynamical characteristics of a system is through the use of *state variables*. The state-variables method is a formulation in the time domain, whereas transfer functions are derived in the frequency domain. At the present time we do not wish to concern ourselves with the definitions and details of these various techniques; they are discussed in detail in subsequent chapters.

In obtaining the necessary quantitative information about a given energy-conversion device, two difficult steps are generally encountered. First, the parameters involved in the equations of motion are difficult to evaulate, and liberal approximations have to be made to obtain models amenable to analysis. To take a specific example, in calculating the inductances of various windings of an electromagnetic device, saturation effects are often neglected as a first approximation. Similarly, in rotary energy converters, realistic considerations such as harmonics, leakages, and effects of slots and teeth cannot conveniently be taken into account, although each of these factors contributes to the performance of the device. As pointed out earlier, in such cases the actual device is replaced by an idealized model, simplifying assumptions are made, and, if necessary, for accuracy the second-order effects are included in the solution by special techniques such as numerical and/or graphical methods. In other words, parameter determination is a difficult, yet important, step in the study of electromagnetic energy-conversion devices.

Once the parameters have been obtained, formulation of the equations of motion is rather a routine matter. A number of methods of obtaining the equations of motion are discussed in the next chapter. The next difficult step is obtainingsolutions for the resulting equations of motion. These equations. are most of ten nonlinear differential equations with time-variable coefficients No general methods are available for solving these nonlinear equations, and approximations have to be made to obtain the end results. For example, in the case of transducers *incremental motion* is assumed (often not too unrealistically) and the resulting nonlinear equations of motion are linearized about a quiescent operating point. After they are linearized, the equations may be solved by any one of the several methods which are extensively available and discussed in the next two chapters. For rotary energy converters such as motors and generators, the procedure for formulating the equations of motion is similar to that for a transducer. The solution of these equations is facilitated either by linear transformations or by some other technique, such as by numerical methods or by using analog or digital computers. These are discussed in detail in Chapters 5 and 9.

Finally, it should be pointed out that classical electromagnetic-field theory has a wide range of applications pertaining to energy-conversion devices. Some of the aspects of the applications of the electromagnetic-field theory are outlined in Chapters 2, 4, 5, and 9.

In summary, the analysis and study of electromagnetic energy converters include the following aspects:

1. *Topological considerations and physical descriptions.* These give the locations of input and output terminals, identify the fixed and moving elements, and specify the structure of the magnetic circuit, winding data, and various physical dimensions.

2. *Choice of a model and simplifying assumptions.* These generally depend on the problem at hand and the degree of refinement desired in the solution. For example, in a magnetic device, as a first approximation, fringing, saturation, and hysteresis are neglected. The permeability of the magnetic material is ideally assumed to be infinitely greater than that of free space; consequently the magnetic energy is assumed to be stored in the airgap alone.

3. *Determination of the system parameters.* This includes the evaluation of resistances, inductances, and capacitances for the electrical portion of the system; and of the mass (or moment of inertia), stiffness, and friction coefficient for the mechanical portion. The parameters are usually obtainable from the physical description of the system together with the simplifying assumptions and choice of model. It may be mentioned here that in the majority of cases of practical importance, parameter determination just includes the calculation of various inductances. As pointed out earlier, this is one of the difficult steps. It is also an important one, as the performance of an electromagnetic device depends a great deal on the various inductances.

4. *Formulation of the electrical and mechanical equations of motion.* Respectively, these turn out to be the volt-ampere equations and the force-balance (or torque-balance) equations. The equations of motion can be derived by one of the methods discussed in the next chapter.

5. *Solution of the equations of motion.* This step is performed after the equations of motion have been formulated. Almost invariably, the resulting differential equations are nonlinear. In simple cases, such as in the case of transducers for small-signal applications, the equations of motion are first linearized and then solved by treating them as linear differential equations with constant coefficients; whereas for complicated cases of rotating electrical machines, certain types of linear transformations become necessary. No sufficiently general method is available which covers all cases. Obtaining the solutions of equations of motion for electromagnetic energy-conversion devices is often the most difficult—yet most important—step in the quantitative study of these devices. Sometimes analog-computer simulation is done; in such instances the block-diagram representation becomes quite useful. In certain other cases, digital-computer solutions are desirable. Often, it is useful to obtain a complete electrical equivalent circuit for the device, the circuit in turn leading to the behavior of the device, and thereby permitting the solution of the equations of motion to be indirectly obtained. Depending

upon the potential application of the device (such as the automatic voltage-regulating system of Fig. 1-3), its transfer function and step or frequency response are determined for a linearized model. In other instances, such as in a multi-input–multi-output device, formulation in terms of state variables is most convenient.

All the above steps, then, constitute the study of the device or system.

REFERENCES

1. Tsu, T. C., "MHD Power Generators in Central Stations," *IEEE Spectrum*, June 1967, p. 59.

2. Lindgren, Nilo, "Electric Cars—Hope Springs Eternal," *IEEE Spectrum*, April 1967, p. 48.

3. Koch, W. H., "The Route to Control," *Electro-Technology*, May 1968, p. 59.

2

Methods of Analysis

In the last chapter we outlined the general problem pertaining to electromagnetic energy-conversion devices. We concluded that the study essentially involves the formulation and solution of dynamical equations of motion. In this chapter we shall consider the various methods of formulation of equations of motion. We shall also discuss some of the analytical techniques useful in the study of energy-conversion devices.

There are three basic methods of formulating the equations of motion and thereby obtaining the characteristics of electromagnetic energy-conversion devices. These are through the applications of:

1. electromagnetic-field theory,
2. the principle of conservation of energy and the principle of virtual work, and
3. variational principles.

Each of the above approaches has its own merits, and each is not entirely independent of the other, although in some cases one cannot be conveniently substituted for the other. In most cases, the choice depends on the problem at hand. We shall examine each of these methods briefly and then explore the scope of each approach. In the equations of motion we must ascertain that electrical forces (such as voltages and currents) of mechanical origin and mechanical forces of electrical origin are properly taken into account.

2.1 A Review of Electromagnetic-Field Theory[1-5]

The basic laws of electricity are governed by a set of equations called *Maxwell's equations*. Naturally, these equations govern the electromagnetic phenomena in energy-conversion devices too. Because of the presence of moving media a direct application of Maxwell's equations to energy-conversion devices is rather subtle. But certain other aspects of applications, such as parameter determination, are straightforward. In this section, we shall briefly review Maxwell's equations, then derive certain other equations from Maxwell's equations, and finally take up a number of examples which will illustrate some of the applications of Maxwell's equations and of other equations derived therefrom.

We know that charged particles are exerted upon by forces when placed in *electric* and *magnetic* fields. In particular, the magnitude and direction of the force $\mathbf{F}$ acting on a charge q moving with a velocity $\mathbf{u}$ in an electric field $\mathbf{E}$ and in a magnetic field $\mathbf{B}$ is given by the *Lorentz force equation*

$$\mathbf{F} = q(\mathbf{E}+\mathbf{u} \times \mathbf{B}) = \mathbf{F}_E+\mathbf{F}_B \tag{2.1}$$

The electric-field intensity $\mathbf{E}$ is thus defined by the following equation:

$$\mathbf{E} = \frac{\mathbf{F}_E}{\Delta q} \tag{2.2}$$

where $\mathbf{F}_E$ is the force on an infinitesimal test charge Δq; the electric field is measured in volts per meter. The magnetic field can also be similarly defined, from Eq. (2.1), as the force on a unit charge moving with unit velocity at right angles to the direction of $\mathbf{B}$. The quantity B is called the *magnetic flux density* and is measured in webers per square meter.

Having defined the electric field, we can now define the *potential difference* dV between two points separated by a distance $d\mathbf{l}$ as

$$dV = -\mathbf{E}\cdot d\mathbf{l} \tag{2.3a}$$

Or, using the vector operator ∇, Eq. (2.3a) is expressed as

$$\mathbf{E} = -\nabla V \tag{2.3b}$$

We also see that, if $\mathbf{B}$ is the magnetic flux density, the total *magnetic flux* ϕ can be expressed as

$$\phi = \int_s \mathbf{B}\cdot d\mathbf{s} \tag{2.4}$$

where the integral is over a surface $\mathbf{s}$.

With the above definitions in mind, we now recall Faraday's law. It states that an electromotive force (EMF) is induced in a closed circuit when the magnetic flux ϕ linking the circuit changes. If the closed circuit consists of an N-turn coil, the induced EMF is given by

$$\text{EMF} = -N\frac{d\phi}{dt} \tag{2.5}$$

The negative sign is introduced to take into account Lenz's law and to be consistent with the positive sense of circulation about a path with respect to the positive direction of flow through the surface (as shown in Fig. 2-1) when Eq. (2.5) is expressed in the integral form as follows.

Because the potential difference has been defined by Eqs. (2.3a,b), it is reasonable to define EMF as

$$\text{EMF} = \oint \mathbf{E}\cdot d\mathbf{l} \tag{2.6}$$

which can be considered as the potential difference about a specific closed path. If we consider "going around the closed path only once", we can put $N = 1$; then from Eqs. (2.4–2.6) it follows that

$$\oint \mathbf{E}\cdot d\mathbf{l} = -\frac{\partial}{\partial t}\int_s \mathbf{B}\cdot d\mathbf{s} \tag{2.7}$$

The partial derivative with respect to time is used to distinguish derivatives with space, since **B** can be a function of both time and space.

Equation (2.7) is an expression of Faraday's law in the integral form. We now recall *Stokes' theorem*, which states that

$$\oint \mathbf{E}\cdot d\mathbf{l} = \int_s (\nabla\times\mathbf{E})\cdot d\mathbf{s} \tag{2.8}$$

where the surface **s** is that enclosed by the closed path, as shown in Fig. 2-1. We notice that the terms on the left-hand sides of Eqs. (2.7) and (2.8) are identical. Equating the terms on the right-hand sides of these equations we have

$$\nabla\times\mathbf{E} = -\frac{\partial\mathbf{B}}{\partial t} \tag{2.9}$$

which is an expression of Faraday's law in the differential form.

A knowledge of vector algebra shows that the divergence of a curl is zero. Expressed mathematically,

$$\nabla\cdot\nabla\times\mathbf{E} = 0 \tag{2.10}$$

From Eqs. (2.9) and (2.10), therefore,

$$\nabla\cdot\mathbf{B} = 0 \tag{2.11}$$

In deriving Eq. (2.9) we assumed that there was no relative motion between the closed circuit and the magnetic field. If, however, there is a relative motion between the circuit and the magnetic field, in addition to the time variation of this field, Eq. (2.9) has to be modified. Referring to the Lorentz force equation, Eq. (2.1), we notice that the fields **E** and **B** are measured in a stationary reference frame. It is only the charge that is moving with a velocity **u**. If, on the other hand, the charge moves with a velocity **u**′ with respect to a moving reference frame having a velocity **u**, the velocity of the charge with respect to the stationary reference frame becomes $(\mathbf{u}+\mathbf{u}')$. The force on the charge is then given by

$$\mathbf{F} = q[\mathbf{E}+(\mathbf{u}+\mathbf{u}')\times\mathbf{B}] = q[(\mathbf{E}+\mathbf{u}\times\mathbf{B})+\mathbf{u}'\times\mathbf{B}] \tag{2.12}$$

Thus the electric field $\mathbf{E}'$ in the moving reference frame becomes

$$\mathbf{E}' = \mathbf{E}+\mathbf{u}\times\mathbf{B} \tag{2.13}$$

where **E** is given by Eq. (2.9). Here we have not taken into account relativistic effects because in electromagnetic energy-conversion devices the velocities involved are considerably small. Using Eqs. (2.9) and (2.13) we see that the electric field **E** induced in a circuit moving with a velocity **u** in a time-varying field **B** is given by

$$\nabla\times\mathbf{E} = -\frac{\partial\mathbf{B}}{\partial t}+\nabla\times\mathbf{u}\times\mathbf{B} \tag{2.14}$$

In Eq. (2.14) we notice that the second term on the right-hand side is introduced to take into account the motion of the circuit. Thus we can identify the first term in Eq. (2.14) as due to a *transformer EMF* and the second as due to a *motional EMF*; that is, the "change in the flux-linkage" rule as well as the "flux-cutting" rule of obtaining the EMF induced in a circuit are both taken into account in Eq. (2.14).

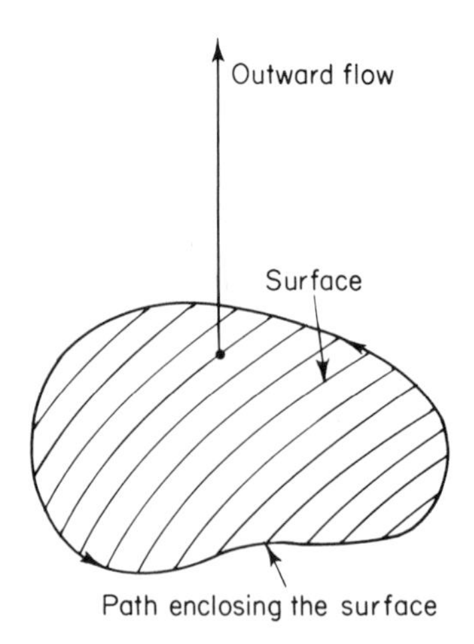

Fig. 2-1 Positive circulation about a path enclosing a surface and positive flow through the surface.

We would like to point out here that Eq. (2.5), expressing Faraday's law, is complete. A slightly more tedious analysis than that given above can be shown to lead to Eq. (2.14); the interested reader should consult Reference six. In practical cases, it is better to identify the transformer and motional voltages separately. A direct application to Eq. (2.5), without extra care, may lead to inconsistent results.

So far we have discussed ways of finding the electric field from given magnetic fields. We shall now consider the relationship between given currents and resulting magnetic fields. In this connection, *Ampere's circuital law* expressed as

$$\oint \mathbf{H}\cdot d\mathbf{l} = I \tag{2.15a}$$

gives the relationship between the *magnetic-field intensity* $\mathbf{H}$ and the current I enclosed by the closed path of integration. If $\mathbf{J}$ is the surface-current density, Eq. (2.15a) can be also written as

$$\oint \mathbf{H}\cdot d\mathbf{l} = \int_s \mathbf{J}\cdot d\mathbf{s} \tag{2.15b}$$

Using Stokes's theorem, Eq. (2.8), the point form of Ampere's law becomes, from Eq. (2.15b),

$$\nabla\times\mathbf{H} = \mathbf{J} \tag{2.15c}$$

The surface-current density $\mathbf{J}$ is related to the volume-charge density ρ through the *continuity equation*

$$\nabla\cdot\mathbf{J} = -\frac{\partial\rho}{\partial t} \tag{2.16}$$

Taking the divergence of Eq. (2.15c) reveals an immediate inconsistency when compared with Eq. (2.16), since $\nabla\cdot\nabla\times\mathbf{H} = 0$. If an additional term $\partial\mathbf{D}/\partial t$ is added to the right-hand side of Eq. (2.15c) such that

$$\nabla\times\mathbf{H} = \mathbf{J}+\frac{\partial\mathbf{D}}{\partial t} \tag{2.17}$$

we find that the continuity equation is satisfied provided that

$$\nabla\cdot\mathbf{D} = \rho \tag{2.18}$$

But Eq. (2.18) is perfectly legitimate, since it expresses *Gauss's law* in the differential form. The quantity $\mathbf{D}$ is called the *electric-flux density* and $\partial\mathbf{D}/\partial t$ is known as the *displacement-current* density.

From the preceding considerations we have the following set of equations relating to the various field quantities:

$$\nabla \times \mathbf{E} = -\frac{\partial \mathbf{B}}{\partial t} \tag{2.19a}$$

$$\nabla \times \mathbf{H} = \mathbf{J} + \frac{\partial \mathbf{D}}{\partial t} \tag{2.19b}$$

$$\nabla \cdot \mathbf{B} = 0 \tag{2.19c}$$

$$\nabla \cdot \mathbf{D} = \rho \tag{2.19d}$$

Equations (2.19a–d) are generally known as *Maxwell's equations.* We can easily verify that Eqs. (2.19c,d) are actually contained in Eqs. (2.19a,b) respectively.

In order to obtain complete information regarding the various field quantities, in addition to Maxwell's equations certain auxiliary relations are also very useful. These relations are as follows:

Ohm's law: For a conductor of conductivity σ,

$$\mathbf{J} = \sigma \mathbf{E} \tag{2.20}$$

where $\mathbf{J}$ = surface-current density and $\mathbf{E}$ = electric-field intensity.

Permittivity: The electric-field intensity and the electric-flux density in a medium are related to each other by

$$\mathbf{D} = \epsilon \mathbf{E} \tag{2.21}$$

where ϵ is called the *permittivity* of the material.

Permeability: The magnetic-field intensity and the magnetic-flux density in a material are related to each other by

$$\mathbf{B} = \mu \mathbf{H} \tag{2.22}$$

where μ is called the *permeability* of the medium.

Finally, we should note that in majority of energy-conversion devices there are no free charges. Consequently, for these cases the Lorentz force equation takes the form

$$\mathbf{F} = \mathbf{J} \times \mathbf{B} \tag{2.23}$$

which follows from Eq. (2.1), since the motion of charges constitutes the flow of current. Because $\mathbf{J}$ and $\mathbf{B}$ are the current and flux densities respectively, Eq. (2.23) determines the force density rather than the total force. We will use the simplified form of the force equation, Eq. (2.23), more often than the original equation, Eq. (2.1).

In summary, so far in this section we have reviewed Maxwell's equations and certain other auxiliary equations useful in the study of electromagnetic problems.

2.1.1 Note on Solving the Field Equations

Maxwell's equations and the auxiliary equations, discussed in the first part of this section, provide complete information about electromagnetic fields due to given currents and charges. There are many methods, ranging from the separation-of-variables methods to the applications of Green's functions and operational methods, which yield solutions to various kinds of field problems.

Beginning with simple cases, it can be shown for a two-dimensional magnetic field (in a rectangular coordinate system) that

$$\nabla^2 B_x = \nabla^2 B_y = 0 \tag{2.24}$$

Eq. (2.24) is of the form of *Laplace's equation* and can be solved by one of the standard methods, such as the product-solution method. From Eq. (2.24) the magnetic field, in a specified region, is directly obtained.

For somewhat complicated cases it is sometimes convenient to introduce new functions, such as scalar and vector potentials, to facilitate the solution. In fact, an *electrostatic scalar potential* V and a vector potential $\mathbf{A}$, the latter called the *magnetic vector potential*, are often used in solving the field equations. These potentials are defined by the following equations:

$$\mathbf{E} = -\nabla V - \frac{\partial \mathbf{A}}{\partial t} + \mathbf{u} \times \mathbf{B} \tag{2.25}$$

$$\mathbf{B} = \nabla \times \mathbf{A} \tag{2.26}$$

A number of useful equations can be derived in terms of scalar and vector potentials. Those pertinent to the present work are:

1. Laplace's equation: $$\nabla^2 V = 0 \tag{2.27a}$$

$$\nabla^2 \mathbf{A} = 0 \tag{2.27b}$$

2. The wave equation: $$\nabla^2 V = \mu\epsilon \frac{\partial^2 V}{\partial t^2} \tag{2.28a}$$

$$\nabla^2 \mathbf{A} = \mu\epsilon \frac{\partial^2 \mathbf{A}}{\partial t^2} \tag{2.28b}$$

3. The diffusion equation: $$\nabla^2 \mathbf{J} = \mu\sigma \frac{\partial \mathbf{J}}{\partial t} \tag{2.29}$$

Once the appropriate equation has been derived, the next and most important step is to solve the equation for specified sources and boundaries. In order to do this, reference frames are chosen and transformations made to account for the relative motion between electromagnetic fields and current-carrying conductors. For the present it is not necessary to go into the details of solutions of the field equations. It must be mentioned, however, that most of the pertinent equations are partial differential equations which have series solutions. The purpose of solving for the fields is that voltages and forces of electromechanical and electromagnetic origin can be obtained either by using the force-density and the **E**-field expressions or by the energy-storage methods eventually leading to the equations of motion.

2.1.2 Range of Application

The scope of the applications of electromagnetic-field methods to electromechanical systems is quite wide. Almost every type of machine can be analyzed starting from the basic field equations. In some cases, such as dc machines and salient-pole synchronous machines, analysis becomes unnecessarily complicated; whereas in certain other instances, such as eddy-current devices and magnetohydrodynamic studies, solution is possible only by field methods. Some of the important aspects of applications of electromagnetic-field theory to electromechanical energy conversion are summarized below.

1. *Parameter determination.* In formulating the equations of motion of electromechanical systems it is necessary to know the parameters of the system. In particular, for the electrical portion, inductances and capacitances must be known. In most cases, the electrostatic field is not of importance (see Problem 2-3) and only inductances need be determined. Field theory is fundamental to the evaluation of inductances. Laws of magnetic circuits yield inductances in simple lumped-parameter transducers. But for many types of rotating machines the problem is much more complicated, and inductances can only be found by solving for the fields in the various zones of the machine. Different methods of obtaining inductances hinge on finding

(a) magnetic stored energy,
(b) flux linkages, or
(c) permeance variations.

These methods differ from each other considerably and are discussed in detail in Chapter 5. In each instance, however, field equations have to be solved.

2. *Energy storage methods.* The fields in various regions of the machine having been determined, the energy stored in the field is evaluated. Equations of motion are then formulated using one of the standard methods, such as the

Lagrangian formulation or the force-balance method. As pointed out in the preceding paragraph, system inductances can also be found by calculating the energy stored in the electromagnetic fields.

It can be shown[1,6] that the energies stored in electric and magnetic fields are respectively given by

$$W_E = \frac{1}{2}\int_v \mathbf{D}\cdot\mathbf{E}\, dv \tag{2.30}$$

and

$$W_m = \frac{1}{2}\int_v \mathbf{B}\cdot\mathbf{H}\, dv \tag{2.31}$$

Again we see that, although the determination of stored energy leads to the equations of motion, the first step is still solving for the fields.

3. *Energy transfer methods.* An alternative method of obtaining the performance characteristics of electromagnetic devices, based on electromagnetic-field theory, consists of calculating the electric and magnetic fields from Maxwell's equations and then obtaining the power density through the use of Poynting's vector:[1,6]

$$\mathbf{P} = \mathbf{E}\times\mathbf{H} \tag{2.32}$$

where **E** and **H** are respectively the electric- and magnetic-field intensities and **P** is the power density. For transient analysis this approach is rather difficult, although it has been applied with success to the steady-state analysis of dc machines and of ac induction and synchronous machines.

4. *Other methods.* Other techniques pertaining to electromagnetic-field theory applicable to electromechanical energy conversion are the method of images, conformal transformations, and graphical and numerical methods. These have a wide range of application to solving problems of practical interest. Some examples are considered in Chapter 9.

It is beyond the scope of the present work to discuss in detail every possible application of electromagnetic-field theory to energy conversion. Only the range of applications is indicated above. A few examples are considered here and certain other applications are considered in Chapters 4, 5, and 9. In summary, for the analysis of devices having continuous media—such as solid-rotor machines and liquid-metal pumps—it is most convenient to use field theory methods. In order to calculate the inductances of electrical machines and to account for leakages, teeth and slots, etc., again the solutions are obtained by the application of field theory.

We shall now consider a number of examples to illustrate some applications of the theory developed in the preceding sections.

EXAMPLE 2–1

First we shall investigate the general form of the **B** field encountered in energy-conversion devices. Because most problems of practical interest can be solved in a rectangular coordinate system, we shall carry out all vector operations in rectangular coordinates unless stated otherwise.

We know from previous considerations, Eq. (2.24), that the x and y components of the **B** field at a point satisfy Laplace's equation. Therefore, it is reasonable to assume that

$$B_y = B_m \cos \beta x \cosh \beta y \cos \omega t \tag{2.33}$$

If a conducting sheet of conductivity σ, extending up to infinity (Fig. 2E-1), is held stationary, what are the instantaneous and time-average force densities on the sheet? Assume current flows in the z direction only.

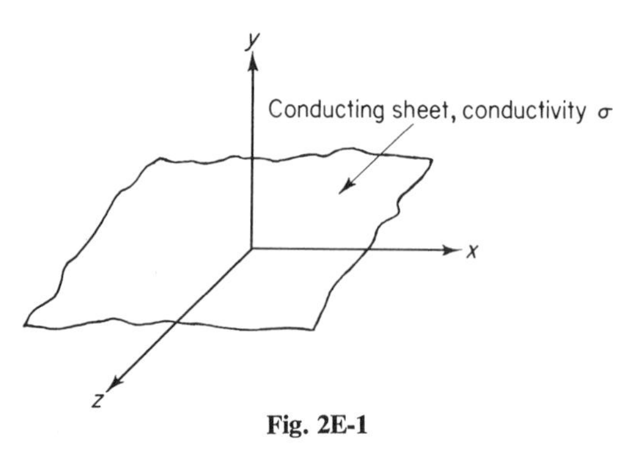

Fig. 2E-1

We recognize that the form of the field given by Eq. (2.33) is reasonable because it satisfies Laplace's equation, Eq. (2.24).

To find the force density **F**, we recall Eqs. (2.23), (2.20), and (2.9), in that order:

$$\mathbf{F} = \mathbf{J} \times \mathbf{B} \tag{2.23}$$

$$\mathbf{J} = \sigma \mathbf{E} \tag{2.20}$$

and

$$\nabla \times \mathbf{E} = -\frac{\partial \mathbf{B}}{\partial t} \tag{2.9}$$

From these equations we see that to determine **F** we have to find **E**, since **E** and **J** are related by Eq. (2.20).

Equations (2.9) and (2.33) yield

$$\frac{\partial E_z}{\partial x} = \frac{\partial B_y}{\partial t} = -\omega B_m \cos \beta x \cosh \beta y \sin \omega t$$

Or, integrating both sides with respect to x,

$$E_z = -\frac{\omega}{\beta} B_m \sin \beta x \cosh \beta y \sin \omega t \tag{2.34}$$

Because of periodicity in x, the constant of integration is zero. From Eqs. (2.20), (2.23), and (2.34) the instantaneous force density is given by

$$F_x = \frac{\sigma\omega}{4\beta} B_m^2 \cosh^2 \beta y \sin 2\beta x \sin 2\omega t \tag{2.35}$$

The only time-varying quantity in Eq. (2.35) is $\sin 2\omega t$, the average value of which is zero. Consequently, the average value of the force density is also zero. We conclude that a conducting sheet, initially at rest, placed in a pulsating field will not move.

In order to obtain some feeling for the magnitude of the force density, the following (reasonable) numerical values are substituted in Eq. (2.35):

$$\sigma = 5.8 \times 10^7 \quad \mho/m$$

$$B_m = 0.1 \text{ Wb/m}^2$$

$$\omega = 377 \text{ rad/sec}$$

and

$$\beta = 1 \text{ m}$$

in which case $(F_x)_{\max} = 5.5 \times 10^7 \text{ N/m}^2$.

EXAMPLE 2–2

We now allow the conducting sheet of Example 2-1 to move with a velocity u_x in the x direction. What are the instantaneous and time-average force densities in this case?

From Example 2-1,

$$B_y = B_m \cos \beta x \cosh \beta y \cos \omega t \tag{2.33}$$

Equation (2.14), when expanded, yields

$$\frac{\partial E_z}{\partial x} = u_x \frac{\partial B_y}{\partial x} + \frac{\partial B_y}{\partial t} = -B_m \cosh \beta y \, (\beta u_x \sin \beta x \cos \omega t + \omega \cos \beta x \sin \omega t)$$

Or, integrating both sides with respect to x,

$$E_z = B_m \cosh \beta y \, (u_x \cos \beta x \cos \omega t - \frac{\omega}{\beta} \sin \beta x \sin \omega t) \tag{2.36}$$

Proceeding as in the last example, the instantaneous-force density is, from Eqs. (2.20), (2.23), and (2.36),

$$F_x = \sigma B_m^2 \cosh^2 \beta y \left(\frac{\omega}{4\beta} \sin 2\beta x \sin 2\omega t - u_x \cos^2 \beta x \cos^2 \omega t \right) \tag{2.37}$$

The time average of the force density, $\langle F_x \rangle$, is found from

$$\langle F_x \rangle = \frac{1}{T} \int_0^T F_x \, dt$$

and Eq. (2.37), from which

$$\langle F_x \rangle = -\frac{\sigma}{2} u_x B_m^2 \cosh^2 \beta y \cos^2 \beta x \tag{2.38}$$

Evidently, when the sheet is held stationary ($u_x = 0$), the average force is zero and the results reduce to those obtained in Example 2-1.

EXAMPLE 2–3

We now wish to determine the density of power flow in the conducting sheet of Examples 2-1 and 2-2.

The power flow in the y direction—that is, into the conducting sheet—is given by Eq. (2.32). Or,

$$P_y = E_z H_x \tag{2.39}$$

Now,

$$B_y = B_m \cos \beta x \cosh \beta y \cos \omega t \tag{2.33}$$

so that from Eqs. (2.19c) and (2.33)

$$\frac{\partial B_x}{\partial x} = -B_m \beta \cos \beta x \sinh \beta y \cos \omega t$$

Or, integrating with respect to x yields

$$B_x = -B_m \sin \beta x \sinh \beta y \cos \omega t \tag{2.40a}$$

Or, from Eqs. (2.22) and (2.40)

$$H_x = \frac{-B_m}{\mu} \sin \beta x \sinh \beta y \cos \omega t \tag{2.40b}$$

For Example 2-1, E_z is given by Eq. (2.34). Therefore, from Eqs. (2.34), (2.39), and (2.40b), the power density is given by

$$P_y = \frac{\omega B_m^2}{4 \mu \beta} \sin^2 \beta x \sinh 2\beta y \sin 2\omega t \tag{2.41a}$$

For Example 2-2, E_z is given by Eq. (2.36) and, in this case, the power density is obtained from Eqs. (2.36), (2.39), and (2.40b), so that

$$P_y = \frac{B_m^2}{4\mu} \sinh 2\beta y \left(\frac{\omega}{\beta} \sin^2 \beta x \sin 2\omega t - u_x \sin 2\beta x \cos^2 \omega t \right) \tag{2.41b}$$

Notice that $\langle P_y \rangle$ is not zero only if $u_x \neq 0$.

2.1.3 Ferromagnetism

Electromagnetic energy-conversion devices invariably have cores made of magnetic materials. It is therefore important that we gain some understanding of ferromagnetism. Magnetic properties of materials can be studied through the use of relativistic quantum mechanics.[7] However, we shall take a much simplified approach here which is sufficient for our purpose.

We know that the magnetic field at a point (r, θ, ϕ), in spherical coordinates, due to a circular loop, of radius a, and carrying a current I is given by[8]

$$B_r = \mu_o \frac{I\pi a^2}{2\pi r^3} \cos \theta$$

$$B_\theta = \mu_o \frac{I\pi a^2}{4\pi r^3} \sin \theta$$

$$B_\phi = 0$$

The quantity $(I\pi a^2)$ is defined as the *magnetic-dipole moment* and acts along $\theta = 0$. Thus, a current-carrying circular loop can be considered as a magnetic dipole. Now, returning to our simplified approach, we consider an atom to consist of a positive nucleus surrounded by electrons which spin about their axes in addition to orbiting around the nucleus. An orbiting electron is similar to a current loop and can thus be considered to have magnetic-dipole moments. The magnetic properties of a material depend on the contributions of the various components of magnetic moments.

Ferromagnetic materials are different from other magnetic (such as para- and diamagnetic) and nonmagnetic materials in two respects; they are characterized by (1) a high permeability (for example, a relative permeability of about 200,000 for iron with 0.05% impurity) and (2) the presence of residual magnetism and occurrence of the *hysteresis* phenomenon. The atoms of a ferromagnetic material have a large magnetic-dipole moment, mainly due to the electron spin moments. Those regions, containing a large number of atoms, in which the spins are aligned are called *ferromagnetic domains*. The domain configuration in a solid determines magnetic-field energy and other associated energies (such as domain-wall energy and anistropy energy). In terms of domain theory, consider a single crystal such as shown in Fig. 2-2. In each domain in Fig. 2-2(a), magnetization is largely due to electron-spin alignment, but because the net magnetic-dipole moment is zero we consider the crystal to be unmagnetized.

Now, if an external magnetic field is applied, the domain walls move and take positions somewhat as shown in Fig. 2-2(b). In this case there is a net magnetic-dipole moment and, macroscopically, we consider the crystal to be magnetized.

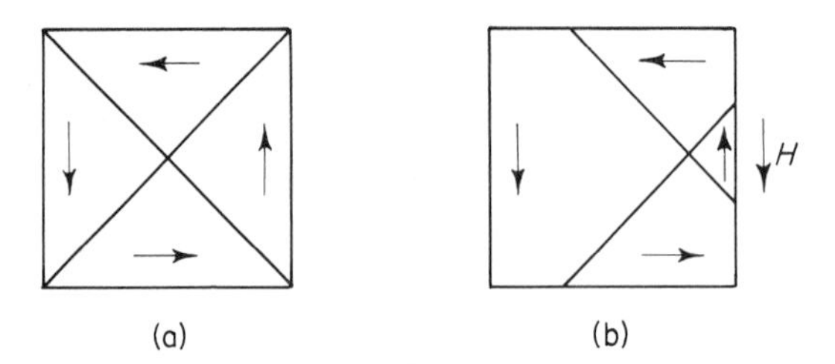

Fig. 2-2 Domains of a ferromagnetic crystal: (a) single crystal, unmagnetized, domains magnetized; (b) single crystal, magnetized, change in domain size due to applied *H*.

We can now define *magnetization* of a material as the magnetic-dipole moment per unit volume. When the external field is removed, a residual magnetic-dipole moment remains in the macroscopic structure of the ferromagnetic material, and the phenomenon of lag of effect (magnetization) behind the cause (magnetizing force) is called *hysteresis*. With reference to the ferromagnetic material, the orientation of domains favorable to the externally applied magnetic field also holds for polycrystalline materials. If the strength of the applied magnetic field H is increased, the domain walls move and thereby increase magnetization and flux density B. However, the

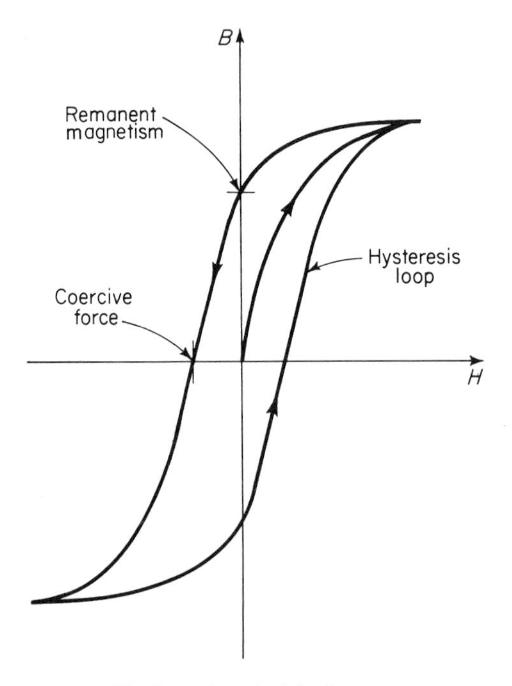

Fig. 2-3 A typical B–H curve.

increase in B with an increase in H does not follow a linear law. Eventually, the material *saturates* with B. A complete cyclic change of H leads to the curve shown in Fig. 2-3, called *hysteresis loop*. Notice that *remanent magnetism* and *coercive force* are defined in Fig. 2-3.

Elements that are ferromagnetic at room temperature are iron, nickel, and cobalt. The hysteresis loop of a ferromagnetic material is a measure of power loss—called *hysteresis loss*— in the material if it is in a magnetic field that goes through cyclic changes. Thus, for ac energy converters, we choose a material for the magnetic core which has a high permeability and a narrow hysteresis loop. For example, high-grade silicon steel is a suitable material for the core of an electrical machine.

So far we have only considered the qualitative properties of ferromagnetic materials. We now wish to introduce certain quantities which measure magnetic properties. We have already seen that magnetic dipoles act as sources of a magnetic field. These dipoles are due to the motion of bound charges, such as orbiting and spinning electrons. The current resulting from the motion of bound charges is called *bound current* or *Amperian current.*[9] Returning now to Eq. (2.15c), we conclude that the bound current $\mathbf{J}_b$ produces a magnetization $\mathbf{M}$ in much the same way as the motion of free charges produces the magnetic field $\mathbf{H}$ in free space. Expressing this statement mathematically, we have

$$\nabla \times \mathbf{H} = \mathbf{J} \tag{2.15c}$$

and

$$\nabla \times \mathbf{M} = \mathbf{J}_b \tag{2.42}$$

The current density $\mathbf{J}$ in Eq. (2.15c) must include all sources (free or bound); that is,

$$\nabla \times \mathbf{H} = \mathbf{J}_f + \mathbf{J}_b \tag{2.43}$$

From Eqs. (2.42) and (2.43)

$$\frac{1}{\mu_o} \nabla \times \mathbf{B} = \mathbf{J}_f + \nabla \times \mathbf{M} \tag{2.44}$$

since, in free space,

$$\mathbf{B} = \mu_o \mathbf{H} \tag{2.45a}$$

Equation (2.44) can be rewritten as

$$\nabla \times \mathbf{H} = \mathbf{J}_f$$

where

$$\mathbf{H} = \frac{\mathbf{B}}{\mu_o} - \mathbf{M}$$

Or, finally,

$$\mathbf{B} = \mu_o(\mathbf{H} + \mathbf{M}) \tag{2.45b}$$

A comparison of Eqs. (2.45a) and (2.45b) indicates how the **B–H** relationship is modified in the presence of a ferromagnetic material. For a linear isotropic medium, a quantity called *magnetic susceptibility* χ_m is defined as

$$\mathbf{M} = \chi_m \mathbf{H} \tag{2.45c}$$

Thus, from Eqs. (2.45a–c) we can write

$$\mathbf{B} = \mu \mathbf{H} \tag{2.22}$$

where the *permeability* μ is

$$\mu = \mu_o \mu_r \tag{2.45d}$$

and μ_r is called *relative permeability* as defined by

$$\mu_r = 1 + \chi_m \tag{2.45e}$$

Relative permeabilities of some ferromagnetic materials are given below.[1]

Material	*Relative Permeability*
Cobalt	250
Nickel	600
Iron (0.2% impurity)	5000
Iron (0.05% impurity)	200,000

2.1.4 The Magnetic Circuit

Having briefly discussed the properties of ferromagnetic materials, we can now consider magnetic circuits consisting of a ferromagnetic core and airgaps. Such a magnetic circuit is shown in Fig. 1-1(a). The laws for magnetic circuits are similar to those for dc electric circuits, except for the phenomena of hysteresis and saturation that occur in the former circuits. A typical saturation curve is shown in Fig. 2-4. We shall illustrate later by means of an example (Example 2-4) how to take into account the effects of saturation in magnetic-circuit calculations. But first of all we shall review briefly the laws of magnetic circuits.

We define a *magnetomotive force* (MMF) $\mathscr{F}$ between two points A and B as

$$\mathscr{F} = \int_A^B \mathbf{H} \cdot d\mathbf{l} \tag{2.46}$$

The flux ϕ has already been defined by Eq. (2.4):

$$\phi = \int_s \mathbf{B} \cdot d\mathbf{s} \tag{2.4}$$

The ratio of MMF to the total flux is called *reluctance* $\mathscr{R}$. The "Ohm's law" for the magnetic circuit can therefore be written as

$$\mathscr{F} = \mathscr{R}\phi \tag{2.47}$$

The reciprocal of reluctance is called *permeance* $\mathscr{P}$. Extending the analogy with the electric circuit a little further, the reluctance of a magnetic circuit of length l and area of cross-section A is given by

$$\mathscr{R} = \frac{l}{\mu A} \tag{2.48a}$$

and the resistance R is given by

$$R = \frac{l}{\sigma A} \tag{2.48b}$$

where μ and σ are permeability and conductivity respectively.

Because we know from Eq. (2.15a) that

$$\oint \mathbf{H} \cdot d\mathbf{l} = I \tag{2.15a}$$

we can express the MMF for an N-turn coil carrying a current I, from Eq. (2.46), as

$$\mathscr{F} = \oint \mathbf{H} \cdot d\mathbf{l} = NI \tag{2.49}$$

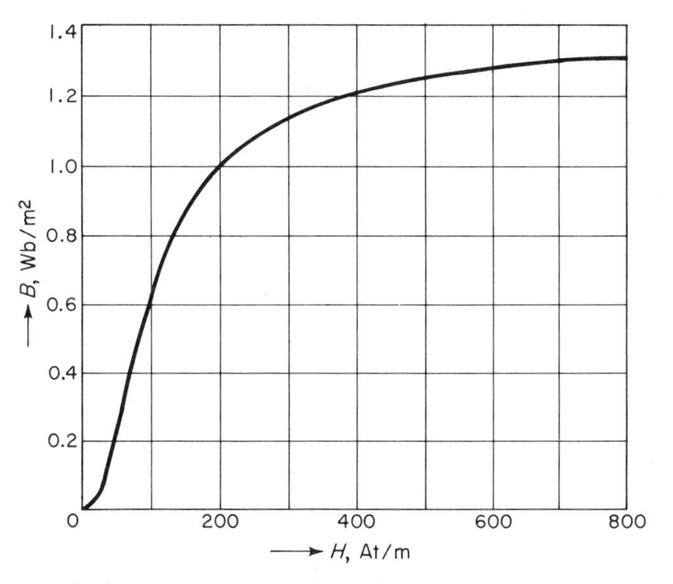

Fig. 2-4 Saturation curve for a silicon sheet steel sample.

EXAMPLE 2-4

A simple magnetic circuit with an airgap is shown in Fig. 2E-4. The mean lengths of the flux path and of the cross-sections are as indicated. How many turns should the exciting coil have in order to establish 1.0 Wb/m^2 (weber per square meter) flux density in the airgap? The maximum allowable current through the coil is 10 A (amperes). The core is of silicon steel.

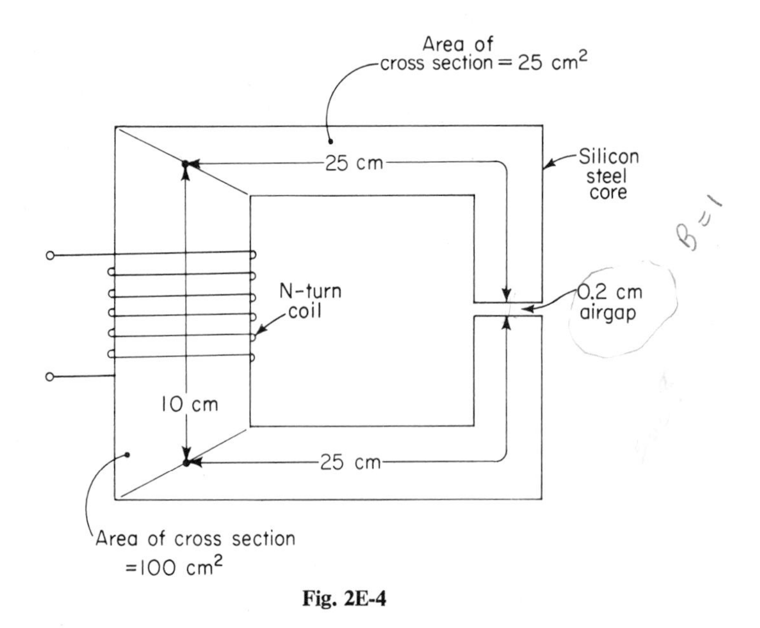

Fig. 2E-4

We see from Fig. 2-4 that at $B = 1.0\ \text{Wb/m}^2$ the B–H relationship is nonlinear. Therefore, it is unwise to begin our reluctance calculations with the iron. The total MMF can be expressed as

$$\mathscr{F} = \mathscr{F}_{\text{steel}} + \mathscr{F}_{\text{air}}$$

The reluctance of the airgap is, from Eq. (2.48a),

$$\mathscr{R}_{\text{air}} = \frac{2 \times 10^{-3}}{4\pi \times 10^{-7} \times 25 \times 10^{-4}} = 6.36 \times 10^5 \quad \text{At/Wb (ampere-turns per weber)}$$

The total flux is

$$\phi = 1.0 \times 25 \times 10^{-4} = 2.5 \times 10^{-3} \quad \text{Wb}$$

which should be the same for the entire magnetic circuit. We have, from Eq. (2.47),

$$\mathscr{F}_{\text{air}} = 6.36 \times 10^5 \times 2.5 \times 10^{-3} = 1590 \text{ At}$$

From Fig. 2-4, for $B = 1.0\ \text{Wb/m}^2$, $H = 200$ At/m. Thus, for the length $(25 + 25)$ cm, $\mathscr{F} = 100$ At.

The total flux in the limb 10 cm long is still 2.5×10^{-3} Wb. Or, flux density $= 2.5 \times 10^{-3}/10^{-2} = 0.25$ Wb/m^2. From Fig. 2-4 again, for $B = 0.25$ Wb/m^2, $H = 70$ At/m. Or, for 10 cm, $\mathscr{F} = 7.0$ At, so that

$$\mathscr{F}_{\text{steel}} = 100 + 7.0 = 107 \text{ At}$$

Or,

$$\mathscr{F} = NI = 107 + 1590 = 1697 \text{ At}$$

The maximum allowable current $I = 10$ A, so that the required number of turns is

$$N = 169.7 = 170 \quad \text{turns}$$

This last example shows that for a given length and cross-section, the MMF for the airgap is a few thousand times more than that for steel, even for saturation flux densities such as 1.0 Wb/m^2. As a first approximation, therefore, we can neglect saturation, in which case the permeability of iron can be considered to be infinity as compared to the permeability of air.

We know that rotating electrical machines consist of a *stator* and *rotor* made of iron. Both stator and rotor have current-carrying windings. For determining the fields in the airgaps of rotating machines, idealized models are sometimes chosen. Such an idealized model and the solution to the field problem is illustrated by the following example.

EXAMPLE 2–5

A rotating machine is ideally represented by the linear structure shown in Fig. 2E-5. The actual windings are assumed to be replaced by current sheets backed

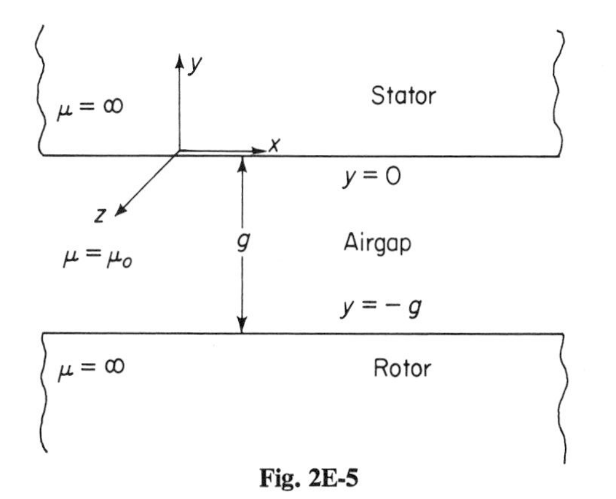

Fig. 2E-5

by iron having infinite permeability. For a single such current sheet on the stator, the magnetic vector potential at the rotor surface is

$$\mathbf{A} = \mathbf{1}_z i \, (\sin x + 0.2 \sin 3x)$$

Find the expression for $\mathbf{A}$ at the center of the airgap and hence determine the B field. ($\mathbf{1}_z$ is the unit vector in the z direction.)

Since $\nabla^2\mathbf{A} = 0$ in the airgap, the solution may be assumed to be:

$$\mathbf{A} = \mathbf{1}_z i\,[\sin x\,(k_1 \cosh y + k_2 \sinh y) + \sin 3x\,(k_3 \cosh 3y + k_4 \sinh 3y)]$$

To evaluate the constants, the given boundary conditions are used. Thus, at $y = -g$, $B_x = 0$ (μ being infinity). Therefore, from Eq. (2.26),

$$B_x = 0 = \frac{\partial A_z}{\partial y} = i\{\sin x\,[k_1 \sinh(-g) + k_2 \cosh(-g)] + \sin 3x\,[3k_3 \sinh(-3g) + 3k_4 \cosh(-3g]\}$$

The coefficients of sin x and sin $3x$ must therefore be zero, that is,

$$k_1 = k_2 \coth g \quad \text{and} \quad k_3 = k_4 \coth 3g$$

From the given expression for **A** at the rotor surface,

$$\begin{aligned} A_z &= i\,(\sin x + 0.2 \sin 3x) \\ &= i\,\{\sin x\,[k_1 \cosh(-g) + k_2 \sinh(-g)] + \sin 3x\,[k_3 \cosh(-3g) + k_4 \sinh(-3g)]\} \end{aligned}$$

Or, equating coefficients and expressing in terms of k_2 and k_4,

$$1 = k_1 \cosh g - k_2 \sinh g = k_2 (\coth g \cosh g - \sinh g)$$

Therefore,

$$k_2 = \sinh g$$

Similarly,

$$k_4 = 0.2 \sinh 3g$$

From which

$$\mathbf{A} = \mathbf{1}_z\, i[\sin x \cosh (y+g) + 0.2 \sin 3x \cosh 3(y+g)]$$

and at $y = -\frac{1}{2}g$,

$$\mathbf{A} = \mathbf{1}_z\, i[\sin x \cosh \tfrac{1}{2} g + 0.2 \sin 3x \cosh (3g/2)]$$

Recalling that $\mathbf{B} = \nabla \times \mathbf{A}$, the **B** field can be determined by taking the curl of the general expression for **A** and then setting $y = -\frac{1}{2}g$.

2.1.5 Inductance and Energy Stored in Inductive Systems

Having considered Faraday's law in Sec. 2.1, Eq. (2.5), we recall that the flux ϕ linking an N-turn coil is expressed as a flux linkage λ such that

$$\lambda = N\phi \tag{2.50}$$

From Eqs. (2.5) and (2.50) we can easily show that the *magnetic stored energy* W_m in an N-turn coil is

$$W_m = \int i\, d\lambda \tag{2.51}$$

This reduces to

$$W_m = \tfrac{1}{2}\lambda i \tag{2.52}$$

or a linear magnetic circuit.

Recalling the definition of inductance as flux linkage per ampere, that is,

$$L = \frac{\lambda}{i} \tag{2.53}$$

the stored energy can also be expressed as

$$W_m = \tfrac{1}{2}Li^2 \tag{2.54}$$

From Eqs. (2.47), (2.49), and (2.50) the alternate expression for the stored energy becomes

$$W_m = \tfrac{1}{2}\mathscr{R}\phi^2 \tag{2.55}$$

In terms of field quantities, stored magnetic energy is given by

$$W_m = \tfrac{1}{2}\int_v \mathbf{B}\cdot\mathbf{H}\,dv \tag{2.31}$$

A comparison of Eqs. (2.31) and (2.54) suggests a method of calculating system inductances. (Similar expressions hold for stored energy in electrostatic fields and capacitive systems).

If the magnetic circuit carries a number of windings, say n windings, as shown in Fig. 2-5, the definition of L as given by Eq. (2.53) still holds, but it

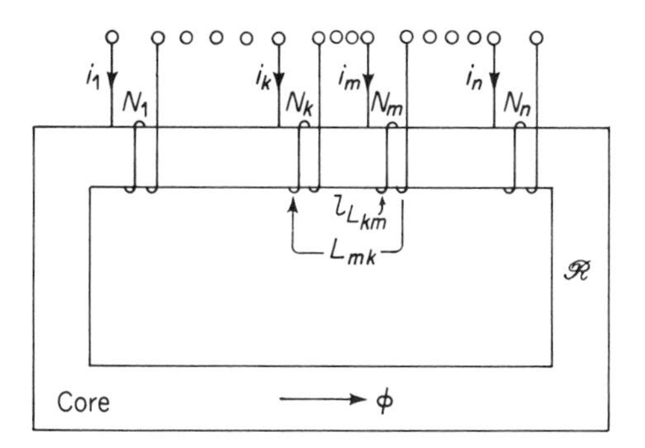

Fig. 2-5 A multiwinding magnetic circuit.

is best to express the L's in the form of a matrix as given below, where the terms along the diagonal denote the self-inductances and the off-diagonal terms are mutual inductances. The L matrix is written as follows:

$$[L] = \begin{bmatrix} L_{11} & L_{12} & \cdots & L_{1n} \\ L_{21} & L_{22} & \cdots & L_{2n} \\ \vdots & \vdots & & \vdots \\ L_{n1} & L_{n2} & \cdots & L_{nn} \end{bmatrix}$$

A typical element of the above L matrix can be determined if the equation defining inductance is recalled, according to which the flux linking a coil per ampere of current through it is the measure of its inductance, so that

$$L_{km} = \frac{\text{flux linking the } k\text{th coil due to the current in the } m\text{th coil}}{\text{current in the } m\text{th coil}} = \frac{\lambda_{km}}{i_m}$$

Or, from Eq. (2.47),

$$L_{km} = \frac{N_k \phi}{i_m} = \frac{N_k N_m}{\mathscr{R}} \tag{2.56a}$$

where N_k = number of turns on the kth coils, and L_{km} is termed as the *mutual inductance* between the kth and mth coils. If $k = m$, L_{mm} is the *self-inductance* of the mth coil.

From Eq. (2.56a) it is evident that $L_{km} = L_{mk}$, that is, that the inductance matrix for the system is *symmetric*. The inductance can also be expressed in terms of the constants of the magnetic circuit. For example, if the given circuit has a length l, and is of uniform area of cross-section A and permeability μ, then from Eq. (2.48a) the inductance of the kth coil with respect to the mth coil becomes

$$L_{km} = \frac{\mu N_k N_m A}{l} \tag{2.56b}$$

When the coils are excited a magnetic field is established. Assuming no losses, all the electrical energy supplied is stored in the magnetic field, the energy stored in the electric field being negligible. Therefore, for small changes in the energies of stationary systems,

$$dW_{\text{electrical}} = dW_{\text{magnetic field}} = dW_m$$

Or,

$$dW_m = \sum_k v_k i_k \, dt = \sum_k i_k \, d\lambda_k \tag{2.56c}$$

since

$$v_k = \frac{d\lambda_k}{dt}$$

where λ_k = flux linking the kth coil.

The total magnetic energy stored can be written as

$$W_m = \int_0^{\lambda_k} \sum_k i_k' \, d\lambda_k' \tag{2.57}$$

where, in the above equations,

$$k = 1, 2, \ldots, n.$$

Now, suppose the flux linkages are expressed in terms of the inductances as follows:

$$\begin{aligned}\lambda_1 &= L_{11}i_1 + L_{12}i_2 + \ldots + L_{1n}i_n\\ \lambda_2 &= L_{12}i_1 + L_{22}i_2 + \ldots + L_{2n}i_n\\ &\vdots\\ \lambda_n &= L_{1n}i_1 + L_{2n}i_2 + \ldots + L_{nn}i_n\end{aligned}$$

These flux linkages, when differentiated and multiplied by proper currents, give

$$\begin{aligned}i_1\,d\lambda_1 &= i_1(L_{11}\,di_1 + L_{12}\,di_2 + \ldots + L_{1n}\,di_n)\\ i_2\,d\lambda_2 &= i_2(L_{12}\,di_1 + L_{22}\,di_2 + \ldots + L_{2n}\,di_n)\\ &\vdots\\ i_n\,d\lambda_n &= i_n(L_{1n}\,di_1 + L_{2n}\,di_2 + \ldots + L_{nn}\,di_n)\end{aligned} \tag{2.58}$$

Therefore, from Eqs. (2.56c) and (2.58),

$$\begin{aligned}dW_m = \sum_k i_k\,d\lambda_k &= L_{11}i_1\,di_1 + L_{22}i_2\,di_2 + \ldots + L_{nn}i_n\,di_n + L_{12}(i_1\,di_2 + i_2\,di_1)\\ &+ L_{13}(i_1\,di_3 + i_3\,di_1) + \ldots + L_{23}(i_2\,di_3 + i_3\,di_2) + L_{24}(i_2\,di_4 + i_4\,di_2) + \ldots\end{aligned}$$

Rearranging terms and expressing under a summation sign, the equation becomes

$$dW_m = \sum_k L_{kk}i_k\,di_k + \tfrac{1}{2}\sum_{\substack{k\\k\neq m}}\sum_{\substack{m\\k\neq m}} L_{km}\,d(i_k i_m) \tag{2.59}$$

Equations (2.57) and (2.59) finally yield, for a linear magnetic circuit, the total magnetic energy stored:

$$W_m = \tfrac{1}{2}\sum_k\sum_m L_{km}i_k i_m \tag{2.60}$$

This result can also be expressed in matrix notation as

$$W_m = \tfrac{1}{2}\,\tilde{i}Li \tag{2.61}$$

where i is a column matrix, $\tilde{i}$ is the transpose of i (that is, $\tilde{i}$ is a row matrix), and L is the inductance matrix of the system.

To summarize the discussions of Sec. 2.1, we reviewed the basic laws of electromagnetic field theory and expressed them in the form of Maxwell's equations. We indicated the usefulness of potential functions in solving field problems and outlined the various aspects of applications of electromagnetic theory to energy-conversion devices. Then we presented a simplified theory of ferromagnetism, discussed the properties of ferromagnetic materials, and considered their use in magnetic circuits. Finally, we introduced the concept of inductance and derived certain expressions for magnetic stored energy.

2.2 Energy-Conservation Principles[10]

It was pointed out in earlier discussions that electromechanical energy conversion is possible if the energy stored in the coupling fields is disturbed. A justification of this is possible from the energy-conservation principles from which the magnitudes of forces can also be calculated. The equations of motion can thus be formulated using the force laws. To be able to use this method, it is necessary to separate the conservative part of the system from its dissipative portion, as shown in Fig. 2-6. Here only a 2-port system is

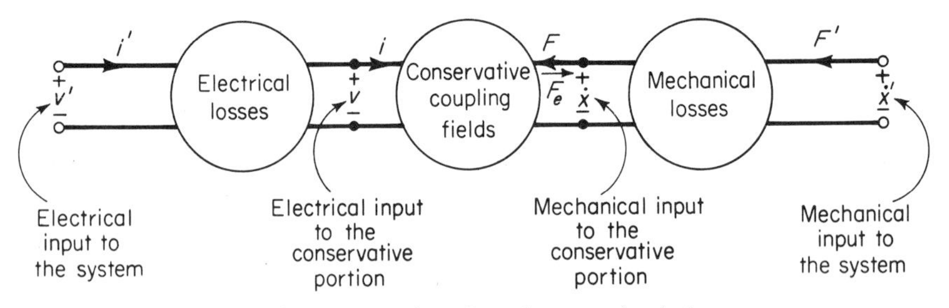

Fig. 2-6 A representation of an electromechanical system.

considered, although the method is applicable to multiport systems also. The total stored energy is the sum of the energy stored in the electric and magnetic fields. In general, this stored energy is a function of six types of variables: λ, F, x, i, q, and v, where λ is flux linkage, F is mechanical-input force, x is mechanical displacement, i is input current, and v is the input voltage, all pertaining to the conservative portion alone. Symbolically, their functional relationship can be written as

$$W = W(F, x, \lambda, i, v, q)$$

where q is the electrical charge. However, there can be only three independent variables, the remaining three governed by the constraint equations

$$v = v(q, x)$$

$$i = i(\lambda, x)$$

$$F = f(\lambda, q, x)$$

Thus the stored energy, expressed as a function of the independent variables (λ, q, x) only, is

$$W = W(\lambda, q, x) \tag{2.62}$$

The voltages and currents obey the relationships

$$v = \frac{d\lambda}{dt} \quad \text{and} \quad i = \frac{dq}{dt}$$

In order to find the force due to the coupling fields, an arbitrary displacement dx is assumed which involves a change in the stored energy as well as changes in the input electrical and mechanical energies. The principle of the conservation of energy requires that

change in stored energy = sum of input energy

Or,

$$dW = F\,dx + vi\,dt$$

If F_e is the force of electromagnetic origin and acts against F (Fig. 2-6), the above equation can be rewritten as

$$dW = -F_e\,dx + vi\,dt \tag{2.63}$$

2.2.1 *Derivation of the Force Equation*

From the preceding remarks it is seen that energy conversion is possible due to the interchange between electrical and mechanical energy via the coupling fields. This fact in turn leads us to the method of calculating mechanical forces of electromagnetic origin. In a majority of electromechanical devices, the energy stored in the electric field is negligible compared with the energy stored in the magnetic field (see Problem 2-3). In such cases, Eqs. (2.62) and (2.63) yield, for an arbitrary displacement dx,

$$F_e\,dx = -dW_m + i\,d\lambda \tag{2.64}$$

If i and x are considered as independent variables, the flux linkage is given by

$$\lambda = \lambda(i, x)$$

$$d\lambda = \frac{\partial\lambda}{\partial i}\,di + \frac{\partial\lambda}{\partial x}\,dx \tag{2.65}$$

And the magnetic energy is given by the single-valued function

$$W_m = W_m(i, x)$$

and

$$dW_m = \frac{\partial W_m}{\partial i}\,di + \frac{\partial W_m}{\partial x}\,dx \tag{2.66}$$

Therefore, Eqs. (2.65) and (2.66) when substituted in Eq. (2.64) give

$$F_e\,dx = -\frac{\partial W_m}{\partial x}\,dx - \frac{\partial W_m}{\partial i}\,di + i\frac{\partial\lambda}{\partial x}\,dx + i\frac{\partial\lambda}{\partial i}\,di$$

Rearranging terms on the right-hand side,

$$F_e\,dx = \left(-\frac{\partial W_m}{\partial x}+i\frac{\partial \lambda}{\partial x}\right)dx+\left(-\frac{\partial W_m}{\partial i}+i\frac{\partial \lambda}{\partial i}\right)di \tag{2.67}$$

In order for the force to be independent of the change in i during the arbitrary displacement, the coefficients of di in Eq. (2.67) must be zero. Consequently, the force is given by

$$F_e = -\frac{\partial W_m}{\partial x}(i, x)+i\frac{\partial \lambda}{\partial x}(i, x) \tag{2.68}$$

This equation holds if i is the independent variable. If, on the other hand, λ is taken as the independent variable, that is

$$W_m = W_m(\lambda, x)$$

and

$$i = i(\lambda, x)$$

then

$$dW_m = \frac{\partial W_m}{\partial \lambda}d\lambda+\frac{\partial W_m}{\partial x}dx$$

which when substituted in Eq. (2.64) yields

$$F_e\,dx = -\frac{\partial W_m}{\partial x}dx-\frac{\partial W_m}{\partial \lambda}d\lambda+i\,d\lambda$$

But $\int i\,d\lambda = W_m$ from Eq. (2.51), and consequently

$$F_e = -\frac{\partial W_m}{\partial x}(\lambda, x) \tag{2.69}$$

Thus the mechanical force of electromagnetic origin is given by Eq. (2.68) if current is an independent variable; if either the flux linkage or voltage is an independent variable, the force is obtained from Eq. (2.69).

EXAMPLE 2–6

An elementary reluctance machine is shown in Fig. 2E-6. The machine is singly-excited, that is, it carries only one winding on the stator. The exciting winding is wound on the stator and the rotor is free to rotate. The rotor and the stator are shaped so that the variation of the inductance of the windings is sinusoidal with respect to the rotor position. It is readily seen that the space variation of the inductance is of double frequency, that is,

$$L(\theta) = L''+L'\cos 2\theta$$

where the symbols are as defined in Fig. 2E-6. For an excitation

$$i = I_m \sin \omega t$$

determine the instantaneous and average torques. The magnetic energy stored is, from Eq. (2.54),

$$W_m = \frac{1}{2} L(\theta)i^2$$

and the flux linkage is, from Eq. (2.53),

$$\lambda(\theta) = L(\theta)i$$

where i is the independent variable. Therefore, from Eq. (2.68),

$$T_e = -\frac{\partial W_m}{\partial \theta} + i\frac{\partial \lambda}{\partial \theta}$$

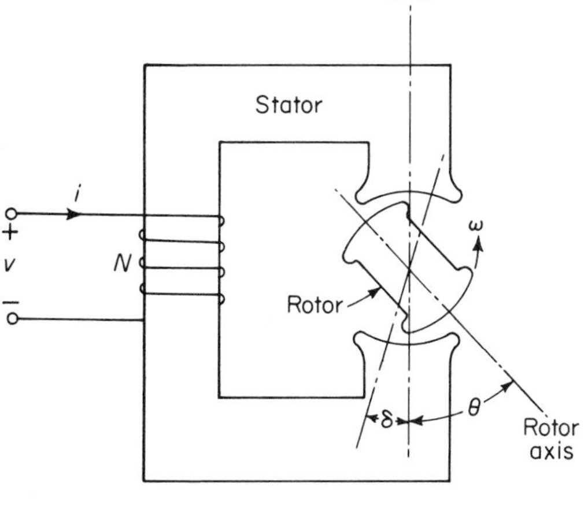

(a)

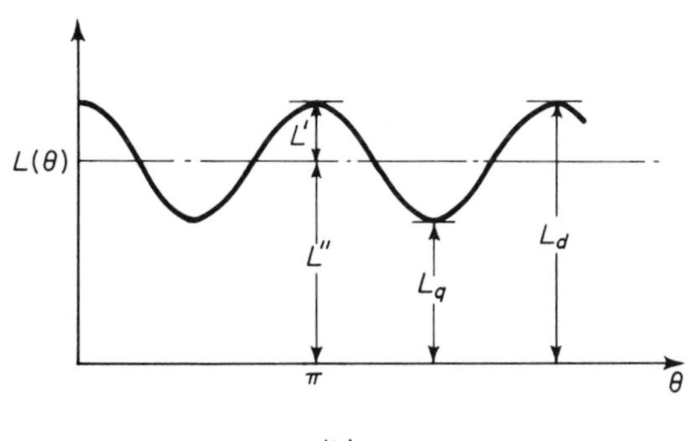

(b)

Fig. 2E-6 (a) An elementary reluctance machine; (b) variation of inductance with rotor position.

Expressing in terms of inductance and current,

$$T_e = -\frac{1}{2} i^2 \frac{\partial L}{\partial \theta} + i^2 \frac{\partial L}{\partial \theta} = \frac{1}{2} i^2 \frac{\partial L}{\partial \theta}$$

For given current and inductance variations,

$$T_e = -I_m^2 L' \sin 2\theta \sin^2 \omega t$$

If the rotor is now allowed to rotate at an angular velocity so that at any instant

$$\theta = \omega' t - \delta$$

(where δ is the rotor position at $t = 0$, when the current i is also zero), then in terms of ω and ω' the expression for instantaneous torque becomes

$$T_e = -\tfrac{1}{2} I_m^2 L' \{\sin 2(\omega' t - \delta) - \tfrac{1}{2}[\sin 2(\omega' t + \omega t - \delta) + \sin 2(\omega' t - \omega t - \delta)]\}$$

In obtaining the above final form we have used the following trigonometric identities:

$$\sin^2 A = \tfrac{1}{2}(1 - \cos 2A)$$

and

$$\sin C \cos D = \tfrac{1}{2} \sin (C + D) + \tfrac{1}{2} \sin (C - D)$$

From the above expression it can be concluded that the time-average torque is zero, since the value of each term integrated over a period is zero. The only case for which the average torque is nonzero is when $\omega = \omega'$ and at this particular frequency the magnitude of the average torque becomes

$$T_{\text{avg}} = \tfrac{1}{4} I_m^2 L' \sin 2\delta$$

Or, from Fig. 2E-6,

$$T_{\text{avg}} = \tfrac{1}{8} I_m^2 (L_d - L_q) \sin 2\delta$$

Thus, for example, at $I_m = 4$ A, $L_d = 0.2$ H (henry), and $L_q = 0.1$ H, the maximum average torque is 0.2 N·m (newton meter).

A number of conclusions can be drawn from the preceding analysis. The machine develops an average torque only at one particular speed, corresponding to the frequency $\omega = \omega'$. This speed is called the *synchronous speed* and the reluctance machine is therefore a synchronous machine. The torque developed by the machine is called the *reluctance torque*, which will evidently be zero if $L_d = L_q$. The torque varies sinusoidally with the angle δ. In other words, δ is a measure of the torque and is called the *torque angle*. The inductances L_d and L_q are the maximum and minimum values of inductance and are called the *direct-axis inductance* and *quadrature-axis inductance* respectively. The maximum torque occurs when $\delta = 45°$, and is called the *pull-out torque*. Any load requiring a torque greater than the maximum torque results in unstable operation of the machine.

EXAMPLE 2–7

In an electromagnetic relay (Fig. 2E-7) the current, flux linkage, and armature position are related to each other by the expression

$$i = \lambda^2 + 2\lambda(x-1)^2, \qquad x<1$$

The force on the plunger is to be evaluated when $x = 0.5$.

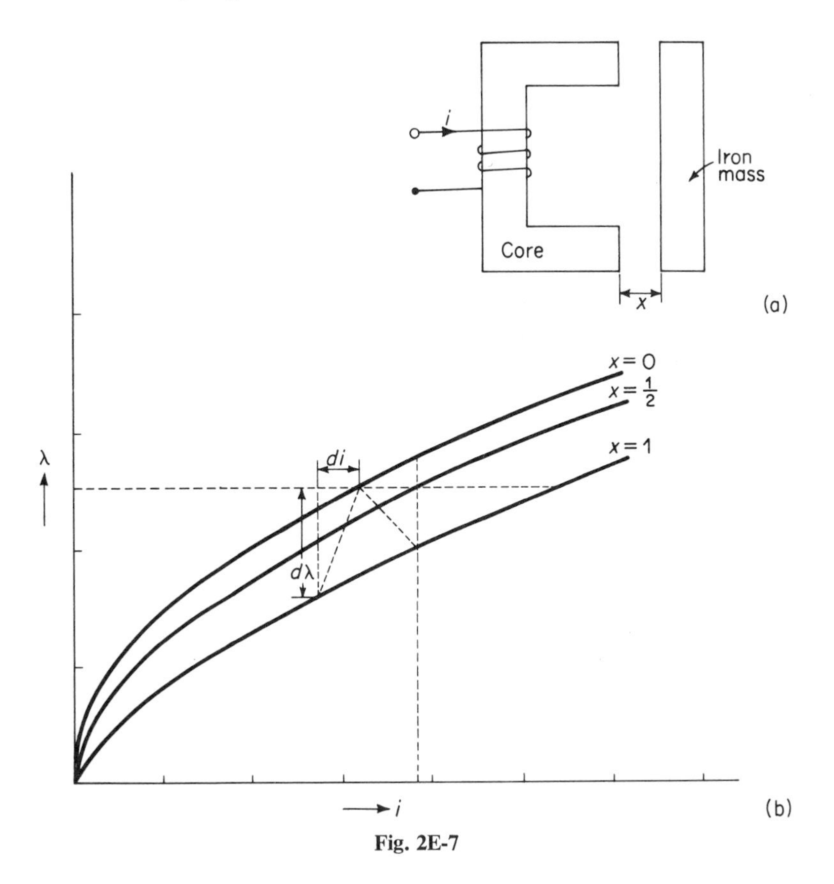

Fig. 2E-7

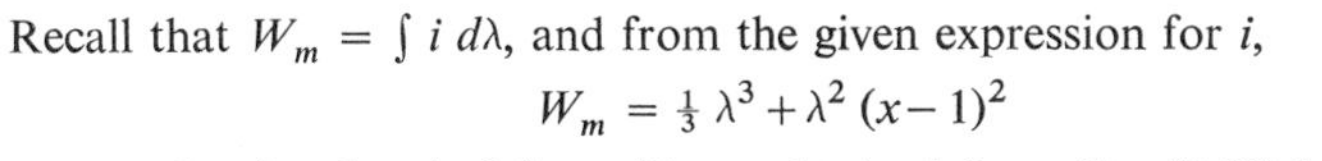

Recall that $W_m = \int i\, d\lambda$, and from the given expression for i,

$$W_m = \tfrac{1}{3}\lambda^3 + \lambda^2 (x-1)^2$$

Consequently, the electrical force F_e, as obtained from Eq. (2.69) becomes

$$F_e = -\frac{\partial W_m}{\partial x} = -2\lambda^2(x-1)$$

At $x = 0.5$ the magnitude of the force is λ^2. If leakages are neglected, this example further illustrates the point that a mechanical force of electrical origin is proportional to the square of applied voltage.

The λ–i relationships for different values of x are shown in Fig. 2E-7(b). Equation (2.70), which is discussed in the next section, is illustrated graphically by Fig. 2E-7(b).

2.2.2 The Concept of Coenergy[11]

The electrical energy supplied at the input terminals of a system is

$$W = \int vi\,dt$$

If the system is conservative and flux linkage is the independent variable, then the energy supplied is stored in the magnetic field and, from Eq. (2.51), this energy is

$$W_m = \int_0^{\lambda} i\,d\lambda' \tag{2.51}$$

When integrated by parts, this yields

$$W_m = i\lambda + \left(-\int_0^{\lambda} \lambda'\,di\right) \tag{2.70}$$

The quantity $\int_0^{\lambda} \lambda'\,di$ is sometimes termed *magnetic coenergy* W_m'. Thus

$$W_m' + W_m = i\lambda = \int_0^{\lambda} \lambda'\,di + \int_0^{\lambda} i\,d\lambda' \tag{2.71}$$

The quantities in Eq. (2.71) are shown graphically in Fig. 2-7. In terms of coenergy the expressions for the force can be written as

$$F_e = \frac{\partial W_m'}{\partial x}(i, x) \tag{2.72}$$

$$F_e = \frac{\partial W_m'}{\partial x}(\lambda, x) - \lambda\frac{\partial i}{\partial x}(\lambda, x) \tag{2.73}$$

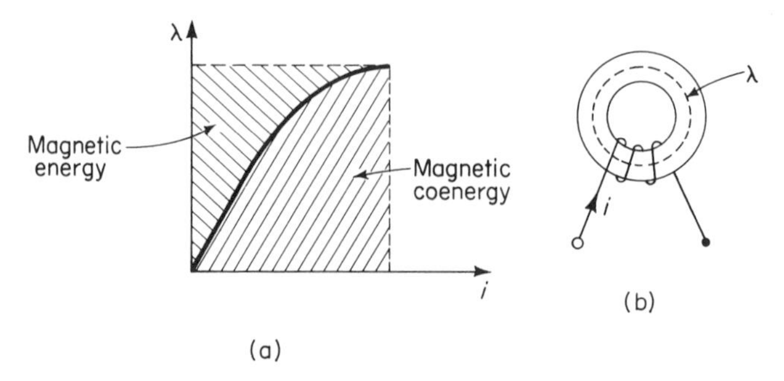

Fig. 2-7 Magnetic energy and coenergy (a) for a circuit (b).

A comparison of Eqs. (2.68) and (2.72) indicates that, with current as the independent variable, the result expressed as Eq. (2.72) is simpler than that given by Eq. (2.68). Evidently, for a linear magnetic circuit energy and coenergy are equal. But if saturation is to be taken into account, it is convenient to use Eq. (2.72) for current-excited systems.

After determining the electrical force F_e, we can write the equation of motion from the force laws of electromechanics. Newton's laws may be used for mechanical motion, while Kirchhoff's laws are applicable to the electrical motion of an electromechanical system.

2.2.3 The Direction of Electrical Force

In preceding sections we derived explicit expressions for a force F_e of electromagnetic origin. For brevity, we shall call this an *electrical force*. In terms of magnetic energy and coenergy the expressions for electrical force are given by Eqs. (2.68), (2.69), (2.72), and (2.73). Particular consideration should be given to Eqs. (2.68) and (2.69):

$$F_e = -\frac{\partial W_m}{\partial x}(i, x) + i\frac{\partial \lambda}{\partial x}(i, x) \tag{2.68}$$

and

$$F_e = -\frac{\partial W_m}{\partial x}(\lambda, x) \tag{2.69}$$

Notice that in Eq. (2.68) the current i is the independent variable, whereas in Eq. (2.69) the linkage λ (or the induced voltage) is the independent variable. By designating i or λ as independent variables we mean that during motion the current or voltage is held constant, as the case may be. We shall illustrate our point by means of the following example, which also indicates the direction of an electrical force and possible motion.

EXAMPLE 2–8

An electromagnetic relay is shown in Fig. 2E-8. If the inductance of the exciting coil is a function of position, that is, if $L = L(x)$, we wish to determine the electrical force for (1) current excitation, and (2) voltage excitation. We assume no saturation of the magnetic circuit and neglect the resistance of the coil.

First, we shall use Eq. (2.68). We know that

$$W_m = \tfrac{1}{2} i^2 L(x)$$

thus

$$\frac{\partial W_m}{\partial x} = \frac{1}{2} i^2 \frac{\partial L}{\partial x} \tag{a}$$

and, since $\lambda = iL(x)$,

$$\frac{\partial \lambda}{\partial x} = i\frac{\partial L}{\partial x} \tag{b}$$

Substituting Eqs. (a) and (b) in Eq. (2.68), we have

$$F_e = -\frac{1}{2} i^2 \frac{\partial L}{\partial x} + i^2 \frac{\partial L}{\partial x} = \frac{1}{2} i^2 \frac{\partial L}{\partial x} \tag{c}$$

In the above we have considered $\partial i/\partial x = 0$, and thereby a current-excited system.

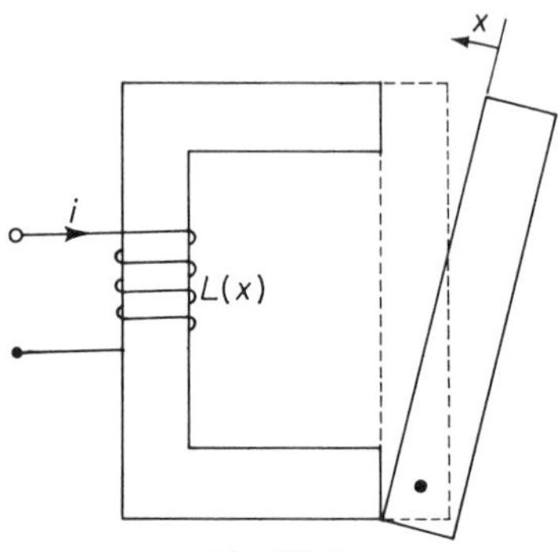

Fig. 2E-8

Next, we consider a voltage-excited system and use Eq. (2.69). In terms of λ and x, the magnetic energy can be expressed as

$$W_m = \frac{1}{2} \frac{\lambda^2}{L(x)} \tag{d}$$

Substituting Eq. (d) in Eq. (2.69), and keeping $\partial\lambda/\partial x = 0$, we have

$$F_e = \frac{1}{2} \frac{\lambda^2}{L^2} \frac{\partial L}{\partial x} \tag{e}$$

Since $\lambda = Li$, Eq. (e) can also be written as

$$F_e = \frac{1}{2} i^2 \frac{\partial L}{\partial x} \tag{f}$$

Notice that Eqs. (c) and (f) are identical. Thus, it must be emphasized that in order to obtain the correct expression for force, Eq. (2.68) must be used in its complete form.

In the study of energy converters, however, it is helpful to use physical reasoning in addition to the mathematical expression. One useful concept is the idea of alignment of magnetic fields—that is, that force or torque is exerted between two current-carrying coupled coils (if there are two coils) so as to align the resulting magnetic fields. From this last example we conclude that the force on the armature is in the direction to reduce net magnetic reluctance, or to minimize stored energy.

EXAMPLE 2–9

We now wish to consider a parallel-plate capacitor filled with a material of dielectric constant ϵ. The area of the plates is A and they are separated by some (variable) distance x. The electrical force exerted between the plates is to be found (1) for a given voltage v, and (2) for a given charge q on the plates.

The results of Sec. 2.2.1 can be easily extended to electrostatic devices in which the energy stored is in the electric field rather than in the magnetic field. In such cases, the stored electrical energy W_e is given by

$$W_e = \int_0^q v\, dq \tag{2.70}$$

If the voltage v is considered the independent variable, the electrical force F_e is given by

$$F_e = -\frac{\partial W_e}{\partial x}(v, x) + v\frac{\partial q}{\partial x}(v, x) \tag{2.71}$$

On the other hand, if the charge q is the independent variable, the electrical force is obtained from

$$F_e = -\frac{\partial W_e}{\partial x}(q, x) \tag{2.72}$$

For the example under consideration,

$$C(x) = \frac{\epsilon A}{x} \tag{a}$$

$$W_e(v, x) = \tfrac{1}{2} Cv^2 \tag{b}$$

and

$$q = Cv \tag{c}$$

When Eqs. (a)–(c) are substituted in Eq. (2.71),

$$F_e = \frac{1}{2} v^2 \frac{\partial C}{\partial x} \tag{d}$$

In terms of q, the stored energy is expressed as

$$W_e = \frac{1}{2}\frac{q^2}{C} \tag{e}$$

Equations (c), (e), and (2.72) yield

$$F_e = \frac{1}{2}\left(\frac{q}{C}\right)^2 \frac{\partial C}{\partial x} = \frac{1}{2} v^2 \frac{\partial C}{\partial x} \tag{f}$$

Notice that Eqs. (d) and (f) are identical, a result which should not be surprising.

2.3 *Variational Principles*[12, 13]

So far we have considered two different methods of obtaining forces of electromagnetic origin in energy-conversion devices. The third method of determining forces—and at the same time formulating the equations of motion—is through the applications of variational principles. This approach has the basic advantage that forces developed internally are automatically taken into account, thus making "bookkeeping" easier. It has the slight disadvantage that we tend to formulate the equations of motion in a routine manner and in so doing are apt to neglect the physical causes and effects of forces.

The key to the development of the variational method, as applied to dynamical systems, is the concept of *generalized coordinates.* We shall first introduce generalized coordinates and then state *Hamilton's principle,* from which we shall derive *Lagrange's equation.* Only the basic results will be given here. Some derivations are given in Appendix A and further details are available in References 12 and 13.

2.3.1 *Generalized Coordinates, Hamilton's Principle, and Lagrange's Equation*

The state of a dynamical system can be specified by means of a finite number of quantities which vary when the state of the system changes. The quantities, denoted by $q_1, q_2, \ldots, q_n$, are called *generalized coordinates.* An additional coordinate of time t has to be added to account for moving constraints. The quantities $\dot{q}_1, \dot{q}_2, \ldots, \dot{q}_n$, which are time derivatives of the generalized coordinates, are called *generalized velocities.* For some cases, it is convenient to introduce a new set of variables $p_1, p_2, \ldots, p_n$, called *generalized momenta.* The variables $q_1, q_2, \ldots, q_n$ and $p_1, p_2, \ldots, p_n$ are called *state variables.* The variables $q_1, q_2, \ldots, q_n$ and $\dot{q}_1, \dot{q}_2, \ldots, \dot{q}_n$ form an alternative set of state variables. It should be noted that the generalized coordinates need not necessarily be mechanical displacements, or any quantity similar to them. A function of state variables (formed from set rules) is known as a *state function.* Having introduced the concept of generalized coordinates and state function, we can now state *Hamilton's principle* as follows:

For a conservative system, the actual motion of a system whose state function is $\mathscr{L}(q_k; \dot{q}_k; t; k = 1, 2, \ldots, n)$ is such as to render the integral

$$\mathscr{I} = \int_{t_1}^{t_2} \mathscr{L}(q_k; \dot{q}_k; t)\, dt, \qquad k = 1, 2, \ldots, n \tag{2.73}$$

between fixed points $q_k(t_1)$ and $q_k(t_2)$, where t_1 and t_2 are two arbitrary instants of time, an extremum (maximum or minimum) with respect to the continuously twice-differentiable function $q_k(t)$, $k = 1, 2, \ldots, n$, for which $q_k(t_1)$ and $q_k(t_2)$ are prescribed for all k's; and $\mathscr{L}$, known as the *Lagrangian*, is defined as $\mathscr{L} = \mathscr{T} - \mathscr{V}$, where $\mathscr{T}$ is the kinetic energy and $\mathscr{V}$ is the potential energy.

For the integral given by Eq. (2.73) to be an extremum, the following necessary condition must be fulfilled:

$$\frac{\partial \mathscr{L}}{\partial q_k} - \frac{d}{dt}\left(\frac{\partial \mathscr{L}}{\partial \dot{q}_k}\right) = 0, \qquad k = 1, 2, \ldots, n \tag{2.74}$$

Equation (2.74) is called *Lagrange's equation*. Therefore, if the Lagrangian for the system is known, Eq. (2.74) gives the dynamical path for the system. (See Appendix A for derivation.)

To include the effects of dissipative forces, a velocity-dependent function $\mathscr{F}$ is used, and the modified Lagrange's equation takes the form

$$\frac{d}{dt}\left(\frac{\partial \mathscr{L}}{\partial \dot{q}_k}\right) - \frac{\partial \mathscr{L}}{\partial q_k} + \frac{\partial \mathscr{F}}{\partial \dot{q}_k} = Q_k, \qquad k = 1, 2, \ldots, n \tag{2.75}$$

where $Q_k = Q_k(t)$ is the nonconservative force acting on a coordinate q_k along with the conservative forces. It should be noted that $\mathscr{F}$ is not needed at all if a part of Q_k is represented in the form $\partial\mathscr{F}/\partial\dot{q}_k$, if so desired. The function $\mathscr{F}$, which may represent dissipative forces, includes the kind introduced by Rayleigh. There are, however, some kinds of velocity-dependent forces which cannot be represented as $\partial\mathscr{F}/\partial\dot{q}_k$.

To summarize, the essential steps involved in using Lagrange's equation to derive the equations of motion of an electromechanical system are: (1) choose the generalized coordinates; (2) determine the nonconservative forces Q_k that arise from sources such as mechanical forces, voltages, or currents, and $\partial\mathscr{F}/\partial\dot{q}_k$ that arise from losses; (3) determine the nonconservative forces associated with dissipation and obtain the dissipation function $\mathscr{F}$; and (4) obtain the conservative Lagrangian $\mathscr{L}$ and substitute in Eq. (2.75).

2.3.2 A Choice of Generalized Coordinates

The first step in the Lagrangian formulation is to choose generalized coordinates. For electromechanical systems a convenient set is as follows:

	Mechanical Coordinates	*Electrical Coordinates*
q_k (generalized displacement)	Mechanical displacement: linear x_k or angular θ_k	Charge q_k
$\dot{q}_k$ (generalized velocity)	Mechanical velocity: linear $\dot{x}_k$ or angular $\dot{\theta}_k$	Current i_k
F_k (generalized force)	Mechanical force F_k or torque T_k	Voltage v_k
p_k (generalized momentum)	Mechanical momentum: linear or angular	Flux linkage λ_k

For the above choice of generalized coordinates the energy stored in the magnetic field W_m is identified as the kinetic energy. This definition of coordinates corresponds to the *force–voltage analogy* of circuit theory. If, on the other hand, the energy stored in the electric field is considered as the kinetic energy, the *force–current analogy* results.

Thus far we have attempted to review Lagrange's equation, choose a set of generalized coordinates, and obtain the Lagrangian state function $\mathscr{L}$. Now these ideas are applied to an example.

EXAMPLE 2–10

An electromagnet (Fig. 2E-10) carrying a current i lifts a mass M. The resistance of the coil is R and its inductance is $L(x)$. The equations of motion are to be written using Lagrangian formulation.

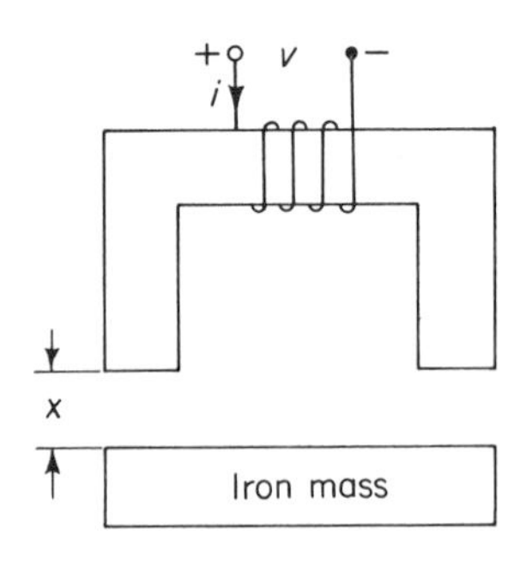

Fig. 2E-10

Using the nomenclature of Sec. 2.3.1,

$$\mathscr{T} = \tfrac{1}{2} Li^2 + \tfrac{1}{2} M\dot{x}^2$$

$$\mathscr{V} = 0$$

$$\mathscr{L} = \mathscr{T} - \mathscr{V} = \tfrac{1}{2} Li^2 + \tfrac{1}{2} M\dot{x}^2$$

$$\mathscr{F} = \tfrac{1}{2} Ri^2$$

The equations of motion are obtained by directly substituting the above quantities in Eq. (2.75). Therefore, for $k = 1$, and $q_1 = q$, the electrical equation is

$$\frac{d}{dt}\frac{\partial \mathscr{L}}{\partial i} + \frac{\partial \mathscr{F}}{\partial i} = v$$

Or

$$\frac{d}{dt}(Li) + Ri = L\frac{di}{dt} + i\frac{dL}{dt} + Ri = v$$

And for $k = 2$, $q_2 = x$, the mechanical equation is

$$\frac{d}{dt}\frac{\partial \mathscr{L}}{\partial \dot{x}} - \frac{\partial \mathscr{L}}{\partial x} + \frac{\partial \mathscr{F}}{\partial \dot{x}} = 0$$

Or,

$$\frac{d}{dt}(M\dot{x})-\frac{1}{2}i^2\frac{\partial L}{\partial x} = 0$$

Or,

$$M\ddot{x} = \frac{1}{2}i^2\frac{\partial L}{\partial x}$$

The term on the right-hand side of this equation is the force of electromagnetic origin. Note that the direction and magnitude of the force are automatically and correctly taken into account in the above procedure, thereby making bookkeeping easier. Notice also that the direction and magnitude of the electrical force in this example are the same as given by Eq. (c), Example 2-8.

2.4 Formulation of Equations of Motion

The equation which yields the electrical behavior of a system is known as an *electrical equation of motion*. This is essentially a voltage-balance or current-balance equation. Thus, for a 2-port system the electrical equation of motion is written as

$$v = Ri+\frac{d}{dt}(Li) \tag{2.76}$$

For a multiport system, the form of this equation remains the same except that the voltages and currents are denoted as column matrices and the system parameters R and L are written down as square matrices. As we shall see later, it is advantageous to express the equations of motion in matrix notation.

The mechanical characteristics of a system are given by a *mechanical equation of motion*, which generally turns out to be a force-balance equation. Thus, for linear motion the mechanical equation of motion is written as

$$M\frac{d^2x}{dt^2}+b\frac{dx}{dt}+kx = F_{\text{applied}} \tag{2.77}$$

In this equation, F_{applied} includes the forces of electromagnetic origin also, and the coefficients b and k are assumed to be constants although they need not necessarily be so.

Consideration of Eqs. (2.76) and (2.77) shows that equations of motion can be obtained by equating the "applied forces" to the "restoring forces". In the electrical equation of motion of an energy-conversion device, electrical forces of mechanical origin (such as a voltage induced by motion) are obtained from Faraday's law. Mechanical forces of electrical origin are determined using the force equations derived earlier. Of course, as Example

2-10 illustrates, application of Lagrange's equation directly yields equations of motion, automatically taking the various forces into account.

Equations of motion generally fall into one of the following categories:

1. linear differential equations with constant coefficients;
2. linear differential equations with variable coefficients;
3. nonlinear differential equations.

To these may be added partial differential equations if field analysis is involved.

There is no general method for solving every type equation of motion. However, special techniques by which solutions are obtained are outlined in the following section.

2.5 Solution of Equations of Motion

The method for solving a dynamic equation of motion depends upon the type equation to be solved. The linear differential equation, being the simplest, is discussed first.

*2.5.1 The Laplace Transform

Besides the classical method of solving linear differential equations with constant coefficients, the other very powerful and useful method is the Laplace transform method.

Laplace transformation is an integral operation. It transforms a linear differential equation into a sort of algebraic equation obeying the general rules of algebra. To obtain the desired solution of the differential equation, the inverse Laplace transform of the algebraic equation is taken. Thus, the Laplace transformation gives the solution of the equation directly. Moreover, it can account for initial conditions very conveniently.

For our needs, only some pertinent results of Laplace transform theory are presented. Details are available in References 14 and 15.

The *Laplace transform* $F(s)$ of a function $f(t)$ is defined as

$$\mathscr{L}[f(t)] = F(s) = \int_0^\infty f(t)e^{-st}\,dt, \qquad t > 0 \tag{2.78}$$

An alternative form of definition is

$$f(t) = \frac{1}{2\pi j}\int_{c-j\infty}^{c+j\infty} F(s)e^{ts}\,ds, \qquad t>0 \tag{2.79}$$

In both equations s is complex (of the form $\sigma+j\omega$) and the integrals are assumed to exist. Equation (2.78) is the direct (unilateral) Laplace transform and expresses $F(s)$ in terms of $f(t)$. In Eq. (2.79) the time function $f(t)$ is expressed in terms of its Laplace transform $F(s)$, and can be called the *inverse transform*. Although Eq. (2.79) is seldom used in the present work, it should be noted that c is the abscissa of absolute convergence and is greater than the real part of all singularities of $F(s)$. (A book on complex variable theory such as Reference 16 may be consulted to obtain definitions of terms unfamiliar to the reader.) Notice that the Laplace transform *transforms* a function of time $f(t)$ to a function of a complex variable $F(s)$. The transformation complex number is generally called a *complex frequency*.

The Laplace Transform of Differential Coefficients. The Laplace transform of the nth derivative of a function $f(t)$ is given by

$$\mathscr{L}[f^n(t)] = s^n F(s) - s^{n-1} f(0+) - s^{n-2} f^1(0+) - \ldots - f^{n-1}(0+) \quad (2.80a)$$

where the superscript on f denotes the order of differentiation, the superscript on s is the power to which it is raised, and $(0+)$ denotes the time at the origin approached through positive values.

From the above theorem it follows that

$$\mathscr{L}[f'(t)] = sF(s) - f(0+) \quad (2.80b)$$

and

$$\mathscr{L}[f''(t)] = s^2 F(s) - sf(0+) - f'(0+) \quad (2.80c)$$

The Laplace transform of an integral is expressed as

$$\mathscr{L}\left[\int_{-\infty}^{t} f(t)\,dt\right] = \frac{1}{s} F(s) + \frac{1}{s} \int_{-\infty}^{t} f(t)\,dt \,\Big|_{t=0+} \quad (2.81)$$

	Time Function $f(t)$	Laplace Transform $F(s)$
Exponential	e^{-at}, $t>0$	$1/(s+a)$
Unit impulse	$\delta(t)$	1
Unit step	$U(t)$	$1/s$
Unit ramp	t	$1/s^2$
Sine	$\sin kt$	$k/(s^2+k^2)$
Cosine	$\cos kt$	$s/(s^2+k^2)$

The above results are now used to illustrate its application to an example.

EXAMPLE 2–11

The following equation is to be solved by the Laplace transform method:

$$f'(t) + f(t) = 100 \quad \text{with} \quad f(0+) = 0$$

Taking the Laplace transform of both sides and rearranging terms gives

$$F(s) = \frac{100}{s(s+1)} = \frac{100}{s} - \frac{100}{s+1}$$

Taking the inverse transform of $F(s)$ gives the required time function

$$f(t) = 100\,(1 - e^{-t})$$

*2.5.2 The Input–output Relationship

A system is often characterized by its input–output relationship. That is, after the input and the output ports have been identified, certain test signals are given at the input port and the output is observed. The output is characteristic of the system. Test inputs need not always be physically realizable and may only be mathematical concepts; nor is it necessary that the test signal actually be given to the system, nor the output actually observed. In any event, the most commonly used and physically realizable test inputs suitable for a large class of electromechanical systems are the step and sinusoidal inputs.

The response of a system to a step input is called the *transient* or *step response*. A step function is sketched in Fig. 2-8(a) and a typical response due to a step input to a system is shown in Fig. 2-8(b), in which the following

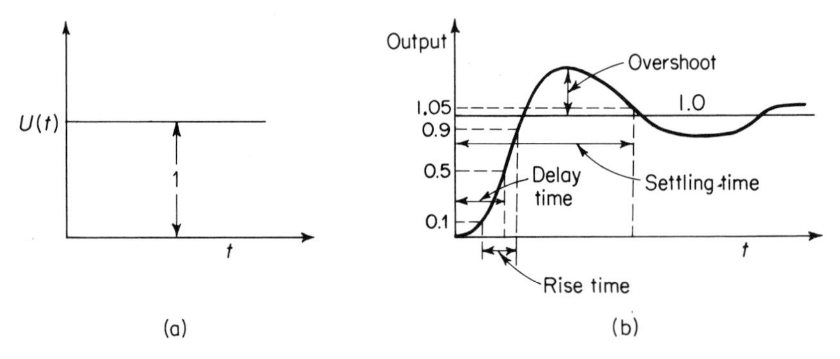

Fig. 2-8 (a) A unit step function; (b) a typical step response.

quantities are indicated: overshoot, delay time, rise time, and settling time.[17] These are some of the basic quantities which indicate the suitability of a given system for a particular job. All these quantities depend on the physical parameters of the system.

The response to sinusoidal inputs is called the *frequency response*. In relation to systems, it refers to the magnitude and phase relationships between the input and output, where the time variation of the input is sinusoidal and the frequency ranges from zero to infinity.

In contrast to step response, frequency response describes the steady-state characteristics of the system. As expected, frequency response also depends upon system parameters, and is of great aid in characterizing a system.

*2.5.3 *Transfer Functions*

Having introduced various methods of characterizing a linear system, we can further extend the technique of studying such systems by investigating the characteristic equation and the *transfer function* of a system. A system transfer function and characteristic equation both describe the natural behavior of that system, and it turns out that the transfer function of a system is the same as the transform of its *impulse response*, that is, its response to an impulse input.

In the simplest case, a linear system can be considered as a black box with an input and an output, as shown in Fig. 2-9. The input and output can be related by means of a linear differential equation. Another way of relating

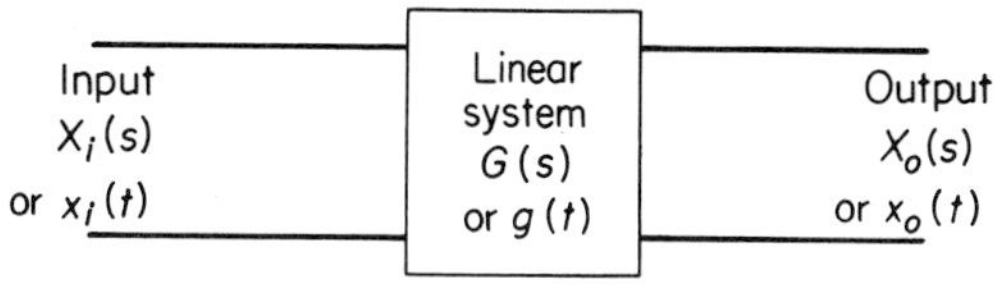

Input $X_i(s)$ or $x_i(t)$

Linear system $G(s)$ or $g(t)$

Output $X_o(s)$ or $x_o(t)$

Fig. 2-9 A linear system as a black box.

them is by taking the Laplace transform of the differential equation and expressing the input–output relationship as a function of complex frequency s. Therefore, a transfer function can also be defined as the ratio of the transform of the output to the transform of the input with all initial conditions zero:

$$\text{transfer function} \quad G(s) = \frac{\mathscr{L}_{\text{output}}}{\mathscr{L}_{\text{input}}} = \frac{X_o(s)}{X_i(s)} \tag{2.82}$$

For a complex but linear system, a transfer function can be defined for each part and the parts can be reconnected in the prescribed manner to make up the entire system. This will be illustrated when complex systems are considered, but for the present rather a simple example is considered.

EXAMPLE 2–12

Consider a series RLC (resistance-inductance-capacitance) circuit such as shown in Fig. 2E-12. Defining the current i as output and the voltage v as input, the transfer function for the system is to be obtained.

The voltage-balance equation is

$$L\frac{di}{dt}+Ri+\frac{1}{C}\int i\,dt = v$$

Taking Laplace transforms of both sides of this equation yields

$$sLI(s)+RI(s)+\frac{1}{Cs}I(s) = V(s) \quad \text{(a)}$$

Notice that in the definition of the transfer function, the initial conditions are not included.

Equation (a) gives the transfer function as

$$G(s) = \frac{I(s)}{V(s)} = \frac{1}{Ls+R+1/Cs} \quad \text{(b)}$$

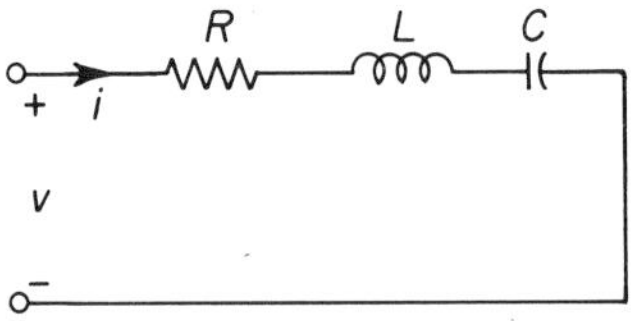

R L C
+ i
v
−

Fig. 2E-12

It is also seen from Eq. (b) that the ratio of voltage to current for sinusoidal excitation is the same as the "impedance" of the circuit, and in this case $s = j\omega$.

Equation (b) can be rewritten as

$$G(s) = \frac{s/L}{s^2+(R/L)s+1/LC} = \frac{s/L}{s^2+2\zeta\omega_n s+\omega_n^2} \quad \text{(c)}$$

where

$$\omega_n = \frac{1}{\sqrt{LC}} = \textit{natural frequency} \text{ of oscillation}$$

$$\zeta = \frac{1}{2}R\sqrt{\frac{C}{L}} = \textit{damping ratio}$$

The parameters ω_n and ζ are often used to replace the actual system parameters, appropriately combined, as the former are more expressive of the system transient behavior. For example, overshoot can be expressed as a function of the damping ratio. For a general second-order system having a transfer function of the form

$$G(s) = \frac{\omega_n^2}{s^2+2\zeta\omega_n s+\omega_n^2}$$

normalized curves are available in standard textbooks[22] which give the quantitative information about the response of the system to various test inputs.

*2.5.4 Asymptotic Approximation of Frequency Response

It was noted in Example 2-12 that for a sinusoidal excitation, the s in the transfer function is replaced by $j\omega$. If $KG(s)$ is the transfer function of a system, then the plot of the log of the magnitude of $KG(j\omega)$, expressed in decibels, dB $= 20 \log_{10}|KG(j\omega)|$, as a function of the log of the frequency ω, is called the *frequency response* or *Bode plot*. Sometimes this plot is also meant to include the phase of $KG(j\omega)$. Only the plots of the amplitude are included here.

The log of the magnitude of $KG(j\omega)$ in dB, also known as the *log modulus* Lm of $KG(j\omega)$, is

$$\text{Lm} = 20 \log_{10}|KG(j\omega)| = 20 \log_{10}K + 20 \log_{10}|G(j\omega)| \tag{2.83a}$$

Or

$$\text{Lm} = \text{Lm } K + \text{Lm } G(j\omega) \tag{2.83b}$$

If the system is composed of several transfer functions such that

$$KG = \frac{K_1 G_1 K_2 G_2}{K_3 G_3}$$

with $j\omega$ understood, then

$$\text{Lm } (KG) = \text{Lm } (K_1) + \text{Lm } (K_2) - \text{Lm } (K_3) + \text{Lm } (G_1) + \text{Lm } (G_2) - \text{Lm } (G_3)$$

It is very convenient to approximate the frequency-response curve by means of asymptotes. It can be shown that a change of ± 20 dB per decade change in the frequency occurs in the Lm versus ω plot, for every time constant present in the system transfer function. This is best illustrated by considering several examples. In each case, the behaviors at extremely low frequencies and at extremely high frequencies are required to determine the entire frequency-response curve.

EXAMPLE 2–13

The asymptotic approximation of frequency response is to be obtained for the following transfer functions: (a) $G(s) = 1/s$; (b) $G(s) = s$; (c) $G(s) = 1/(s+1)$; (d) $G(s) = (s+1)$; (e) $G(s) = (s+1)/(s+10)$.

$$G(s) = \frac{1}{s} \tag{a}$$

Or,

$$|G(j\omega)| = \left|\frac{1}{j\omega}\right| = \frac{1}{\omega}$$

Or,

$$\text{Lm}(G) = 20 \log_{10} 1 - 20 \log_{10} \omega$$
$$= -20 \log_{10} \omega$$

Now the following table is made:

ω	Lm $G(j\omega)$
0	0
10	−20
100	−40

and so on...

Thus the plot shown in Fig. 2E-13(a) is obtained hereby.

$$G(s) = s \tag{b}$$

Or

$$|G(j\omega)| = |j\omega| = \omega$$

Following exactly the same procedure as for transfer function (a), the curve for (b), shown in Fig. 2E-13(b), is obtained.

$$G(s) = \frac{1}{s+1} \tag{c}$$

Or

$$G(j\omega) = \frac{1}{j\omega+1} \quad \text{or} \quad \text{Lm}\, G = -\text{Lm}\,(j\omega+1)$$

The two extremes of frequencies are considered. When ω is small compared with unity, the modulus of $G(j\omega)$ is unity, for which the Lm is zero [Fig. 2E-13(c)]. When ω is large compared with unity, $G(j\omega)$ approaches $1/j\omega$, for which the Lm has the same form as for transfer function (a). In this way the slopes of the two asymptotes are determined. Evidently, the two intersect at a frequency for which $1 = 1/|j\omega| = 1/\omega$. This frequency is known as the *break* or *corner frequency*. Thus the locations of the two asymptotes are uniquely determined. These asymptotes are shown in Fig. 2E-13(c).

$$G(s) = (s+1) \tag{d}$$

Or

$$G(j\omega) = j\omega+1$$

Or

$$\text{Lm}\,(G) = \text{Lm}\,(j\omega+1)$$

Following exactly the same procedure as for transfer function (c), the curve shown in Fig. 2E-13(d) is obtained.

$$G(s) = \frac{s+1}{s+10} \tag{e}$$

This is composed of two transfer functions: $G(s) = G_1(s)/G_2(s)$, where $G_1(s) = (s+1)$ and $G_2(s) = 1/(s+10)$. The plot of $G_1(s)$ has been discussed in part (d), whereas the plot of $G_2(s)$ is similar to that in part (c), except that in this case the break frequency is 10. The plots for $G_1(s)$ and $G_2(s)$ are then combined and the curve shown in Fig. 2E-13(e) is obtained.

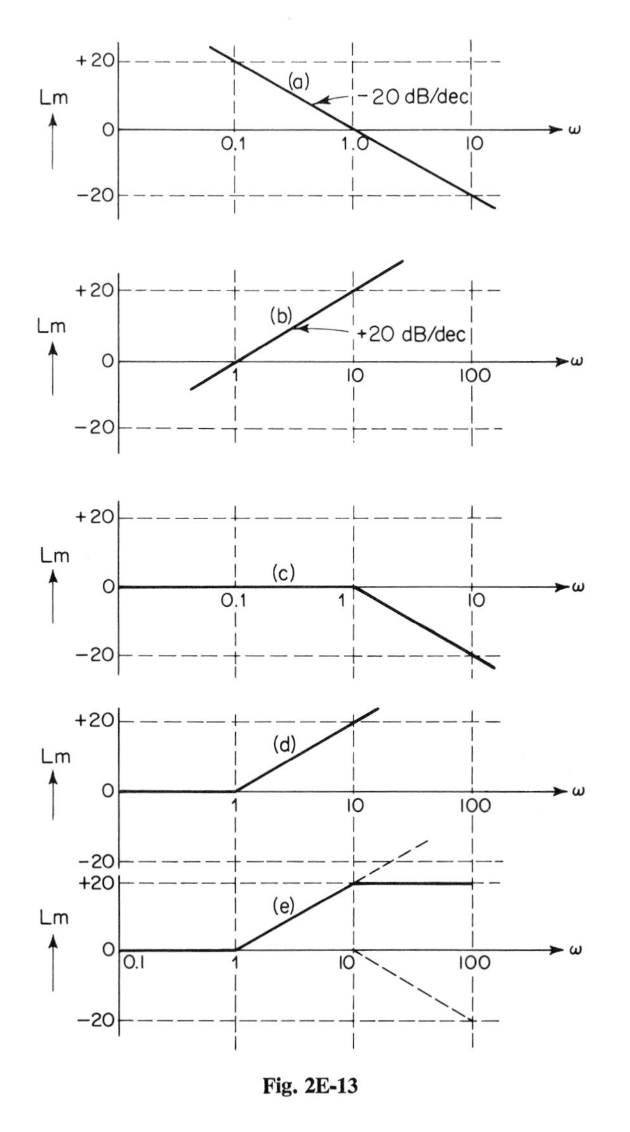

Fig. 2E-13

The difference between the actual curve and its asymptotic approximation can be seen by calculating the Lm at the corner frequency. For example, in part (c), when $\omega = 1$, the magnitude of $G(j\omega) = (\sqrt{2})^{-1}$, which when expressed as a log modulus value equals -3 dB. The deviation at other frequencies can be calculated similarly.

2.5.5 Block Diagrams

An alternative way of representing the equations describing a system is by means of block diagrams. Block-diagram algebra has been developed[18] to facilitate the simplification of complex systems of block diagrams. They can be used to represent pictorially the signal flow in a linear system, as well as the location of various elements, inputs, disturbances, and outputs. With some modifications, certain types of nonlinearities can also be represented. Indeed, their use in analog-computer representation of systems is one of the most important applications of block diagrams.

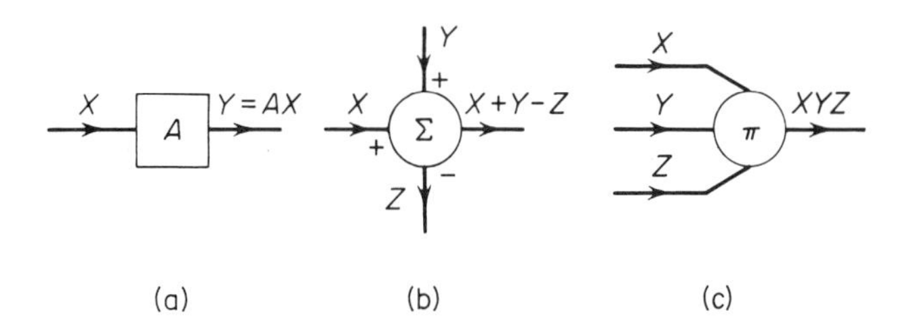

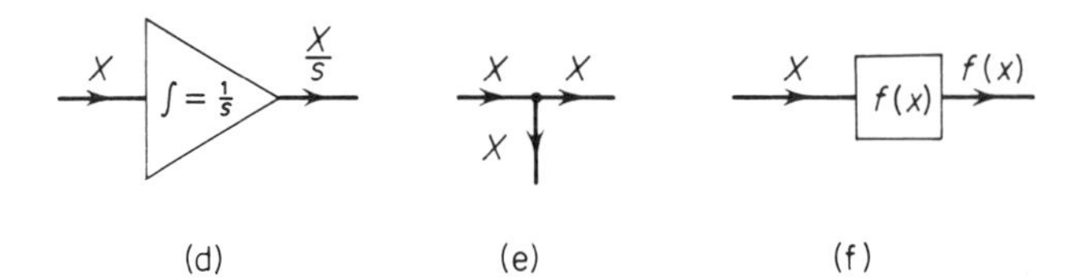

Fig. 2-10 Symbolic representation of various circuits: (a) multiplication by a constant; (b) addition and subtraction; (c) multiplication by a constant; (d) integration; (e) pick-off (or identity); (f) function generation.

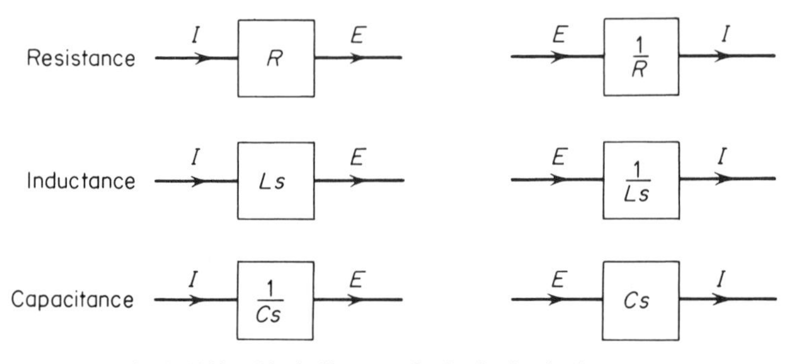

Fig. 2-11(a) Block diagrams for basic circuit elements.

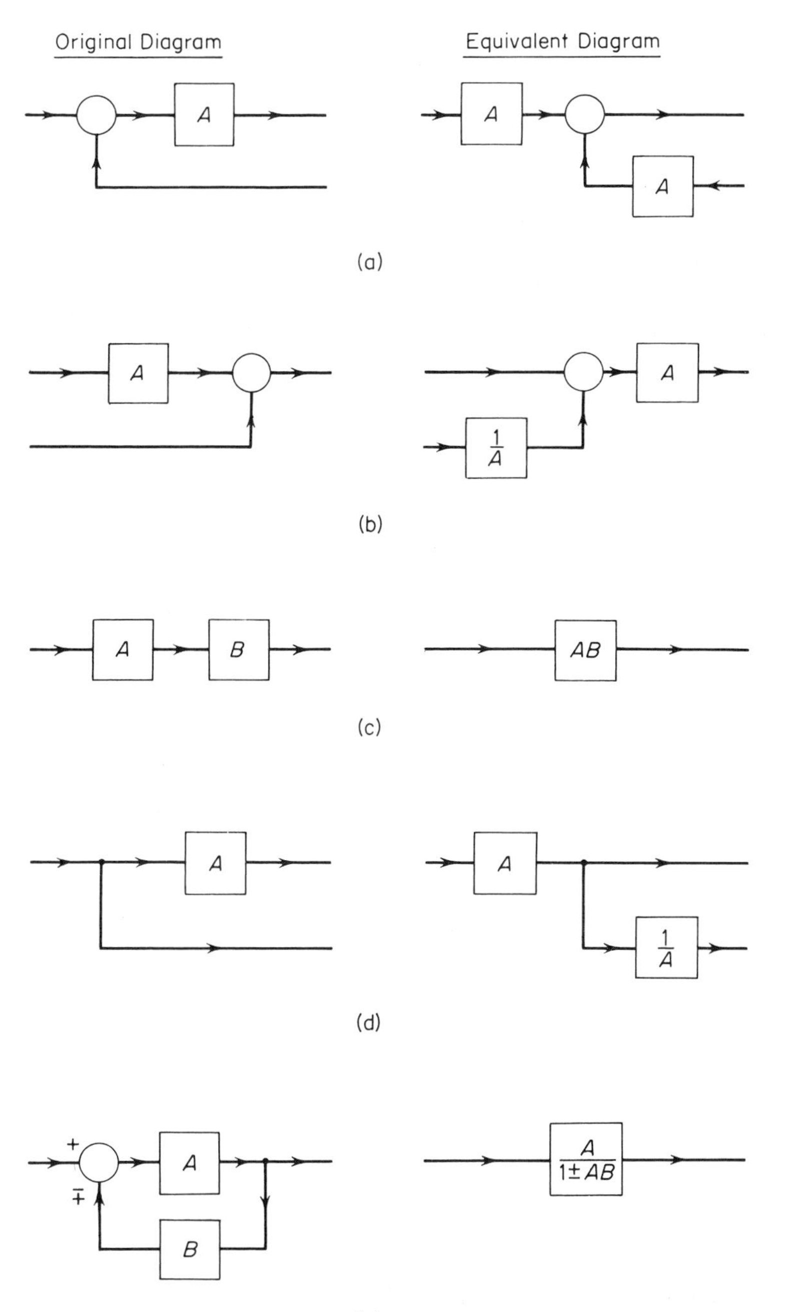

Fig. 2-11(b) Rules for block diagram transformations.

With one restriction in mind—namely, the block diagrams must be considered unilateral—a summary of nomenclature and block-diagram algebra is given on p. 55. Figure 2-10 explains the essential symbols. A set of basic relationships involving the fundamental circuit elements is shown in Fig. 2-11 in block-diagram notation.

EXAMPLE 2–14

The transfer function for the network shown in Fig. 2E-14(a) is to be obtained using the block-diagram method and block-diagram algebra.

The block diagram in Fig. 2E-14(b) is developed from the following equations:

$$i = \frac{e_1 - e_o}{R_1} \quad \text{and} \quad e_o = R_2 i$$

The diagram thus obtained is reduced to the final form of Fig. 2E-14(c) by using the rules tabulated in Fig. 2-11(b).

It can be easily seen that a block diagram for a given circuit is not unique.

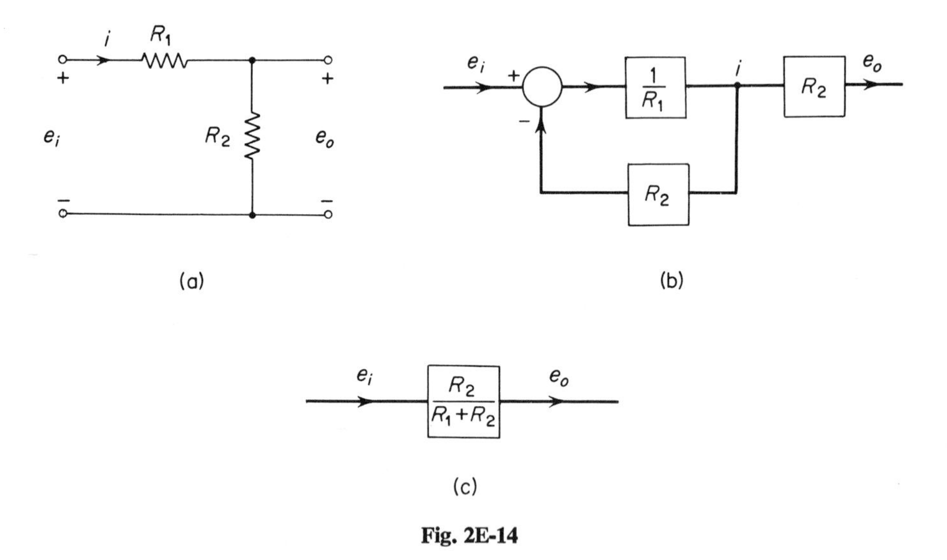

Fig. 2E-14

2.5.6 *Equivalent Circuits*

Electromechanical systems can be represented purely by electrical circuits called *equivalent circuits*. Equivalent-circuit representation is extremely useful in analyzing a system. The approach for obtaining the equivalent circuit of a physical system is best understood by means of an example.

EXAMPLE 2–15

A purely mechanical system consisting of a spring, a mass, and a damper [Fig. 2E-15(a)] is considered, where the damping force is assumed to be directly proportional to the velocity. If F_{ext} is an applied external force, then from Newton's laws the force-balance equation is written as

$$M\frac{d^2x}{dt^2} + b\frac{dx}{dt} + kx = F_{ext}$$

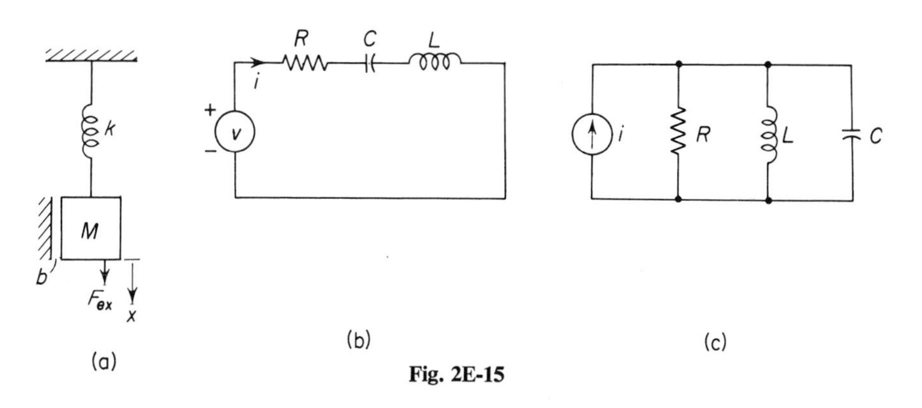

Fig. 2E-15

If the force F_{ext} is assumed to be analogous to voltage v, then the network of Fig. 2E-15(b) can be drawn from the following correspondences:

Force–Voltage Analogy

Force F	Voltage V
Velocity $\dot{x}$	Current i
Damping b	Resistance R
Mass M	Inductance L
Spring constant k	Elastance $1/C$ = Reciprocal of capacitance

On the basis of the principle of *duality*, if force is taken to be similar to current, the following analogy can also be drawn:

Force–Current Analogy

Force F	Current i
Velocity $\dot{x}$	Voltage v
Mass M	Capacitance C
Damping b	Conductance G (= $1/R$)
Spring constant k	$1/L$ = Reciprocal of inductance

This results in the circuit shown in Fig. 2E-15(c).

The preceding example involved a very simple, purely mechanical system. But its method is applicable to more complicated electromechanical systems. Several examples will be considered in the next chapter.

2.5.7 *Linearization Techniques*

The preceding methods and techniques relate to linear systems, the dynamics of which is given by linear differential equations. Electromechanical energy conversion is a nonlinear process, however, and the equations pertaining to it are invariably nonlinear. Nevertheless, for small signals and incremental motion, useful information about the system can be obtained by solving linearized differential equations of motion.

The process of linearization is similar to that used in the incremental analysis of vacuum-tube characteristics by Taylor's series approximation, where a quiescent operating point is defined by a given load line, and the excursions about the operating point are small enough so that the second- and higher-order terms are negligible. This method is also suitable for linearizing the equations of motion of electromechanical systems. These equations generally contain the product type of nonlinearity which can be removed by assuming small perturbations about an operating point.

In general terms, the time-dependent functions are written as follows:

$$\left.\begin{aligned} v(t) &= V_o + v_1(t) \\ i(t) &= I_o + i_1(t) \\ q(t) &= Q_o + q_1(t) \\ f(t) &= F_o + f_1(t) \\ x(t) &= X_o + x_1(t) \end{aligned}\right\} \tag{2.84}$$

where the capital letters denote the dc or steady-state operating point and the lower-case letters with subscripts are small variations about the operating point. Thus, an equation containing a nonlinear term of the form i^2 can be linearized to $I_o^2 + 2I_o i_1$ and so on.

Having linearized the variables, the next step is to linearize the parameters if these happen to be nonlinear. In this case a Taylor or binomial expansion can be made and second- and higher-order terms neglected. For instance, a commonly encountered nonlinear parameter is the inductance expressed as

$$L(x) = \frac{A}{x} \tag{2.85}$$

where A is some constant. This can be linearized by using Eq. (2.84) and writing

$$L(x_1) = \frac{A}{x_1 + X_o} = \frac{A}{X_o}\left(1 + \frac{x_1}{X_o}\right)^{-1}$$

or

$$L(x_1) = \frac{A}{X_o}\left(1 - \frac{x_1}{X_o} + \frac{x_1^2}{X_o^2} - \dots\right)$$

Neglecting the second- and higher-order terms, we may write the linearized expression for inductances as

$$L(x_1) = \frac{A}{X_o}\left(1 - \frac{x_1}{X_o}\right) \tag{2.86}$$

The above example merely illustrates the procedure, which may have to be slightly modified for complicated cases (Chapter 3). In short, in studying a system we must determine whether the quiescent operating point exists; if so, the dynamic characteristics of the system may be obtained by solving the resulting linear equations of motion.

2.6 State Variables and Equations of Motion

In Secs. 2.4 and 2.5 we discussed the formulation of equations of motion and various methods of solving them. In obtaining these solutions we emphasized methods pertaining to linear differential equations. Even if the resulting equations of motion are nonlinear, we are justified in linearizing the equations for incremental motion. In any event, equations of motion yielding the dynamical behavior of a system involve *state variables* (see, for example, Sec. 2.3.1). Because state variables are now so extensively used in system studies,[14,19] we shall briefly discuss linearized equations of motion of electromechanical systems in terms of state variables. This approach broadens our scope and has a very definite advantage in obtaining solutions of equations of motion using matrix algebra.

For the sake of illustration we reconsider Example 2-10. The electrical and mechanical equations of motion are, respectively,

$$\frac{d}{dt}(Li) + Ri = v \tag{2.87a}$$

and

$$M\ddot{x} = \frac{1}{2}i^2\frac{dL}{dx}$$

If a spring of constant k and natural length l_o is attached to a mass M (Fig. 2-12), the mechanical equation of motion is modified to

$$M\ddot{x} + k(x - l_o) = \frac{1}{2}i^2\frac{dL}{dx} \tag{2.87b}$$

Notice that Eqs. (2.87a,b) are both nonlinear. In order to linearize these equations, we assume (arbitrarily) the quiescent values as defined by Eq. (2.84) to be $X_o = 1$, $I_o = 1$. The inductance of this system is of the form given to Eq. (2.85). We further assume, for simplicity's sake, that $A = 1$,

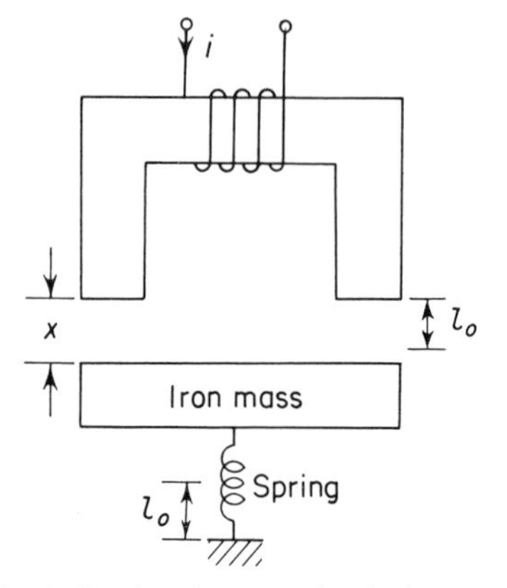

Fig. 2-12 An electromechanical system.

$R = \frac{5}{2}$, $M = \frac{1}{5}$ and $k = \frac{6}{5}$. The expression for inductance in terms of x_1 becomes

$$L(x_1) = (1 - x_1 + x_1^2 \ldots)$$

from which

$$\frac{dL}{dx_1} = -1 + 2x_1$$

and the linearized inductance is

$$L(x_1) = (1 - x_1)$$

Substituting these and Eq. (2.84) in Eqs. (2.87a,b), we have the following linearized dynamical equations of incremental motion:

$$\frac{di_1}{dt} = -\frac{5}{2} i_1 + \dot{x}_1 + v_1 \tag{2.88a}$$

and

$$\ddot{x}_1 = -5i_1 - x_1 \tag{2.88b}$$

Now, we introduce a change of variables defined by

$$\left.\begin{aligned} y_1 &= i_1 \\ y_2 &= x_1 \\ y_3 &= \dot{x}_1 = \dot{y}_2 \end{aligned}\right\} \tag{2.89}$$

and

When Eqs. (2.89) are substituted in Eqs. (2.88a,b) we have

$$\left.\begin{aligned} \dot{y}_1 &= -\tfrac{5}{2}y_1+y_3+v_1 \\ \dot{y}_2 &= y_3 \\ \dot{y}_3 &= -5y_1-y_2 \end{aligned}\right\} \tag{2.90a}$$

Equations (2.90a) can be rewritten in matrix notation (for a review of matrix algebra see Appendix B) as follows:

$$\begin{bmatrix} \dot{y}_1 \\ \dot{y}_2 \\ \dot{y}_3 \end{bmatrix} = \begin{bmatrix} -\frac{2}{5} & 0 & 1 \\ 0 & 0 & 1 \\ -5 & -1 & 0 \end{bmatrix} \begin{bmatrix} y_1 \\ y_2 \\ y_3 \end{bmatrix} + \begin{bmatrix} 1 \\ 0 \\ 0 \end{bmatrix} v_1 \tag{2.90b}$$

Or

$$\dot{\mathbf{y}} = \mathbf{A}\mathbf{y}+\mathbf{B}\mathbf{v} \tag{2.90c}$$

where each term in Eq. (2.90c) is a matrix corresponding to the terms in Eq. (2.90b) (with subscripts dropped). The variables (y_1, y_2, y_3) are the state variables and the vector $\mathbf{y}$ is called the *state vector*. With the passing of time the state vector will trace a trajectory in the *state space*. A familiar trajectory of the state vector in two dimensions is the *phase-plane trajectory*.

It is now appropriate to comment briefly on the choice of state variables. First of all, state variables comprise a set of a minimum number of variables which contain enough information about the past history of the system such that all future states of the system can be computed from the state differential equations. However, the choice of state variables is not unique. This implies that equations of motion are not unique, although solutions for a given system are unique. Regarding the number of state variables for an electromechanical system, it is convenient to assign a state variable to each energy-storage element. In the example discussed above, there are three energy-storage elements—inductance, spring, and mass—and therefore three state variables.

Having formulated the state vector-differential equation, Eq. (2.90b), we can proceed to solve it for given input and initial conditions. The method for solving Eq. (2.90b) is similar to that which we use for a first-order differential equation such as

$$\dot{y} = ay+bv \tag{2.91a}$$

where y, $\dot{y}$, and v are all scalars. Taking Laplace transforms of both sides yields, after simplification,

$$Y(s) = \frac{y(0)}{s-a} + \frac{b}{s-a}V(s) \tag{2.91b}$$

The solution is, then,

$$\mathscr{L}^{-1}[Y(s)] = y(t) = e^{at}y(0) + \int_0^t e^{a(t-\tau)}bv(\tau)\, d\tau \tag{2.91c}$$

The solution of Eq. (2.90b) is, similarly,

$$\mathbf{y}(t) = e^{\mathbf{A}t}\mathbf{y}(0) + \int_0^t e^{\mathbf{A}(t-\tau)}\mathbf{B}v(\tau)\, d\tau \tag{2.92a}$$

where the exponential of a matrix is defined as

$$e^{\mathbf{A}t} = 1 + \mathbf{A}t + \frac{\mathbf{A}^2t^2}{2!} + \ldots + \frac{\mathbf{A}^nt^n}{n!} + \ldots \tag{2.92b}$$

which converges for finite t.

Because Eq. (2.90b) is in the matrix form, the transformed equation, corresponding to Eq. (2.91b), is written as

$$\mathbf{Y}(s) = [s\mathbf{I} - \mathbf{A}]^{-1}\mathbf{y}(0) + [s\mathbf{I} - \mathbf{A}]^{-1}\mathbf{B}V(s) \tag{2.92c}$$

where

$$[s\mathbf{I} - \mathbf{A}]^{-1} = \boldsymbol{\phi}(s) \tag{2.92d}$$

and

$$\boldsymbol{\phi}(t) = e^{\mathbf{A}t} \tag{2.92e}$$

is called the *state-transition matrix*. We conclude that to find the solution for Eq. (2.90b) we have to find the transition matrix.

To illustrate the necessary steps, we assume $\mathbf{y}(0) = 0$ and $v(t) = U(t)$, the unit step function. For the example under consideration

$$\mathbf{A} = \begin{bmatrix} -\frac{5}{2} & 0 & 1 \\ 0 & 0 & 1 \\ -5 & -1 & 0 \end{bmatrix}, \qquad V(s) = \frac{1}{s}$$

and

$$[s\mathbf{I} - \mathbf{A}]^{-1} = \frac{1}{s^3 + \frac{5}{2}s^2 + 6s + \frac{5}{2}} \begin{bmatrix} s^2+1 & -1 & s \\ -5 & s^2 + \frac{5}{2}s + 5 & s + \frac{5}{2} \\ -5s & -s - \frac{5}{2} & s^2 + \frac{5}{2}s \end{bmatrix} \tag{2.92f}$$

which is the transform of the state-transition matrix. From Eqs. (2.92c,f) we have

$$Y_1(s) = \frac{s^2+1}{s(s+\frac{1}{2})(s+1+j2)(s+1-j2)} \tag{2.93a}$$

$$Y_2(s) = \frac{-5}{s(s+\frac{1}{2})(s+1+j2)(s+1-j2)} \tag{2.93b}$$

$$Y_3(s) = \frac{-5s}{s(s+\frac{1}{2})(s+1+j2)(s+1-j2)} \tag{2.93c}$$

We consider only Eq. (2.93a) and express the right-hand side in terms of its partial fractions as

$$Y_1(s) = \frac{2}{5}\cdot\frac{1}{s} - \frac{10}{17}\frac{1}{(s+\frac{1}{2})} + (8+j19)\frac{1/85}{(s+1+j2)} + (8-j19)\frac{1/85}{(s+1-j2)}$$

Taking the inverse transform of this expression yields

$$\begin{aligned} y_1(t) &= \tfrac{2}{5} - \tfrac{10}{17}e^{-(1/2)t} + \tfrac{1}{85}[(8+j19)e^{-(1+j2)t} + (8-j19)e^{-(1-j2)t}] \\ &= \tfrac{2}{5} - \tfrac{10}{17}e^{-0.5t} + \tfrac{2}{85}e^{-t}(19\sin 2t + 8\cos 2t) \end{aligned}$$

Recalling that $y_1(t) = i_1(t)$ we have, finally,

$$i_1(t) = \tfrac{2}{5} - \tfrac{10}{17}e^{-0.5t} + C_1 e^{-t}\sin(2t+\theta)$$

where

$$C_1 = \tfrac{2}{85}\sqrt{425} \quad \text{and} \quad \tan\theta = \tfrac{8}{19}$$

Solutions of Eqs. (2.93b,c) can be obtained by a similar procedure. This example demonstrates the advantage of the state-variable formulation in obtaining explicit solutions of equations of motion.

2.7 Signal-Flow Graphs

In Sec. 2.5.5 we discussed block-diagram representation of linear systems. For complex systems the block-diagram algebra is quite involved and the chances of making a mistake in block-diagram reductions increase. An alternative, more simplified representation of a system is accomplished by means of a *signal-flow graph*.[20,21]

A signal-flow graph represents a set of simultaneous equations. This set of equations may represent the dynamics of a system. The signal-flow graph consists of *nodes*, which represent system variables (such as state variables) and which are connected by directed *branches*. A branch indicates the direction of signal flow and the *transmittance* or transfer function. Thus, Fig. 2-13(a) represents the following equations:

$$y_1 = ax_1 + bx_2$$

$$y_2 = cy_1$$

Like the algebra of block diagrams, signal-flow graphs also follow a set of rules. These are summarized in Fig. 2-13(b). In addition to these rules, the *Mason transmittance formula* is most useful in obtaining the overall transmittance of the system. The formula is given by[22]

$$T = \frac{\Sigma\, T_n \Delta_n}{\Delta} \tag{2.94}$$

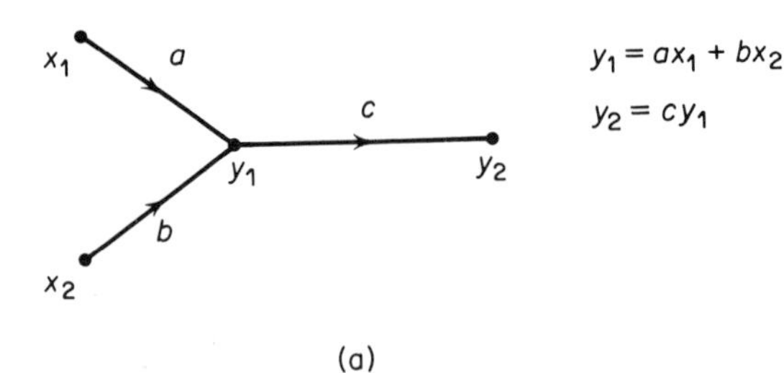

(a)

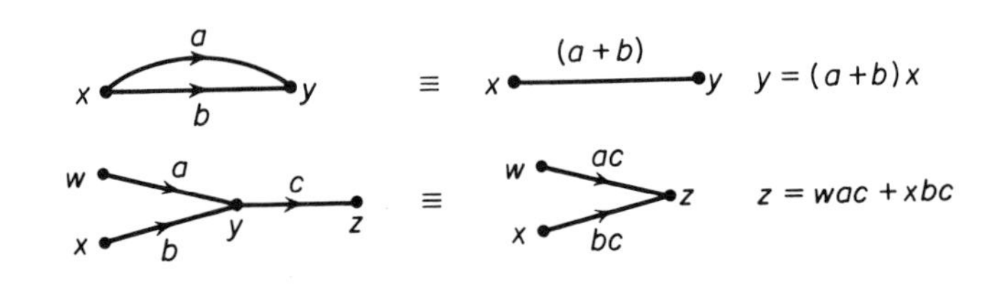

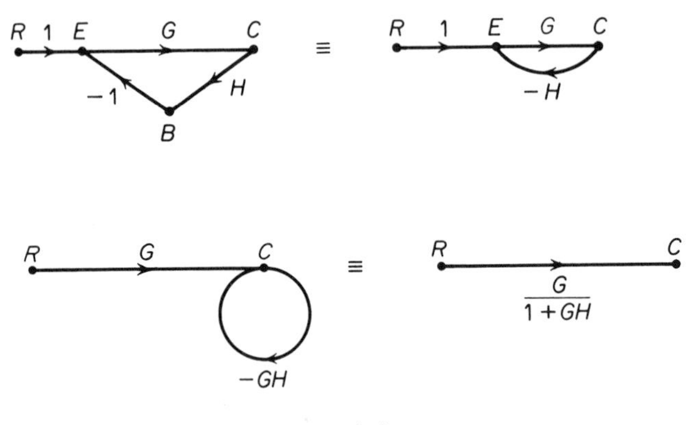

(b)

Fig. 2-13 Rules for signal-flow graphs.

where T_n = transmittance of each forward path between two nodes; $\Delta = 1 - \Sigma\, L_1 + \Sigma\, L_2 - \Sigma\, L_3 + \ldots$; L_1 = transmittance of each closed path and $\Sigma\, L_1$ = sum of transmittances of all closed paths; L_2 = product of transmittances of all possible combinations of nontouching loops taken two at a time; L_3 = product of transmittances of three nontouching loops, and so on; Δ_n = cofactor of T_n. The application of this rule is illustrated by the following example.

EXAMPLE 2–16

A system is represented by the block diagram shown in Fig. 2E-16(a). Its overall transfer function is to be found by the Mason formula.

The signal-flow graph is shown in Fig. 2E-16(b), from which it is seen that the system has two loops whose transmittances are $-G_2H_1$ and $-G_1G_2G_3H_2$. Thus,

$$\Sigma L_1 = -G_2H_1 - G_1G_2G_3H_2$$

Notice that there is no nontouching loop. Hence $L_2 = L_3 = \ldots = 0$.

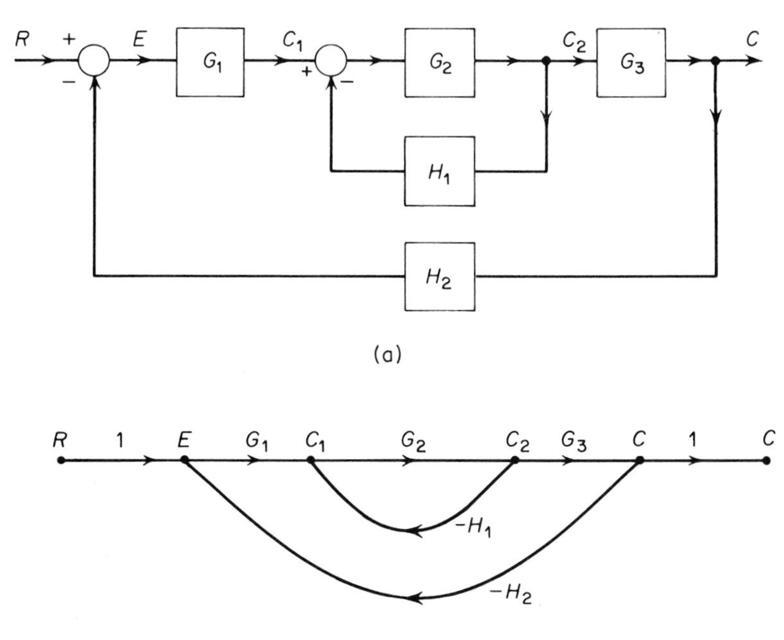

Fig. 2E-16 (a) Block diagram; (b) signal-flow graph.

Now, considering the forward path, we see that the transmittance is $G_1G_2G_3$. The cofactor is unity, because when the nodes from the forward path are removed, the remaining flow graph has no loops. Therefore, from Eq. (2.94), the required transmittance is

$$T = \frac{G_1G_2G_3}{1 + G_2H_1 + G_1G_2G_3H_2}$$

For this example it can be easily verified that the same result can be obtained by using block-diagram algebra.

From this brief discussion we see that signal-flow graphs can be used to represent a set of simultaneous equations. Recall that the state vector differential equation is a set of equations involving state variables. We can now use Eq. (2.90b) (Sec. 2.6), to show that a state-transition matrix can be obtained from a signal-flow graph. At present, we wish to determine $y_1(t)$. From Eq. (2.90b) we have (for zero initial conditions)

$$Y_1(s) = -\frac{5}{2s}Y_1(s) + \frac{1}{s}Y_3(s) + \frac{1}{s}V_1(s)$$

$$Y_2(s) = \frac{1}{s}Y_3(s)$$

$$Y_3(s) = -\frac{5}{s}Y_1(s) - \frac{1}{s}Y_2(s)$$

These equations can be represented by the signal-flow graph shown in Fig. 2-14. To relate $Y_1(s)$ and $V_1(s)$ we use the Mason transmittance formula,

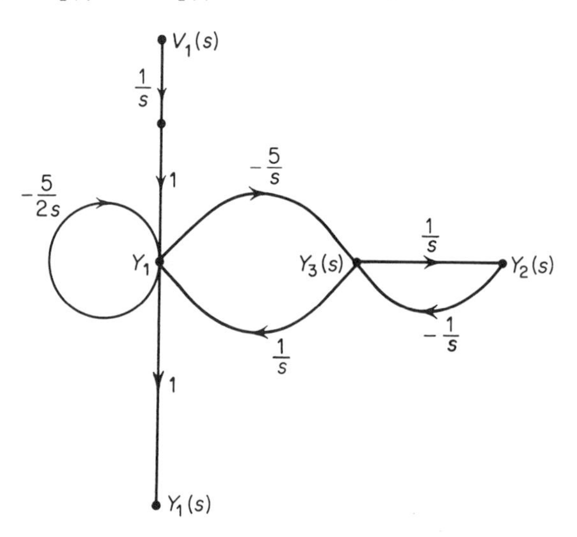

Fig. 2-14 Signal-flow graph.

Eq. (2.94). Here we have three loops with transmittances $-(5/2s)$, $-(5/s^2)$, and $-(1/s^2)$. Therefore,

$$\Sigma L_1 = -\frac{6}{s^2} - \frac{5}{2s}$$

The loops which do not touch each other have transmittances $-(5/2s)$ and $-(1/s^2)$, so that

$$\Sigma L_2 = \frac{5}{2s^3}$$

Hence

$$\Delta = 1 + \frac{6}{s^2} + \frac{5}{2s} + \frac{5}{2s^3}$$

When the forward path together with the node Y_1 are taken off, the loop with transmittance $-(1/s^2)$ remains. Consequently

$$\Delta_1 = 1+\frac{1}{s^2}$$

From Fig. 2-14 we see that $T_1 = 1/s$. Substituting these in Eq. (2.94) we have

$$T = \frac{(1/s)(1+1/s^2)}{1+6/s^2+5/2s+5/2s^3} = \frac{s^2+1}{s^3+\frac{5}{2}s^2+6s+\frac{5}{2}}$$

Or,

$$Y_1(s) = \frac{s^2+1}{s(s+\frac{1}{2})(s+1+j2)(s+1-j2)} \tag{2.95}$$

We see that Eqs. (2.93a) and (2.95) are identical. This example, therefore, illustrates the usefulness of signal-flow graphs.

2.8 Summary

The methods of analysis and the objectives of the study of electromechanical systems were discussed in this chapter. The quantitative behavior of the system can be obtained from equations of motion, which can be formulated from (1) field analysis, (2) energy-conservation principles, or (3) variational principles. The scope of the various methods was outlined.

It was seen that equations relating to electromechanical energy-conversion processes are nonlinear. Linearization techniques were discussed and a number of methods suggested for solving linearized equations and thereby obtaining the useful information contained in the equations.

A brief discussion on the state-variables method was included. These equations were solved for a particular example to show some of the advantages in this formulation. Finally, signal-flow graphs were discussed and their usefulness in obtaining solutions for equations of motion illustrated.

PROBLEMS

2-1. Show that Eqs. (2.19c) and (2.19d) are contained in Eqs. (2.19a) and (2.19b).

2-2. Show that the expression

$$B_y = B_m \sin(\omega t - \beta x)$$

represents a traveling wave. What is the velocity of the wave? Determine the electric field induced in a conducting sheet (a) held stationary in the above field, (b) moving with a velocity U_x in the x direction in the given B field. Assume z-directed currents only.

2–3. The y component of the B field in the airgap of a machine due to the stator currents constrained to flow in z directions only is (for a right-hand co-ordinate system)

$$B_y = B_m \cos(\omega t - \beta x)$$

The rotor is unexcited and is open-circuited. (a) Determine the ratio of the stored-energy density in the electric field to the energy density in the magnetic field and hence demonstrate that the energy stored in the electric field is negligible in conventional machines at power frequencies. (b) Show that

$$2\frac{\partial W_m}{\partial t} + \frac{\partial P_x}{\partial x} = 0$$

where W_m is the energy per unit volume in the magnetic field and P_x is the power density in the x direction. (See Fig. 2E-5.)

2–4. The radial component of the airgap field referred to the rotor (Fig. 2P-4) at the rotor surface ($y = 0$) of a sleeve rotor is

$$B_y = B_m \sin(\omega t - \beta x)$$

Determine the average force density on the rotor, taking into account the thickness of the sleeve but neglecting the end effects.

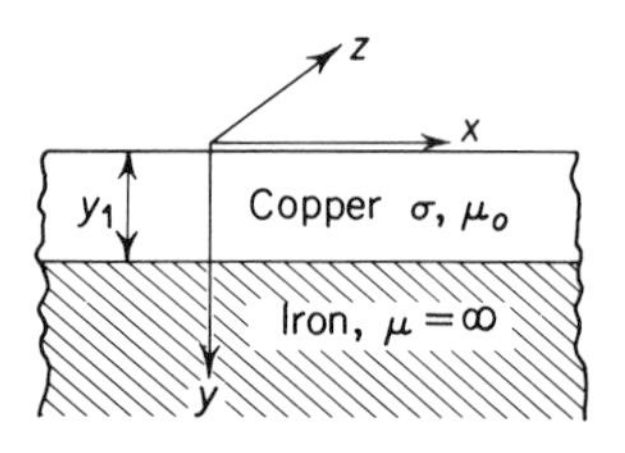

Fig. 2P-4

2–5. If for a two-winding transformer it is postulated that the Poynting vector yields the correct value of power density everywhere, how does energy get from the source to load? Why does this point of view make the concept of leakage reactance of fundamental significance?

2–6. Derive the expressions for the energy stored in a capacitive system consisting of

(a) three capacitances in series,
(b) three capacitances in parallel.

Generalize the results, if possible.

2–7. Show that, for a linear magnetic circuit, the magnetic energy is equal to the coenergy. Using a power-series approximation to the saturation curve, obtain the ratio of energy to coenergy in a nonlinear magnetic circuit having the characteristics shown in Fig. 2-4.

2–8. An electromechanical system is described by the following equation:

$$\frac{A}{x}+Bxi+Cx\frac{di}{dt}+Di^2 = E$$

where A, B, C, D, and E are constants. For the case of small variations $x_1(t)$, $i_1(t)$ about the steady-state operating point X_o, I_o, linearize the equation.

2–9. The singly-excited electromechanical system shown in Fig. 2P-9 is constrained to move only horizontally. The pertinent dimensions are shown in the diagram. Calculate the electrical force exerted on the movable iron member for

(a) current excitation:

$i = I \cos t$

(b) voltage excitation:

$v = V \cos t$

For parts (a) and (b) neglect the winding resistance, leakage fields, and fringing. Assume all energy to be stored in the airgaps, that is, the permeability of iron is very large as compared with that of free space. What modifications have to be made if the winding resistance is not negligible?

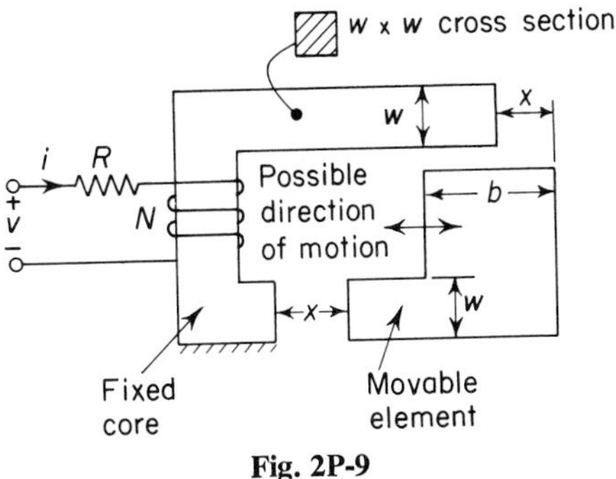

Fig. 2P-9

2–10. The electromagnetic structure shown in Fig. 2P-10 is characterized by the following θ-dependent inductances

$$L_{11} = A+B \cos 2\theta = L_{22}$$

$$L_{12} = E+F \cos \theta = L_{21}$$

Assume winding resistance to be zero. Find the torque as a function of θ, V, and t, when both windings are connected to the same source, so that $v_1 = v_2 = V \sin \omega t$.

2–11. The system shown in Fig. 2P-11 carries two coils, having self- and mutual inductances L_{11}, L_{12}, and L_{22}. Coil no. 1 carries a current $i_1 = I_1 \sin \omega_1 t$, while coil no. 2 carries a current $i_2 = I_2 \sin \omega_2 t$. The inductances are: L_{11}

$= k_1/x$, $L_{22} = k_2/x$, and $L_{12} = k_3/x$, where k_1, k_2, and k_3 are constants. Derive an expression for the instantaneous force on the armature. Give an expression (in integral form) for the average force. Find a relation between ω_1 and ω_2 for (i) maximum average force, and (ii) minimum average force. Calculate the maximum, minimum, and intermediate values of the average force. Give a physical reasoning to your answers.

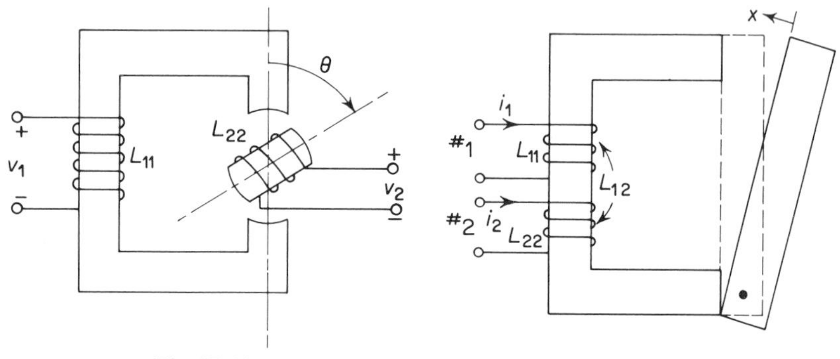

Fig. 2P-10 Fig. 2P-11

2–12. Using Laplace transforms, obtain the transfer function for the networks shown in Fig. 2P-12(a–f). The resistances are in megohms, inductances in millihenries, and capacitances in microfarads.

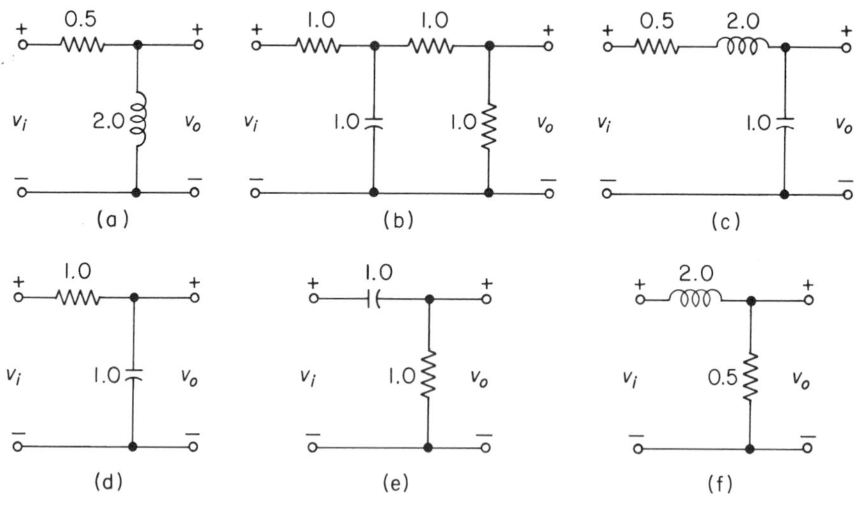

Fig. 2P-12

2–13. Obtain the step response for the networks (a), (b), (d), and (f) in Fig. 2P-12, if a unit step is given as the input in each case.

2–14. Plot the exact frequency-response curves for the networks of Problem 2-13. On each curve, also show the asymptotic approximation.

2–15. Give an asymptotic approximation to the frequency-response curves for the following transfer functions: (a) $(s+10)/(s+1)$; (b) $(s+1)\ (s+10)/(s+5)$; (c) $(s+1)\ (s+5)/(s+10)$.

2–16. Develop the block diagram for the networks of Problem 2–13. Reduce the block diagrams, using block-diagram algebra, to obtain the transfer function in each case.

2–17. Obtain suitable electrical networks which have the transfer functions given in Problem 2-15. This is a synthesis problem. Use can be made, however, of the transfer functions of the networks in Problem 2-12. Otherwise, the student is urged to consult an introductory book in network synthesis.

2–18. A mechanical system consisting of spring, mass, and damping is shown in Fig. 2P-18. Choose a suitable coordinate system and write down the equations of motion from Newton's laws. Obtain its electrical analog using

(a) the force–voltage analogy,
(b) the force–current analogy.

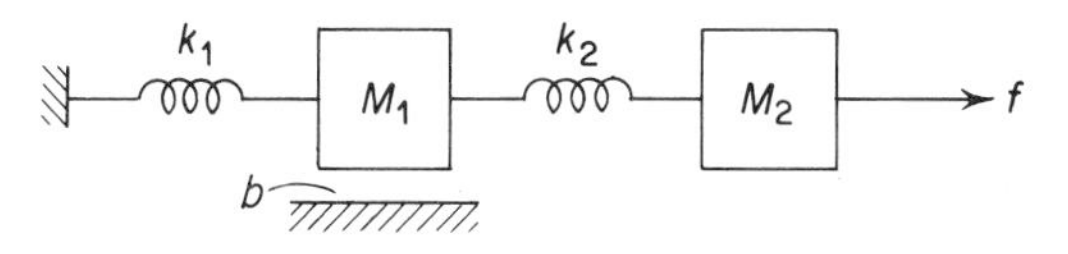

Fig. 2P-18

Observe if there is duality between the networks obtained in parts a and b. Using block diagrams, give a suitable analog-computer representation for the system. The damping force is directly proportional to the velocity.

2–19. For the network shown in Fig. 2P-12(c) choose a suitable set of state variables and write down the state vector differential equation. Solve for all state variables by finding the state-transition matrix. Check that the step response thus obtained agrees with the result of Problem 2-13(c). Assume zero initial conditions.

2–20. Obtain the state vector differential equation for the system shown in Fig. 2P-18. Draw the signal-flow graph and thereby obtain the transmittance (ratio of the applied force F to the displacement x_1 of mass M_1).

REFERENCES

1. Bradshaw, Martin D., and William J. Byatt, *Introductory Engineering Field Theory*. Englewood Cliffs, N. J.: Prentice-Hall, Inc., 1967.

2. Hammond, P., "A Short Modern Review of Fundamental Electromagnetic Theory," *Proc. IEE* (London), Vol. 101, Part I, 1954, p. 147.

3. Cullen, A. L., and T. H. Barton, "A Simplified Electromagnetic Theory of the Induction Motor Using the Concept of Wave Impedance," *Proc. IEE* (London), Vol. 105(c), 1958, p. 331.
4. Nasar, S. A., "Electromagnetic Theory of Electrical Machines," *Proc. IEE* (London), Vol. 111, 1964, p. 1123.
5. Barlow, H. E. M., "Simplified Treatment of Mechanical Forces on Materials in an Electromagnetic Field," *Proc. IEE* (London), Vol. 113, 1966, p. 373.
6. Nussbaum, Allen, *Electromagnetic Theory for Engineers and Scientists*, pp. 214–216. Englewood Cliffs, N.J.: Prentice-Hall, Inc., 1965.
7. Nussbaum, Allen, *Electromagnetic and Quantum Properties of Materials*, Chapter 6. Englewood Cliffs, N.J.: Prentice-Hall, Inc., 1966.
8. Ramo, S., J. R. Whinnery, and T. Van Duzer, *Fields and Waves in Communication Electronics*, pp. 122–124. New York: John Wiley & Sons, Inc., 1965.
9. Hayt, William H., Jr., *Engineering Electromagnetics*, 2nd ed., Chapter 9. New York: McGraw-Hill Book Company, 1967.
10. White, D. C., and H. H. Woodson, *Electromechanical Energy Conversion*. New York: John Wiley & Sons, Inc., 1959.
11. Mawardi, O. K., "On the Concept of Coenergy," *Jour. Franklin Inst.*, Vol. 264, 1957, p. 313.
12. Goldstein, H., *Classical Mechanics*. Reading, Mass.: Addison-Wesley Publishing Company, Inc., 1959.
13. Weinstock, R., *Calculus of Variations*. New York: McGraw-Hill Book Company, 1952.
14. Gupta, S. C., *Transform and State Variable Methods in Linear Systems*. New York: John Wiley & Sons, Inc., 1966.
15. Thompson, W. T., *Laplace Transforms*. Englewood Cliffs, N.J.: Prentice-Hall, Inc., 1960.
16. Guillemin, E. A., *Mathematics of Circuit Analysis*. New York: John Wiley & Sons, Inc., 1959.
17. Truxal, John G., *Automatic Feedback Control Systems Synthesis*. New York: McGraw-Hill Book Company, 1955.
18. Graybeal, T. D., "Block Diagram Network Transformation," *Electrical Engineering*, Vol. 70, 1951, p. 985.
19. Ogata, Katsuhiko, *State Space Analysis of Control Systems*. Englewood Cliffs, N.J.: Prentice-Hall, Inc., 1967.
20. Mason, S. J., "Feedback Theory: Further Properties of Signal Flow Graphs," *Proc. IRE*, Vol. 44, 1956, p. 920.
21. Robichaud, L. P. A., M. Boisvert, and J. Robert, *Signal Flow Graphs and Applications*. Englewood Cliffs, N.J.: Prentice-Hall, Inc., 1962.
22. D'Azzo, John J., and Constantine H. Houpis, *Feedback Control System Analysis and Synthesis*, 2nd ed. New York: McGraw-Hill Book Company, 1966.

3

Incremental-Motion Electromechanical Systems

The methods of formulating equations of motion of energy-conversion devices, developed in the last chapter, can be classified as suitable for (1) lumped-parameter devices and (2) distributed-parameter devices. In the present chapter we shall concern ourselves with the analysis of lumped-parameter incremental-motion electromechanical systems. The incremental-motion constraint suggests that equations of motion can be linearized about an operating point by the method developed in Sec. 2.5.7. Some examples of incremental-motion energy processors are moving-coil microphones, telephone receivers, electromagnetic relays, electrostatic microphones, and numerous other transducers.

We shall now consider a number of examples to illustrate the applications of some of the techniques developed in the last chapter. We shall formulate equations of motion for a given system, and shall solve the resulting equations for different operating conditions.

Equations of motion are formulated, either by using force laws or by Lagrangian formulation. Examples are considered to illustrate both the methods. In both cases the parameters of the system must be known. For a lumped-parameter electromagnetic transducer, for example, the inductance can be determined by the method given in Sec. 2.1.5. Other parameters are found from the physical description and dimensions of the system.

Solutions to the equations of motion are given in terms of the linearized behavior of the system, that is, in terms of transfer functions, equivalent circuits, state equations, and so forth. The examples discussed in subsequent sections illustrate the details of the various steps involved in the formulation of equations of motion and in their solutions.

Current-excited as well as voltage-excited systems are considered, as it is useful to distinguish between the two forms of excitations.

3.1 An Angular-Motion Electromechanical System

3.1.1 Description of the System

The electromechanical system shown in Fig. 3-1 is a composite of two interacting subsystems, one electromagnetic and the other mechanical. The mechanical subsystem is a torsion pendulum consisting of a flywheel, driving and damping blades, the torsion spring, and inherent frictional damping. The electromagnetic subsystem consists of a driving coil wound on a laminated core. The driving blade is placed in the airgap (Fig. 3-1). The current path is through the driving coil and the flux path is through the iron core, airgap, and driving blade. The airgap thus acts as a link between the magnetic and mechanical parts of the system. The driving blade, which is made of iron,

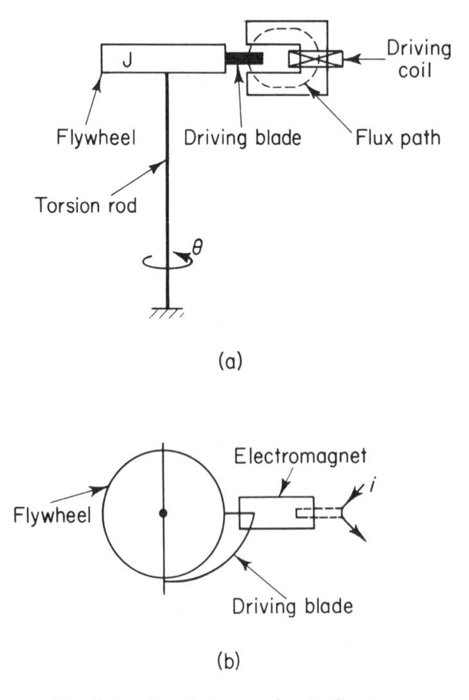

Fig. 3-1 An electromechanical system.

is so shaped that the variation of inductance of the coil is approximately linear with respect to mechanical position.

3.1.2 The Equations of Motion

The physical description of the system and topological considerations lead to the following parameters:

Mechanical: J = moment of inertia of the rotating parts, N·m/rad/sec²
b = velocity-dependent friction coefficient, N·m/rad/sec
k = stiffness or spring constant, N·m/rad
Electrical: R_1 = resistance of the coil, ohm
$L = L' + L''\theta$ = coil inductance, henry

The generalized coordinates, discussed in Sec. 2.3.2, are used to describe the system. Therefore,

$$\begin{aligned}
\mathscr{T} &= \text{kinetic energy} = \tfrac{1}{2}(L'+L''\theta)i^2 + \tfrac{1}{2}J\dot{\theta}^2 \\
\mathscr{V} &= \text{potential energy} = \tfrac{1}{2}k\theta^2 \\
\mathscr{F} &= \text{dissipation function} = \tfrac{1}{2}R_1 i^2 + \tfrac{1}{2}b\dot{\theta}^2 \\
\mathscr{L} &= \text{Lagrangian function} = \tfrac{1}{2}(L'+L''\theta)\,i^2 + \tfrac{1}{2}J\dot{\theta}^2 - \tfrac{1}{2}k\theta^2
\end{aligned}$$

Substituting the above quantities in Eq. (2.75) yields the following equations of motion:

$$\text{Mechanical: } k = 1; \quad q_1 = \theta; \quad J\ddot{\theta} + b\dot{\theta} + k\theta - \tfrac{1}{2}L''i^2 = 0 \tag{3.1}$$

$$\text{Electrical: } k = 2; \quad q_2 = \int i\,dt$$

$$(L'+L''\theta)\frac{di}{dt} + L''i\dot{\theta} + R_1 i = v \tag{3.2}$$

where v = voltage applied across the driving coil.

3.1.3 Equations for Incremental Motion

The presence of the terms containing i^2 and $i\dot{\theta}$ makes Eqs. (3.1) and (3.2) nonlinear. As such, it is rather difficult to obtain their solutions. However, for practical purposes, the equations of motion can be linearized about the equilibrium point, which exists, for small perturbations about the quiescent operating point. To do so, let

$$\left.\begin{aligned}
\theta &= \Theta_o + \theta_1 \\
v &= V_o + v_1 \\
i &= I_o + i_1
\end{aligned}\right\} \tag{3.3}$$

where (Θ_o, V_o, I_o) denotes the steady-state equilibrium point and (θ_1, v_1, i_1) is the small time-dependent variation such that the second-order terms, such as i_1^2, $i_1\theta_1$, etc., may be neglected. Equations (3.1–3.3), therefore, yield the following linearized equations of motion describing the dynamical behavior of the system:

$$\text{Mechanical:} \quad J\ddot{\theta}_1 + b\dot{\theta}_1 + k\theta_1 - I_o L'' i_1 = 0 \tag{3.4}$$

$$\text{Electrical:} \quad (L' + L''\Theta_o)\frac{di_1}{dt} + R_1 i_1 + I_o L''\dot{\theta}_1 = v_1 \tag{3.5}$$

And the mechanical and electrical equilibrium points are respectively given by

$$k\Theta_o = \tfrac{1}{2} L'' I_o^2 \tag{3.6}$$

$$R_1 I_o = V_o \tag{3.7}$$

3.1.4 *Equivalent Circuits*

In order to develop equivalent circuits for the given system, Eqs. (3.4) and (3.5) are rewritten as

$$C_2 \frac{dv_2}{dt} + G_2 v_2 + \frac{1}{L_2}\int v_2\, dt = i_1 \tag{3.8}$$

$$L_1 \frac{di_1}{dt} + R_1 i_1 = v_1 - v_2 \tag{3.9}$$

where

$$\left.\begin{aligned} L_1 &= L' + L''\Theta_o \\ L_2 &= \frac{(L''I_o)^2}{k} \\ C_2 &= \frac{J}{(L''I_o)^2} \\ G_2 &= \frac{b}{(L''I_o)^2} \end{aligned}\right\} \tag{3.10}$$

and $(L''I_o)\dot{\theta}_1 = v_2$.

The circuit representing Eqs. (3.8) and (3.9), with the constants defined by Eq. (3.10), is given in Fig. 3-2(a).

Equations (3.4) and (3.5) can also be written as

$$L_2 \frac{di_2}{dt} + R_2 i_2 + \frac{1}{C_2}\int i_2\, dt - Ri_1 = 0 \tag{3.11}$$

$$L_1 \frac{di_1}{dt} + R_1 i_1 + Ri_2 = v_1 \tag{3.12}$$

where

$$\left.\begin{aligned} L_2 &= J \\ R_2 &= b \\ C_2 &= \frac{1}{k} \\ L''I_o &= R \\ \dot{\theta} &= i_2 \end{aligned}\right\} \tag{3.13}$$

Note that reciprocity of mutual interaction does not hold because the coefficients of the coupling terms in the two equations do not appear with the same sign. This can be taken into account by the use of a *gyrator*, as shown in Fig. 3-2(b).

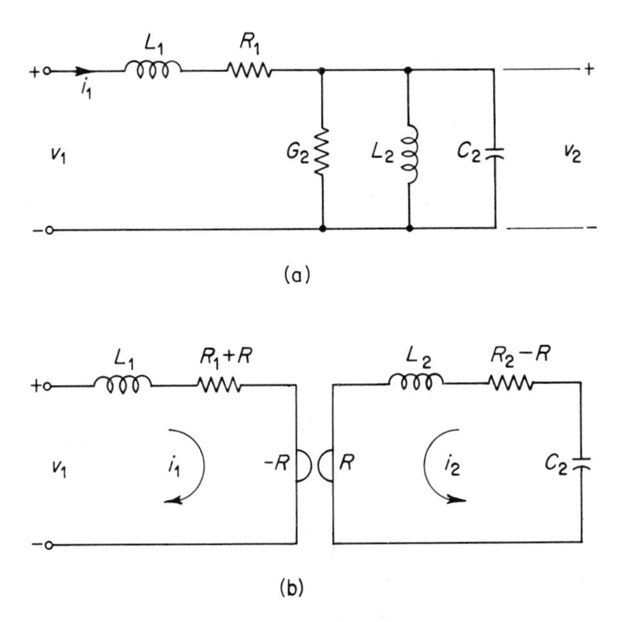

Fig. 3-2 Equivalent circuits.

3.1.5 *Block Diagrams and Transfer Functions*

Taking Laplace transforms of the linearized equations, Eqs. (3.11) and (3.12), yields

$$L_2sI_2(s)+R_2I_2(s)+\frac{1}{C_2s}I_2(s)-RI_1(s) = 0$$

$$L_1sI_1(s)+R_1I_1(s)+RI_2(s) = V_1(s)$$

Equations (3.8) and (3.9) can be represented by the block diagram shown in Fig. 3-3. Such a representation facilitates analog-computer studies of the system and determination of the transfer function.

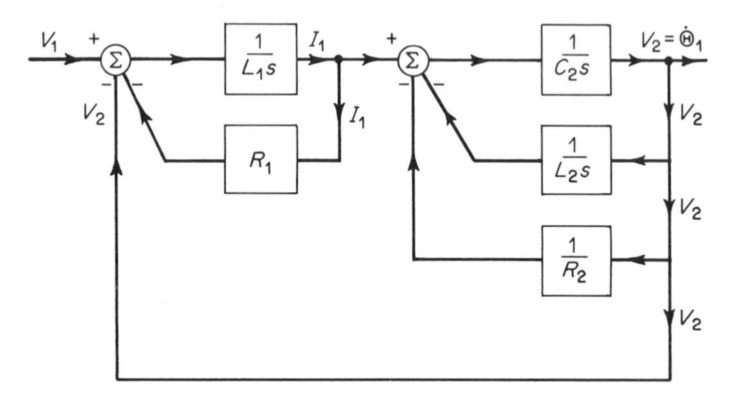

Fig. 3-3 Block diagram for linearized system.

The transfer function of the linearized system can be obtained also by taking Laplace transforms of Eqs. (3.4) and (3.5) such that

$$Js^2\theta_1(s)+bs\theta_1(s)+k\theta_1(s)-\beta I_1(s) = 0 \tag{3.14}$$

$$L_1sI_1(s)+R_1I_1(s)+\beta\theta_1(s) = V_1(s) \tag{3.15}$$

where $\beta = L''I_o$.

For a constant current excitation the system evidently behaves as a perfect second-order system and the transfer function is

$$G_i(s) = \frac{\theta_1(s)}{I_1(s)} = \frac{\beta}{Js^2+bs+k} = \frac{\beta/J}{s^2+(b/J)s+k/J} \tag{3.16}$$

For a constant-voltage input the transfer function is, from Eqs. (3.14) and (3.15),

$$G_v(s) = \frac{\theta_1(s)}{V_1(s)} = \frac{\beta}{(Js^2+bs+k)\,(L_1s+R_1)+\beta^2} \tag{3.17a}$$

If, however, the mechanical time constant J/b is considerably greater than the electrical time constant L_1/R_1, the approximate transfer function for voltage excitation becomes (assuming $L_1 \simeq 0$)

$$G_v(s)|_{\text{approx}} = \frac{\beta}{(Js^2+bs+k)R_1+\beta^2} \tag{3.17b}$$

3.1.6 Step and Frequency Response

A comparison of Eqs. (3.16) and (3.17b) indicates that the system approximately behaves as a second-order system. The step and frequency responses of the system are shown in Figs. 3-4(a) and (b).

To summarize, an angular-motion electromechanical system is analyzed. Its equations of motion are derived from Lagrange's equation. Equivalent circuits, block diagram, transfer functions, step response, and frequency response are obtained for incremental motion.

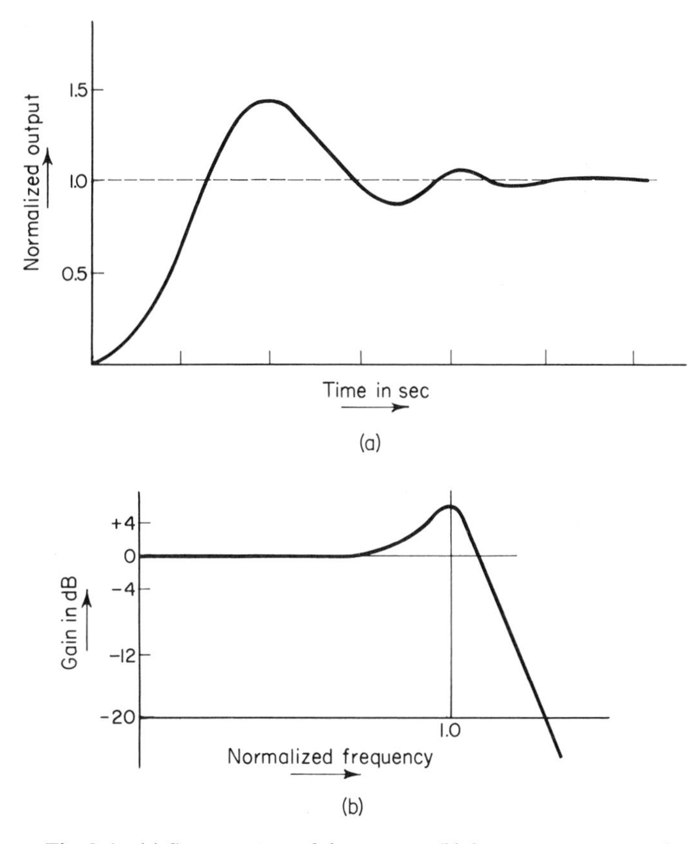

Fig. 3-4 (a) Step response of the system; (b) frequency response of the system.

3.1.7 State Equation and Signal-Flow Graph

To complete the study we shall now attempt to obtain the state vector differential equation. Using the signal-flow graph we shall then obtain an explicit solution for the position θ for a unit-step current input. For this case, we need to consider only Eq. (3.4), which is rewritten as

$$\ddot{\theta}_1 = -\frac{b}{J}\dot{\theta}_1 - \frac{k}{J}\theta_1 + \frac{I_o L''}{J} i_1 \tag{3.4a}$$

For the sake of illustration we choose $J = 1$, $I_o L'' = 1$, $b = 2$, and $k = 5$, so that Eq. (3.4a) becomes

$$\ddot{\theta}_1 = -2\dot{\theta}_1 - 5\theta_1 + i_1 \tag{3.4b}$$

Now putting $y_1 = \theta_1$ and $y_2 = \dot{y}_1$ in Eq. (3.4b), and proceeding as in Sec. 2.6, we have

$$\begin{bmatrix} \dot{y}_1 \\ \dot{y}_2 \end{bmatrix} = \begin{bmatrix} 0 & 1 \\ -5 & -2 \end{bmatrix} \begin{bmatrix} y_1 \\ y_2 \end{bmatrix} + \begin{bmatrix} 0 \\ 1 \end{bmatrix} i_1 \tag{3.18}$$

which is the required state vector-differential equation. The state variables are y_1 and y_2. The **A** matrix is

$$\mathbf{A} = \begin{bmatrix} 0 & 1 \\ -5 & -2 \end{bmatrix}$$

so that

$$[s\mathbf{I} - \mathbf{A}]^{-1} = \frac{1}{s^2 + 2s + 5} \begin{bmatrix} s+2 & 1 \\ -5 & s \end{bmatrix} \tag{3.19}$$

Therefore, from Eq. (2.92c), assuming zero initial conditions, the desired solutions can be obtained, as in Sec. 2.6.

If, on the other hand, we draw the signal-flow graph as shown in Fig. 3-5 and use the transmittance formula, we directly obtain the transfer function as

$$G_i(s) = \frac{\theta_1(s)}{I_1(s)} = \frac{1}{s^2 + 2s + 5} \tag{3.20}$$

which is the same as Eq. (3.16).

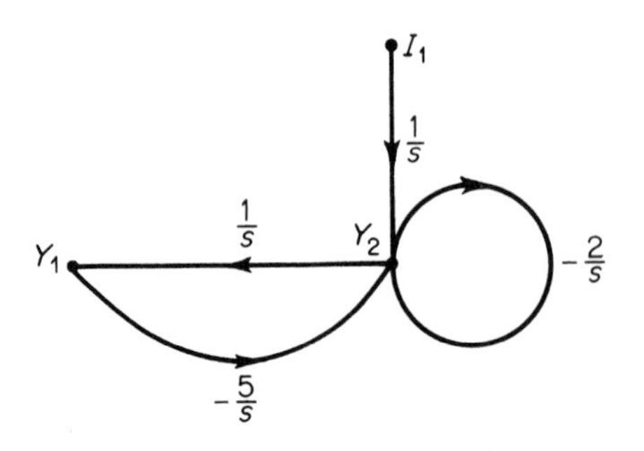

Fig. 3-5 Signal flow graph.

From this example, we note the usefulness of the various methods developed in the last chapter. In particular, we have formulated the equations of motion by the Lagrangian method; linearized the equations for incremental motion; developed block diagrams, equivalent circuits, and transfer functions; obtained the step and frequency response; and, finally, obtained the state equations and used the signal-flow graph to obtain the transfer function by an alternative method.

3.2 An Electrostatic Transducer

We shall next study the electromechanical system with electric coupling field shown in Fig. 3-6.

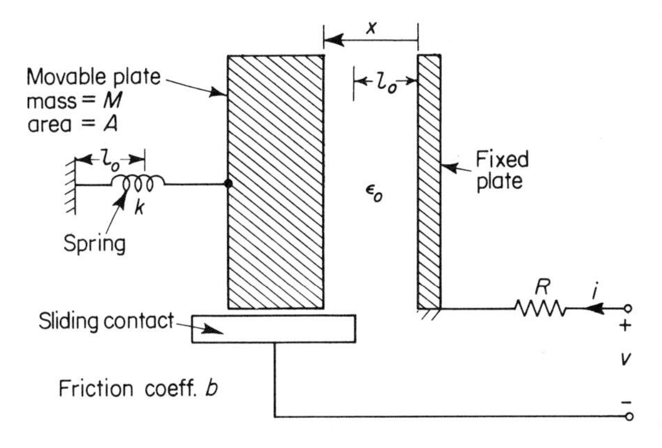

Fig. 3-6 An electrostatic transducer.

Assumptions: Fringing of the electric lines of force is ignored and it is assumed that the coupling field is purely electrostatic. The friction force is directly proportional to the velocity.

Parameters: The mass M, friction coefficient b, and spring constant k are given. The natural length of the spring, when the applied voltage is zero, is l_o. The resistance R is also specified. If ϵ_o is the permittivity, the capacitance C as a function of position x is

$$C(x) = \frac{\epsilon_o A}{x}$$

3.2.1 Equations of Motion

If q is the charge on the capacitor plates, the equations of motion are obtained from Lagrangian formulation as follows:

$$\text{Kinetic energy} \quad \mathscr{T} = \tfrac{1}{2} M\dot{x}^2$$

$$\text{Potential energy} \quad \mathscr{V} = \tfrac{1}{2} k(x-l_o)^2 + \frac{1}{2}\frac{q^2 x}{\epsilon_o A}$$

$$\text{Lagrangian} \quad \mathscr{L} = \mathscr{T} - \mathscr{V}$$

$$\text{Dissipation} \quad \mathscr{F} = \tfrac{1}{2} R\dot{q}^2 + \tfrac{1}{2} b\dot{x}^2$$

Therefore, from Eq. (2.75), the equations of motion are

$$\text{Mechanical:} \quad M\ddot{x} + b\dot{x} + k(x-l_o) + \frac{1}{2}\frac{q^2}{\epsilon_o A} = 0 \tag{3.21}$$

$$\text{Electrical:} \quad R\dot{q} + \frac{qx}{\epsilon_o A} = v \tag{3.22}$$

The electrical force is

$$F_e = -\frac{1}{2}\frac{q^2}{\epsilon_o A}$$

The equations of motion are *nonlinear* due to the presence of terms such as q^2 and qx.

Linearization is accomplished by assuming that a steady-state operating point (Q_o, X_o, V_o) exists such that

$$v = V_o + v_1$$

$$x = X_o + x_1$$

and

$$q = Q_o + q_1$$

which when substituted in Eqs. (3.21) and (3.22) yield

$$V_o + v_1 = R\dot{q}_1 + \frac{Q_o X_o}{\epsilon_o A} + \frac{X_o q_1}{\epsilon_o A} + \frac{Q_o x_1}{\epsilon_o A}$$

and

$$0 = M\ddot{x}_1 + b\dot{x}_1 + k(X_o - l_o) + kx_1 + \frac{1}{2}\frac{Q_o^2}{\epsilon_o A} + \frac{Q_o q_1}{\epsilon_o A}$$

The *steady-state operating point* is given by

$$V_o = \frac{Q_o X_o}{\epsilon_o A}$$

and

$$\frac{1}{2}\frac{Q_o^2}{\epsilon_o A} = -k(X_o - l_o)$$

The *dynamics about the operating point* is described by

$$v_1 = R\dot{q}_1 + S_o q_1 + \beta x_1 \tag{3.23}$$

and

$$0 = M\ddot{x}_1 + b\dot{x}_1 + kx_1 + \beta q_1 \tag{3.24}$$

where

$$S_o = \frac{1}{C_o} = \frac{X_o}{\epsilon_o A} \quad \text{and} \quad \beta = \frac{Q_o}{\epsilon_o A} = \frac{Q_o S_o}{X_o}$$

An equivalent circuit for the system can be developed if $\dot{x}_1 = i_2$ and $\dot{q}_1 = i_1$ are substituted in Eqs. (3.23) and (3.24) to obtain

$$v_1 = Ri_1 + S_o \int i_1\, dt + \beta \int i_2\, dt$$

$$0 = M\frac{di_2}{dt} + bi_2 + k\int i_2\, dt + \beta \int i_1\, dt$$

The equivalent circuit for this transducer is shown in Fig. 3-7.

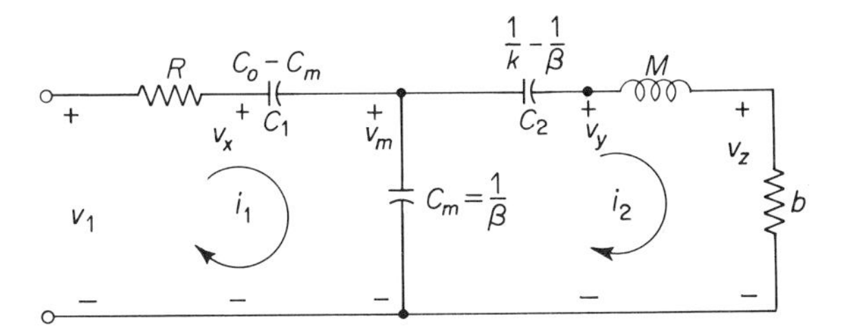

Fig. 3-7 Equivalent circuit for the electrostatic transducer.

The block diagram for the system is shown in Fig. 3-8.

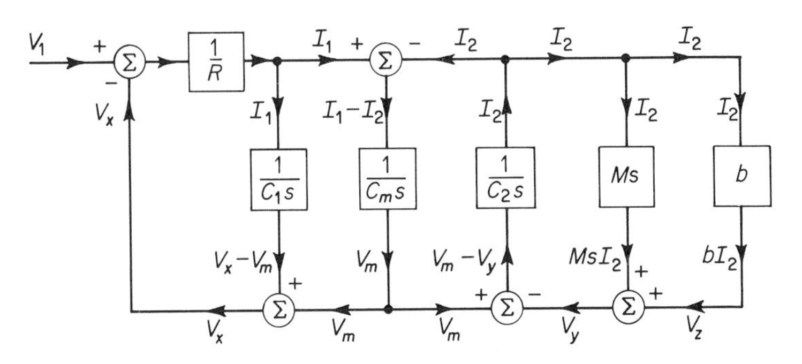

Fig. 3-8 Block diagram for the electrostatic transducer.

Realizability conditions are

$$C_o \geqslant C_m$$

and

$$\beta \geqslant k$$

If V_1 is defined as the input and X_1 as the output, the *transfer function* for the system can be obtained from Eqs. (3.23) and (3.24):

$$G(s) = \frac{X_1(s)}{V_1(s)} = \frac{-\beta}{(Ms^2+bs+k)(Rs+S_o)-\beta^2}$$

The *step* and *frequency response* of the system can be easily obtained once the transfer function has been determined. Notice that, for the example under consideration, the polynomial in the denominator of $G(s)$ is a cubic in s. It has three roots; at least one of the roots must be real, and the other two roots may be either real or a pair of complex conjugate roots. For the response of the system to be bounded, the real root and the real part of the complex root must be negative. Furthermore, there are three energy-storage elements—spring, mass, and capacitance—in the system. Consequently, there must be three time constants. This also corresponds to the three roots of the polynomial just mentioned.

3.3. An Electromagnetic Transducer

In the examples of the preceding two sections the equations of motion were derived from the Lagrangian formulation. An electromagnetic transducer is now considered, for which the equations of motion are obtained using the force laws. The system is shown in Fig. 3-9. The area of cross-sections of the magnetic circuit is A, the coil has N turns, and the magnetic energy is assumed to be stored in the airgap. The friction coefficient b is a (linear) constant. Fringing is ignored.

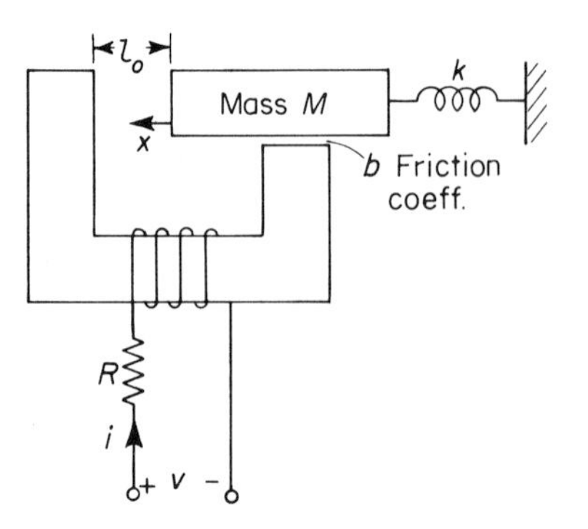

Fig. 3-9 An electromagnetic transducer.

The mechanical parameters are M, k, and b, and the electrical constants are R and L, where L is obtained from Eq. (2.56b) as

$$L(x) = \frac{\mu_o N^2 A}{x}$$

From Eq. (c), Example 2-8, the force of electrical origin is given by

$$F_e = \frac{1}{2} i^2 \frac{\partial L}{\partial x}$$

since the current i is considered as the independent variable. This force is taken as the externally applied force and the equations of motion are written as

$$L\frac{di}{dt} + i\frac{dL}{dx}\cdot\frac{dx}{dt} + Ri = v \tag{3.25}$$

$$M\frac{d^2x}{dt^2} + b\frac{dx}{dt} + k(x - l_o) = \frac{1}{2} i^2 \frac{dL}{dx} \tag{3.26}$$

Both equations are nonlinear. The nonlinear terms are $i(dx/dt)$ in Eq. (3.25) and i^2 in Eq. (3.26). Notice that v, i, and x are time-dependent quantities.

If it so happens that a quiescent operating point exists and that the deviation from this point is small, the equations of motion can be linearized. In such a case let

$$\left.\begin{aligned} v &= V_o + v_1 \\ i &= I_o + i_1 \\ x &= X_o + x_1 \end{aligned}\right\} \tag{3.27}$$

where the quantities with subscripts o denote the steady-state operating point and those with subscript 1 are small variations about this point.

Now recall that the inductance L is

$$L = \mu_o \frac{AN^2}{x} \tag{3.28}$$

From Eqs. (3.27) and (3.28) we have

$$L = \mu_o \frac{AN^2}{(X_o + x_1)} = \mu_o \frac{AN^2}{X_o}\left(1 + \frac{x_1}{X_o}\right)^{-1} = L_o\left(1 - \frac{x_1}{X_o} + \frac{x_1^2}{X_o^2} - \cdots\right) \tag{3.29}$$

Or,

$$L \simeq L_o\left(1 - \frac{x_1}{X_o}\right) \tag{3.30}$$

neglecting higher-order terms, where

$$L_o = \mu_o \frac{AN^2}{X_o}$$

From Eq. (3.29),

$$\frac{dL}{dx_1} \simeq L_o \left(-\frac{1}{X_o} + \frac{2x_1}{X_o^2} \right) \tag{3.31}$$

From Eq. (3.27),

$$i^2 \simeq I_o^2 + 2i_1 I_o \tag{3.32}$$

From Eqs. (3.31) and (3.32),

$$\frac{1}{2} i^2 \frac{dL}{dx_1} = -\frac{1}{2}(I_o^2 + 2i_1 I_o) \frac{L_o}{X_o} + I_o^2 \frac{L_o}{X_o^2} x_1 \tag{3.33}$$

Now, Eqs. (3.25–3.33), when simplified, and with second- and higher-order terms neglected, give

$$L_o \frac{di_1}{dt} - \frac{L_o I_o}{X_o} \frac{dx_1}{dt} + Ri_1 + RI_o = V_o + v_1 \tag{3.34}$$

and

$$M \frac{d^2x_1}{dt^2} + b \frac{dx_1}{dt} + kx_1 + k(X_o - l_o) = -\frac{1}{2} \frac{I_o^2 L_o}{X_o} - \frac{I_o L_o}{X_o} i_1 + \frac{I_o^2 L_o}{X_o^2} x_1 \tag{3.35}$$

The quiescent operating point can now be obtained by separating the time-dependent parts, and putting $x_1 = v_1 = i_1 = 0$, to yield

$$V_o = RI_o \tag{3.36}$$

and

$$-k(X_o - l_o) = \frac{1}{2} \frac{I_o^2 L_o}{X_o} \tag{3.37}$$

The equilibrium point exists only if Eqs. (3.36) and (3.37) hold for realistic cases. This is illustrated in Fig. 3-10, where the spring force and the electrical force (for three different currents) are plotted as a function of the distance x. (Notice that the spring force is opposite to the electrical force.) An equilibrium point exists if the line denoting the spring force intersects the curve for the electrical force. Thus, for a current I_3 there is no equilibrium point, and the mass M in Fig. 3-10 will hit the core. For the current I_2, the point C indicates that a slight disturbance from this point will cause the mass to be pulled down by the electrical force and that the mass will hit the core. For the current I_1, there are two equilibrium points A and B. If the mass is at point B, a slight disturbance in the upward direction increases the spring force and consequently the mass tends to be pulled down to position B. Similarly, a slight disturbance in the downward direction from point B will result in a decreased spring force and the mass will return to point B again. Therefore, point B represents stable equilibrium. Similar reasoning will indicate that point A represents unstable equilibrium.

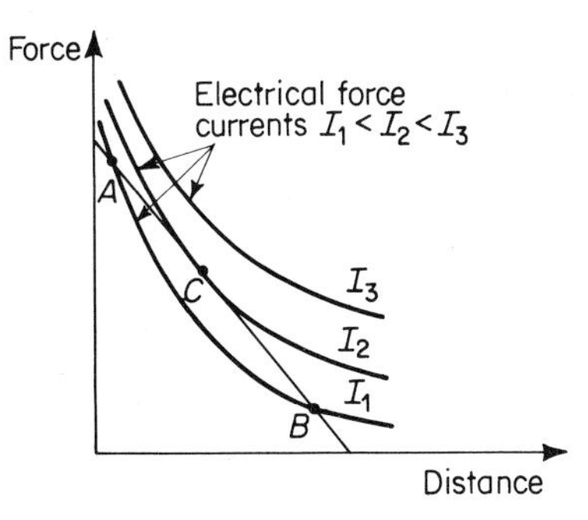

Fig. 3-10 Determination of the equilibrium point.

The behavior for small disturbances about the equilibrium point is described by the following equations:

$$L_o \frac{di_1}{dt} + Ri_1 - \gamma \frac{dx_1}{dt} = v_1 \tag{3.38}$$

$$M \frac{d^2x_1}{dt^2} + b \frac{dx_1}{dt} + \beta x_1 + \gamma i_1 = 0 \tag{3.39}$$

where

$$\gamma = \frac{L_o I_o}{X_o} \quad \text{and} \quad \beta = k - \frac{L_o I_o^2}{X_o^2}$$

The remaining steps in the study of this system are identical to those for the preceding two examples. The particular purpose of this example is to determine the existence of the stable equilibrium point.

3.4 *Instantaneous, Average, and RMS Values*

The instantaneous and average values of a force of electromagnetic origin are important to the study of electromechanical systems. For instance, to start the mechanical motion the instantaneous force must be nonzero.

Similarly, for continuous motion the average force must be nonzero. The average force is defined as

$$\langle F \rangle = \frac{1}{T}\int_0^T F_{\text{inst}}\, dt \tag{3.40}$$

where T is a suitable period and F_{inst} is the instantaneous force.

From the definition of the average force a new interpretation of the RMS (root-mean-square) value emerges. This is illustrated by the following equations. For a current-excited inductive system the instantaneous force is

$$F_i = \frac{1}{2} i^2 \frac{\partial L}{\partial x} \tag{3.41}$$

where the current i may have any waveform. The average force is, from Eqs. (3.40) and (3.41),

$$\langle F \rangle = \frac{1}{2}\frac{\partial L}{\partial x}\left[\frac{1}{T}\int_0^T i^2\, dt\right] \tag{3.42}$$

If a direct current I_{dc} excites the system, the average force is

$$\langle F' \rangle = \frac{1}{2}\frac{\partial L}{\partial x} I_{\text{dc}}^2 \tag{3.43}$$

The value of the direct current which produces the same average force is, therefore, from Eqs. (3.42) and (3.43),

$$I_{\text{dc}} = \left(\frac{1}{T}\int_0^T i^2\, dt\right)^{1/2} \tag{3.44}$$

Recall that Eq. (3.44) is identical to the definition of the RMS or effective value. Thus, in order that the same average force be exerted for both ac and dc excitations, it is necessary that the RMS value of the alternating current be equal to the magnitude of the direct current.

3.5 A Voltage-Excited System

The average force exerted on a movable iron is to be found for a voltage-excited system. The coil has N turns and its resistance is R ohms. The leakage fields are neglected (Fig. 3-11).

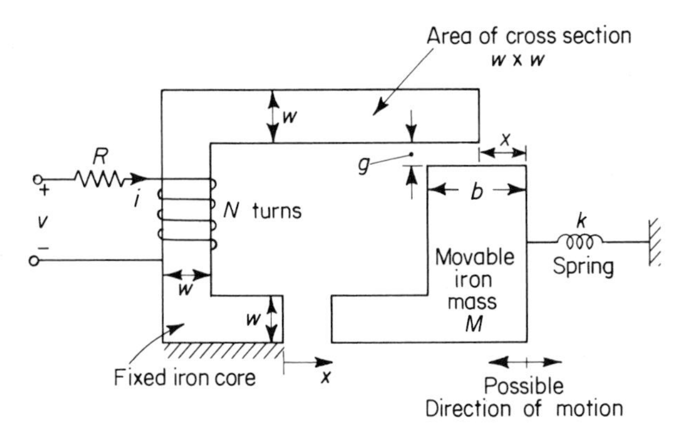

Fig. 3-11 A voltage-excited system.

This system is different from those of the preceding examples in that it is a voltage-excited system. In this case the instantaneous force (see Problem 3-13) is given by

$$F_e = -\frac{1}{2}\phi^2\frac{\partial \mathscr{R}}{\partial x}$$

Now, the reluctance $\mathscr{R}$ is given by

$$\mathscr{R} = \frac{1}{\mu_o w}\left[\frac{bx-x^2+gw}{w(b-x)}\right]$$

and the flux ϕ is obtained from the voltage-balance equation

$$v = Ri+N\frac{d\phi}{dt}, \quad \text{where } \phi = \frac{Ni}{\mathscr{R}}$$

Or, for sinusoidal excitation, where $v = V_m \sin \omega t$, the solution to this equation is

$$\phi = \phi_m \sin(\omega t+\psi)$$

where

$$\phi_m = \frac{NV_m}{\sqrt{(\mathscr{R}R)^2+N^4\omega^2}}$$

and

$$\psi = \tan^{-1}\frac{\omega N^2}{R\mathscr{R}}$$

Consequently, the instantaneous force is

$$F_e = -\frac{1}{2}\frac{1}{\mu_o w}\frac{N^2V_m^2}{[(\mathscr{R}R)^2+N^4\omega^2]}\left[\frac{1}{w}+\frac{g}{(b-x)^2}\right]\sin^2\left[\omega t+\tan^{-1}\frac{\omega N^2}{\mathscr{R}R}\right]$$

And the average force is

$$\langle F_e \rangle = \frac{V_m^2}{4\mu_o w}\left[\frac{1}{w}+\frac{g}{(b-x)^2}\right]\frac{N^2}{(\mathscr{R}R)^2+(N^2\omega)^2}$$

Two cases are of interest: (1) when the coil resistance is negligible, the average force is

$$\langle F_e \rangle = \frac{1}{4\mu_o w(\omega N)^2} V_m^2 \left(\frac{1}{w}+\frac{g}{(b-x)^2}\right)$$

(2) when $x = 0$ this force becomes

$$\langle F_e \rangle\big|_{x=R=0} = \frac{V_m^2}{4\mu_o w\omega^2 N^2}\left(\frac{1}{w}+\frac{g}{b^2}\right)$$

3.6 Multiwinding Systems

In the preceding examples, we considered devices each with one input and one output. However, there exist a large number of energy conversion devices which carry more than one winding for input and/or output. Some common examples of such devices are: polyphase rotating machines and dc machines used for control purposes such as the amplidyne or other multiply-excited dc machines. In general, we call such devices multiwinding systems. The theory developed for singly-excited systems is also applicable to multi-winding systems.

As pointed out in Sec. 2.4, for a multiwinding system the equations of motion are best expressed in matrix notation. For an n-winding angular-motion system these equations can be written as

Electrical:

$$\begin{bmatrix} v_1 \\ v_2 \\ \cdot \\ \cdot \\ \cdot \\ v_n \end{bmatrix} = \begin{bmatrix} R_1 & & & & \\ & R_2 & & & \\ & & \cdot & & \\ & & & \cdot & \\ & & & & \cdot & \\ & & & & & R_n \end{bmatrix} \begin{bmatrix} i_1 \\ i_2 \\ \cdot \\ \cdot \\ \cdot \\ i_n \end{bmatrix} + \frac{d}{dt}\begin{bmatrix} L_{11} & L_{12} & \cdots & L_{1n} \\ L_{12} & L_{22} & \cdots & L_{2n} \\ \cdot & & & \cdot \\ \cdot & & & \cdot \\ \cdot & & & \cdot \\ L_{1n} & L_{2n} & \cdots & L_{nn} \end{bmatrix} \begin{bmatrix} i_1 \\ i_2 \\ \cdot \\ \cdot \\ \cdot \\ i_n \end{bmatrix} \tag{3.45}$$

Mechanical:

$$J\ddot{\theta}+b\dot{\theta}+k\theta = \frac{1}{2}[\tilde{i}]\frac{\partial}{\partial\theta}[L][i] \tag{3.46}$$

Here an angular-motion electromechanical system is assumed. In Eq. (3.46) $[\tilde{i}]$ is a row matrix corresponding to the transpose of the current column matrix $[i]$. The L matrix is the same as that given in Eq. (3.45). The remaining details are identical to those for singly-excited systems previously discussed.

3.7 Current and Flux Variations

In a voltage-excited system it is interesting to investigate the variation of the input current as a function of time. The following discussion is meant to be of qualitative nature only.

For the sake of illustration, consider the system shown in Fig. 3-11. With no applied voltage the movable iron is, say, at a distance x_o from the core. The corresponding inductance is L_o, the minimum value of the inductance, and the time constant τ_o is L_o/R. If the arm is held in the original position and a step voltage is applied, the circuit behaves like an RL circuit, with time constant τ_o, and final current is V/R where V is the applied voltage. Now, if the iron is allowed to move and its final position is x_f, the inductance of the circuit increases to L_f and the corresponding time constant is $\tau_f = L_f/R$. Evidently $\tau_f > \tau_o$. For the initial and final positions of the movable iron the currents are respectively shown by curves (a) and (b) in Fig. 3-12.

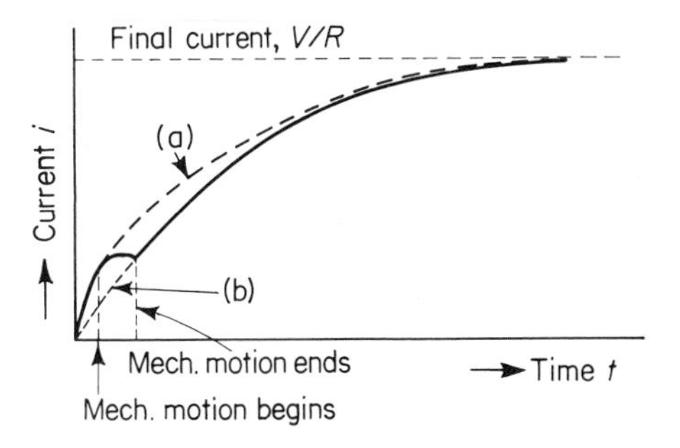

Fig. 3-12 Variation of i with t.

However, the transition from (a) to (b) is not smooth because as soon as the iron starts moving, the mechanical time constant τ_m of the system comes into play. It should be recognized that $\tau_m > \tau_f > \tau_o$. This explains the nature of the current variation when the movable iron is allowed to move at constant voltage.

In case of constant current excitation, the variations of flux are of interest. For initial position x_o the reluctance is maximum and the corresponding flux ϕ_o is a minimum. The flux reaches its maximum value ϕ_f when the motion is complete. The change of flux from ϕ_o to ϕ_f follows a path similar to that indicated in Fig. 3-13. The variation of the flux is governed by the mechanical as well as by the electrical time constants, as in the preceding case. It may be of interest to obtain the variation of flux as a function of time (Problem 3-11).

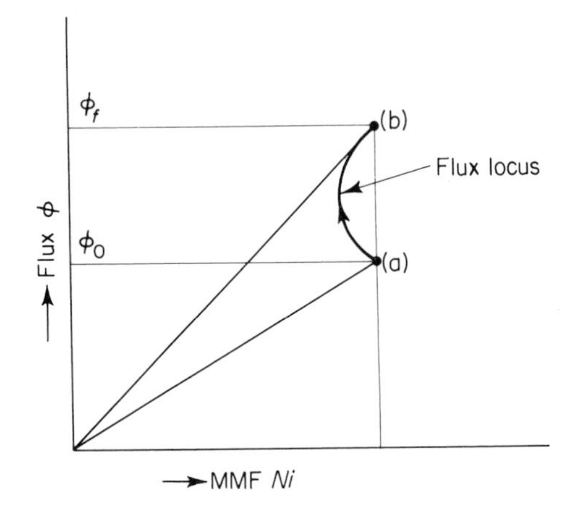

Fig. 3-13 Flux–MMF variation.

3.8 Effects of Saturation

The effect of saturation on stored magnetic energy is illustrated in Fig. 2-7. It can be readily verified that the end result in the form of the force equation $F_e = \frac{1}{2} i^2 \; (\partial L/\partial x)$ is not directly applicable. Nevertheless, the energy-conservation principles are still valid and, if saturation has to be taken into account, it is best to begin with the energy-balance equation

$$dW_{\text{elec}} = dW_{\text{mag}} + dW_{\text{mech}}$$

for a conservative system. This leads to Eq. (2.68), which is the most general equation. The following example illustrates the point that energy-conservation principles and a direct application of Eq. (2.68) yield identical results. This is not surprising, because Eq. (2.68) has been derived from the energy-balance equation without imposing the constraint of linear relationship between current and flux linkage.

EXAMPLE 3–1

The flux linkage and current for the magnetic circuit of an iron plunger are related by the equation

$$\lambda = \frac{2(i^{1/2} + i^{1/3})}{x+1}$$

where x is the position of the plunger. It is desired to find the force of electromagnetic origin F_e when the plunger is at $x = 0$.
From Eq. (2.68),

$$F_e = -\frac{\partial W_m}{\partial x} + i\frac{\partial \lambda}{\partial x}$$

and from the given λ-i-relationship (holding x constant),

$$d\lambda = \frac{2}{(x+1)}\left(\frac{1}{2}i^{-1/2} + \frac{1}{3}i^{-2/3}\right)di$$

and

$$W_m = \int_0^i i\,d\lambda = \frac{2}{x+1}\int_0^i i\left(\frac{1}{2}i^{-1/2} + \frac{1}{3}i^{-2/3}\right)di = \frac{2}{x+1}\left(\frac{1}{3}i^{3/2} + \frac{1}{4}i^{4/3}\right)$$

Thus

$$-\frac{\partial W_m}{\partial x} = \frac{2}{(x+1)^2}\left(\frac{1}{3}i^{3/2} + \frac{1}{4}i^{4/3}\right)$$

and

$$i\frac{\partial \lambda}{\partial x} = \frac{-2}{(x+1)^2}(2i^{3/2} + i^{4/3})$$

Hence

$$F_e = \frac{-1}{(x+1)^2}\left(\frac{4}{3}i^{3/2} + \frac{3}{2}i^{4/3}\right) \quad \text{newton}$$

Or, at $x = 0$,

$$F_e = -\left(\frac{4}{3}i^{3/2} + \frac{3}{2}i^{4/3}\right) \quad \text{newton}$$

It can easily be verified that the same result is obtained starting from the energy-balance equation.

3.9 Summary

In this chapter a number of electromechanical systems were analyzed. Forces of electromagnetic origin were evaluated for voltage-excited and current-excited systems. The application of linearization technique was illustrated. The dynamical behavior of linearized systems was obtained from the equations of motion and from equivalent circuits. An example was considered to illustrate the method of accounting for saturation of the magnetic circuit.

PROBLEMS

3–1. An equivalent circuit of an electromechanical system is shown in Fig. 3P-1. Compare this circuit with that shown in Fig. 3-7, and obtain the electromechanical system which can be represented by this circuit. What are the realizability conditions for this circuit? Obtain the dual of the circuit. Does the dual also represent the same electromechanical system?

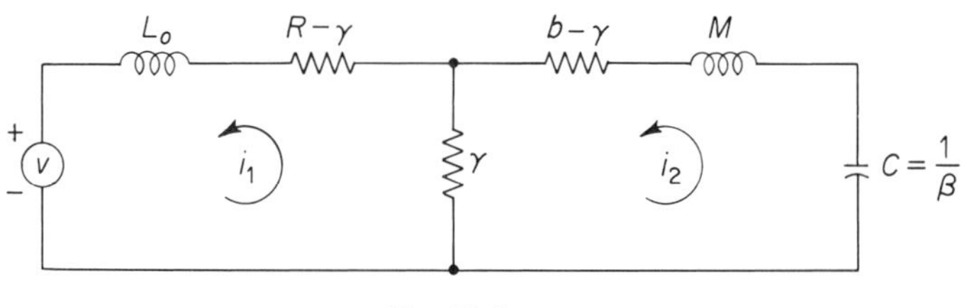

Fig. 3P-1

3–2. a. A permanent-magnet moving-coil meter has an inertia J, spring constant k, and friction coefficient b. The resistance of the coil is R and its inductance is negligible. Defining the position of the pointer as the output and the current (or voltage), which the meter is designed to measure as input, obtain the transfer function for the meter. Assume that the electrical torque exerted on the moving system is proportional to the current through the coil and that the friction force is proportional to the velocity.

b. If the moment of inertia of the moving system of the meter is 3×10^{-8} kg-m^2, the spring constant is 2×10^{-8} kg-m/rad, and the friction coefficient is 1×10^{-8} kg-m/rad/sec. Calculate the damping ratio and the natural frequency of undamped oscillations. Plot the exact frequency response, showing clearly the behavior near the break frequency. Give an asymptotic approximation to the frequency response, and indicate the frequency range for which the meter is suitable.

c. Obtain the step response of the meter given in part b. This gives an indication of the behavior of the meter on dc. Is the performance satisfactory? If not, name the undesirable characteristics, and state what parameters need be changed to improve the performance. It is preferred that some numerical values be suggested. Is the system under-, over-, or critically-damped?

3–3. The following data relate to the system shown in Fig. 3P-3: R_1, R_2 = resistances of the two coils; i_1, i_2 = specified currents; N_1, N_2 = number of turns; μ_o = permeability of free space; A = area of cross-section of the center limb and of the outer limbs; M = mass of the armature; k = spring constant; b = friction coefficient; l_o = length of the spring when $i = 0$. Assume all energy to be stored in the airgap.

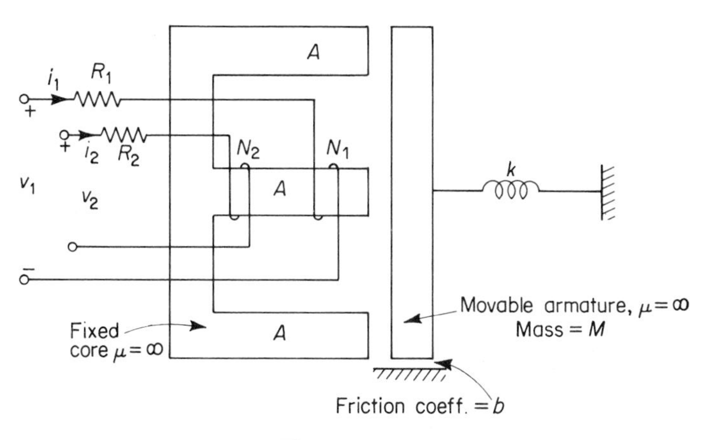

Fig. 3P-3

a. If $i_1 = I_{dc}$ and $i_2 = I_m \sin t$, write
   i. the electrical equations of motion in matrix notation;
   ii. the mechanical equation of motion.
   iii. If the above equations are nonlinear, identify the nonlinear terms.

b. If $i_1 = I_{dc}$ and $i_2 = 0$,
   i. obtain the quiescent operating point and linearize the equation about this point;
   ii. determine the transfer function of the system.

3–4. The plunger of infinite permeability shown in Fig. 3P-4 is free to move into the coil along a frictionless path.

a. If $L(x) = k/(x+a)$ where a is a constant, write the mechanical equation of motion after the switch has been closed. The current I is maintained constant. Solve the equation of motion.

b. Explain physically the behavior of the system in terms of mechanical motion.

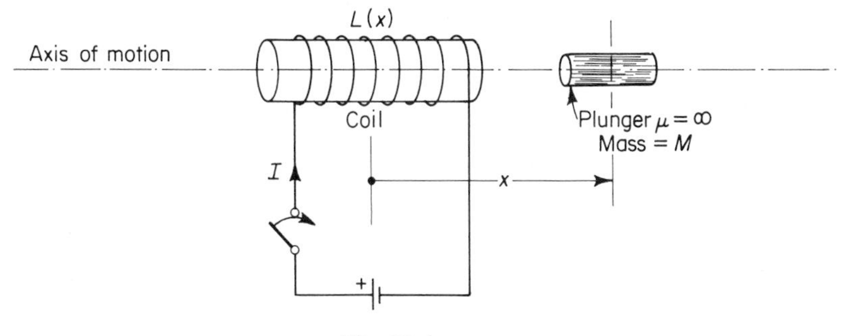

Fig. 3P-4

3–5. Given the electromechanical system shown in Fig. 3P-5, derive an expression for the electrical force developed. State all the simplifying assumptions made in this derivation. Write down the electrical and mechanical equations of motion. Find the equilibrium position. What is direction of the electrical force? Next assume that the spring is detached and friction ignored. An external force is applied to pull the plunger through a distance d. Calculate the change in the stored energy in the coupling field. If there is a change in the stored energy, where does it go? Or, where does it come from, as the case may be?

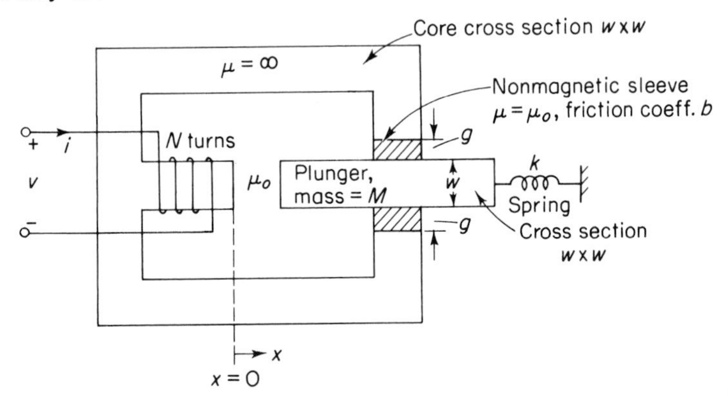

Fig. 3P-5

3–6. The cross-sectional area of the magnetic circuit shown in Fig. 3P-6 is A. The slide has a viscous damping force $F_{\text{fric}} = b\dot{x}$ opposing motion of the slide and acting at the support. Position $x = x_o$ due to the spring with spring constant k is obtained with no electrical excitation. With $e = E_m \cos \omega t$ and negligible winding resistance, determine the following in terms of the given symbols:

a. the magnetic force acting on the slide;
b. the distance x as a function of time in the steady state, and the condition for the validity of this expression;
c. the average power supplied by the voltage source if the magnetic material is lossless.

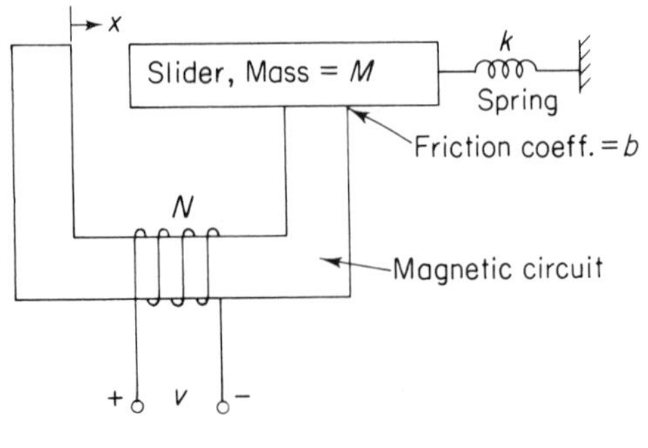

Fig. 3P-6

3–7. Draw the dual of the circuit shown in Fig. 3P-1.

3–8. Reduce the block diagram shown in Fig. 3-8 to obtain the transfer function $G(s) = X(s)/V_1(s)$. Check that the same result is obtained from the signal-flow graph.

3–9. Draw a block diagram for the meter of Problem 3-2. Simplify the block diagram to obtain the transfer function. Give an equivalent circuit for the meter, and label all parameters involved in the circuit.

3–10. An electromechanical system is shown in Fig. 3P-10. The armature is supported by three springs, each of spring constant k. The area of cross-section of each limb of the core is A m^2 and the limbs I and III each carry N_1 turns, as shown. A second coil of N_2 turns is wound on limb II. The mass of the armature is M. The permeability of free space is μ_o and that of steel infinite. A current i flows in the first coil. Making any simplifying assumptions if necessary:

a. Calculate e_2.
b. Calculate the force exerted on the armature.
c. Write the electrical and mechanical equations of motion.
d. Repeat part b if the terminals 22′ are short-circuited.
e. Express the inductance of the system in matrix notation.

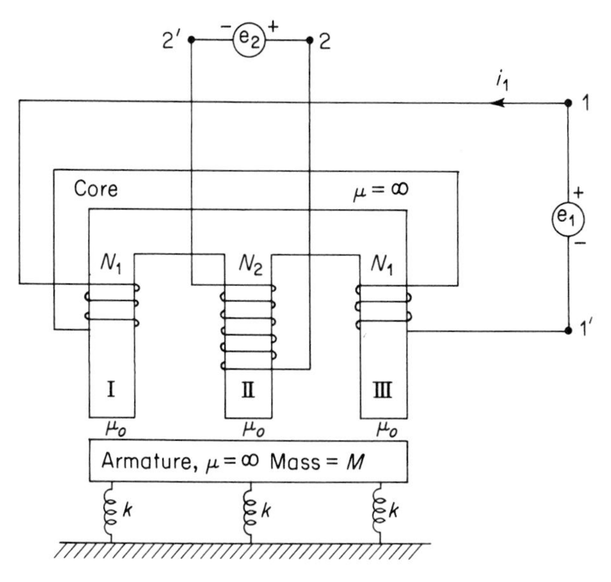

Fig. 3P-10

3–11. a. For the system shown in Fig. 3-11 obtain an explicit expression for the input current as a function of time. Choose suitable numerical values of the system parameters and plot the current–time relationship.

b. If the system of Fig. 3-11 is current-excited, repeat part a for flux variations.

3–12. For the system shown in Fig. 3P-12, choose a suitable coordinate system.

a. Using Lagrangian formulation, obtain the equations of motion.
b. Verify that the same equations are obtained from the force-balance equations.
c. If the applied voltage is $v = V_m \sin \omega t$, find the frequency ω so that the iron mass may be in an equilibrium position.
d. The capacitance may be considered to act as a spring. Determine the equivalent spring constant.
e. A mechanical spring may undergo a permanent set. Is it possible for this to happen in an "equivalent electrical spring"?

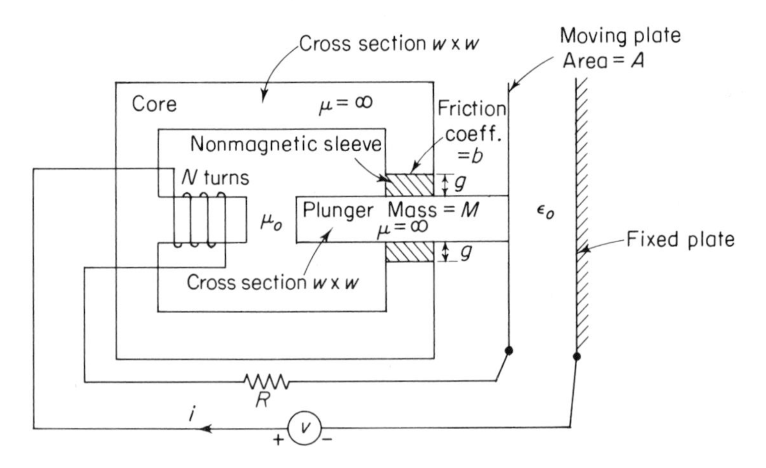

Fig. 3P-12

3–13. Derive the force equation

$$F_e = -\frac{1}{2}\phi^2 \frac{\partial \mathscr{R}}{\partial x}$$

of Sec. 3.5.

REFERENCES

1. White, D. C., and H. H. Woodson, *Electromechanical Energy Conversion*. New York: John Wiley & Sons, Inc., 1959.
2. Seely, S., *Electromechanical Energy Conversion*. New York: McGraw-Hill Book Company, 1962.
3. Nasar, S. A., "A Theoretical and Experimental Study of an Electromechanical System," *Inter. Jour. E. E. Education*, July 1965, p. 225. Oxford: Pergamon Press, Ltd.

4

Energy Conversion in Continuous Media

4.1 General Remarks

In the last chapter we studied certain examples of lumped-parameter incremental-motion electromechanical systems. These examples illustrated the applications of some of the techniques presented in Chapter 2. In the present chapter we shall consider some distributed-parameter devices and study the energy-conversion process in continuous media. Some of the outstanding examples of energy conversion in continuous media are magnetohydrodynamic (MHD) power converters, liquid-metal pumps, conventional eddy-current motors, and homopolar or acyclic machines.

Energy-conversion processes involve macroscopic interactions between current-carrying conductors and electromagnetic fields, with relative motion between the conductors and the fields. In practice, in dc devices the fields (excited by voltage or current sources) are generally stationary and the conductors are allowed to move, whereas in ac devices the field and the conductor may both be moving but with different velocities. In an MHD converter, the moving conductor may be an ionized gas (or plasma); in a liquid-metal pump, a molten metal may constitute the moving conductor; and in a conventional eddy-current motor, a copper disc or conducting-sleeve rotor may play the part of moving conductor.

Such moving conductors consist of continuous media, such as an ionized gas or a molten metal or a copper plate, and in the analysis of such electromagnetic devices field equations should be used. Starting with Maxwell's equations (outlined in Chapter 2), a number of equations will be derived in this chapter. These equations can be conveniently applied to problems involving distributed parameters or continuous media. In this chapter a simplified approach to such problems will be presented.

4.2 Simplifying Assumptions

A number of simplifying assumptions are made for the following discussions, and it is best to state these assumptions at the outset. These are:

1. Electric and magnetic circuits are linear; and for all practical purposes $\mu_{air} \ll \mu_{iron}$, so that all magnetic energy may be assumed to be stored in the airgap.
2. Space and time variations of electromagnetic fields are sinusoidal unless stated otherwise.
3. The relative velocity between the conductor and the field is small compared to the velocity of light, so that relativistic effects are negligible.
4. Current flows in one direction only (z in a rectangular coordinate system) and the effects of circumferential currents (in the x direction) are not considered.
5. Edge effects, fringing, and leakage fields are neglected.
6. Idealized models are chosen to replace actual devices.
7. Velocities involved are considered to be uniform, and acceleration is not taken into account.

A few words of justification of the above assumptions are perhaps in order. We assume the permeability of iron to be much greater than that of air. This is a realistic assumption; for instance, the relative permeability of iron (with 0.2% impurity) is about 5000. However, magnetic circuits cannot be assumed to be linear under all circumstances. We assume such a linearity as a first-order approximation. Refinements in solutions can be made to take into account the saturation of a magnetic circuit by a number of methods. We shall consider an example in Chapter 9. For the present we neglect saturation. We assume the space variation of the magnetic field in a given device to be sinusoidal because in actual machines every attempt is made to dispose the current-carrying windings (called *exciting windings*) such that the resulting magnetic fields are sinusoidal. In some cases, other considerations lead to nonsinusoidal winding distributions. The magnetic field produced by such windings can be resolved into a series of *harmonics* by Fourier analysis. For the present we assume sinusoidal space variations. The results thus derived can easily be extended to take into account harmonic fields, as will be shown subsequently.

Assumption 3 is obviously a realistic assumption. Assumption 4 implies that current-carrying conductors are embedded in slots which are axially directed. In many machines the slots are slightly twisted, or *skewed.* We shall not consider these for the present. Likewise, it is extremely difficult to take into account the effects mentioned in assumption 5, and we shall neglect these for the sake of simplicity, at least for the time being. Assumption 6 requires us to choose models amenable to analysis. This is a very practical step. But care must be exercised to choose models which take into account as many of the features of the actual devices as possible. Assumption 7 covers a wide range of operation of many devices. Consequently, we shall assume uniform velocities in the following discussions.

4.3 Production of a Traveling Field

It was pointed out earlier that the energy-conversion process results from the interaction of current-carrying conductors and electromagnetic fields. There is always relative motion between the conductor and the field, and often (in most ac devices) the field also travels, although with a different velocity than that of the moving conductor. In fact, the operation of a large class of ac electromagnetic energy-conversion devices, such as induction and synchronous machines, is based on the existence of a traveling magnetic field. A number of questions might be asked at this point. Why is it at all necessary to have a traveling magnetic field? Where do these fields exist and how are they produced? How are the fields evaluated quantitatively? What are the conditions that must be fulfilled to establish a traveling field? And so on. These questions shall now be considered, but only partial answers can be given at this stage.

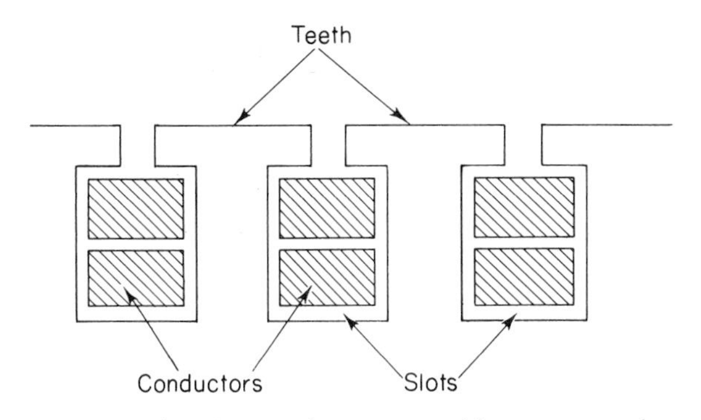

Fig. 4-1 A slotted magnetic structure with current-carrying conductors.

Recall from Example 2-1 that the time-average force on a conducting sheet initially at rest in a time-varying magnetic field is zero. Consequently, the sheet remains stationary unless acted upon by some other force. On the other hand, from Problem 2-2 it follows that the time-average force on the stationary conducting sheet placed in a traveling magnetic field is nonzero, and thus the traveling field will "drag" the sheet along and impart continuous motion to it as long as the velocity of the conducting sheet is less than the velocity of the traveling field. Therefore, for starting the motion of the conductor, the existence of the traveling field is necessary (such as in an induction motor). Once the conductor begins to move, even a pulsating field will sustain the motion (as shown in Example 2-2), although it might be preferable to have a traveling field for continuous motion also, because the force in this case is more uniform than in the case of a purely pulsating field.

Having discussed the need for a traveling field, we next consider its production. In electromagnetic devices, the sources of electromagnetic fields are current-carrying windings embedded in slotted magnetic structures, as shown in Fig. 4-1. These windings are generally on the stationary member, or stator, of the device. The windings are so disposed that the resulting magnetic flux-density distribution has ideally a sinusoidal variation in space. In reality, however, harmonic fields are often present, although their presence is undesirable. From the viewpoint of energy conversion, the region in which such fields are of greatest interest is the airgap. Fields in other regions cause secondary and undesired effects.

Quantitatively, the fields are determined either by using Ampere's law or by solving Laplace's equation for the fields in the airgap for given sources and boundaries. For such solutions, an idealized model is chosen in which the actual slotted structure is replaced by a smooth surface and the current-carrying windings are replaced by fictitious infinitely thin current elements, called *current sheets*, having linear current densities. The procedure for the determination of electromagnetic fields is discussed in detail in later sections. For the present, we assume that these fields are known and, ideally, that their space and time variations are sinusoidal.

Turning now to the determination of the conditions necessary for the production of a traveling field, we consider two B fields which are displaced from each other by 90° in space and in time. Explicitly, let these flux densities be

$$B_1 = B_m \sin \beta x \sin \omega t \tag{4.1}$$

and

$$B_2 = B_m \sin (\beta x + 90^\circ) \sin (\omega t + 90^\circ) \tag{4.2}$$

as shown in Fig. 4-2. The resultant flux density is obtained by adding Eqs. (4.1) and (4.2), so that

$$B = B_1 + B_2 = B_m \cos (\omega t - \beta x) \tag{4.3}$$

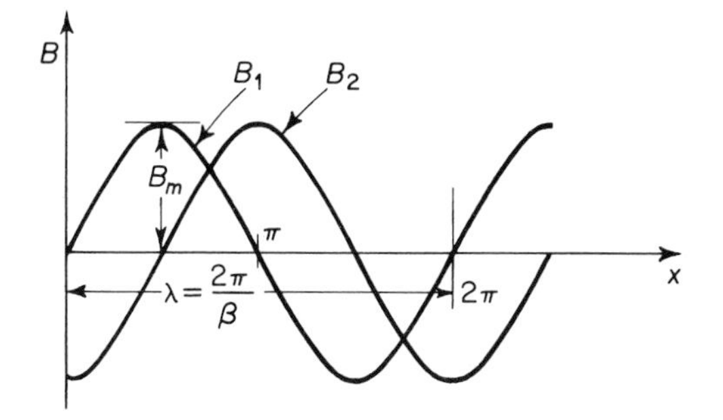

Fig. 4-2 Sinusoidal flux-density distribution resulting from a 2-phase excitation.

It can be readily verified that Eq. (4.3) is a solution to the wave equation, Eq. (2.28), and represents a traveling wave (Problem 2-2) which travels with a velocity $dx/dt = \omega/\beta$. The flux densities of the form given by Eqs. (4.1) and (4.2) are obtained from a set of two independent windings, displaced from each other by 90° in space and carrying currents having a phase displacement of 90° in time. For simplicity, we impose the further constraint that the windings must be sinusoidally distributed in space. This form of excitation, yielding the fields given by Eqs. (4.1) and (4.2), is termed a 2-*phase excitation.*

Therefore, for the production of a traveling field it is necessary to have a polyphase excitation, and there must be some overlap in space and time between the fields resulting from this excitation. (A general case of traveling-field production is given in Problem 4-1.)

Having discussed the production of a traveling field, we can now consider in some detail the different viewpoints and approaches leading to the mechanism of force production and its quantitative evaluation in eddy-current devices.

4.4 Traveling-Field Theory

In order fully to appreciate traveling-field theory, we shall first introduce the concept of *electromagnetic momentum.*[3,4]

We know from the laws of mechanics that a mechanical force F acting on a mass m and moving with a velocity u is given by

$$F = \frac{dG_m}{dt} = \frac{d}{dt}(mu) \tag{4.4a}$$

where

$$G_m = mu \tag{4.4b}$$

is the *mechanical momentum*. Equation (4.4a) can be rewritten as

$$F = \frac{d}{dt}\int_v (\rho_m u)\, dv \tag{4.4c}$$

where ρ_m is the mass density of the material. The principle of *conservation of momentum* is a basic concept in mechanics. Thus, for an isolated system with no forces acting on it we have

$$\frac{dG_m}{dt} = \frac{d}{dt}(mu) = \frac{d}{dt}\int_v (\rho_m u)\, dv = 0 \tag{4.5a}$$

In Eqs. (4.4c) and (4.5) the integration is over a volume v occupied by the mass.

Now, the Lorentz force, Eq. (2.1), arising from the presence of moving charges in electromagnetic fields is given by

$$\text{force} = \int_v (\rho\mathbf{E}+\mathbf{J}\times\mathbf{B})\, dv = \frac{dG_m}{dt} \tag{4.5b}$$

Substituting $\rho = \nabla\cdot\mathbf{D}$, from Eq. (2.19d), and $\mathbf{J} = (1/\mu)\,\nabla\times\mathbf{B} - \partial\mathbf{D}/\partial t$,

$$\frac{dG_m}{dt} = \int_v \left[\mathbf{E}(\nabla\cdot\mathbf{D}) + \mathbf{B}\times\frac{\partial\mathbf{D}}{\partial t} - \frac{\mathbf{B}}{\mu}\times\nabla\times\mathbf{B}\right] dv \tag{4.5c}$$

But

$$\mathbf{B}\times\frac{\partial\mathbf{D}}{\partial t} = -\frac{\partial}{\partial t}(\mathbf{D}\times\mathbf{B}) + \mathbf{D}\times\frac{\partial\mathbf{B}}{\partial t} \tag{4.5d}$$

so that Eq. (4.5c) can be rewritten, using Eqs. (2.19a,c), as

$$\frac{dG_m}{dt} = \frac{-\partial}{\partial t}\int_v (\mathbf{D}\times\mathbf{B})\, dv$$

$$+\int_v [\mathbf{E}\,(\nabla\cdot\mathbf{D}) + \mathbf{H}(\nabla\cdot\mathbf{B}) - (\mathbf{H}\times\nabla\times\mathbf{B} + \mathbf{D}\times\nabla\times\mathbf{E})]\, dv \tag{4.5e}$$

In the absence of free charges, and using Eq. (4.4c), Eq. (4.5e) can be expressed as

$$\frac{d}{dt}\int_v (\rho_m\mathbf{u} + \mathbf{D}\times\mathbf{B})\, dv = -\int (\mathbf{H}\times\nabla\times\mathbf{B} + \mathbf{D}\times\nabla\times\mathbf{E})\, dv \tag{4.5f}$$

The quantity $\rho_m\mathbf{u}$ is the mechanical-momentum density and $(\mathbf{D}\times\mathbf{B})$ can be termed the *electromagnetic-momentum* density q_n.[4,5] We know, from

Poynting's theorem (Sec. 2.1.2), that the power density **P** of an electromagnetic wave is given by

$$\mathbf{P} = \mathbf{E} \times \mathbf{H} \tag{4.6}$$

Thus, if the wave travels with a velocity **u** carrying an energy density W, from Eq. (4.6) we have

$$\mathbf{P} = \mathbf{E} \times \mathbf{H} = W\mathbf{u} \tag{4.7}$$

The electromagnetic-momentum density q_n associated with the moving energy density W is given by

$$q_n = \frac{P}{u_o^2} = \frac{Wu}{u_o^2} \tag{4.8}$$

where $u_o = 1/\sqrt{\mu\epsilon}$ = velocity of the wave in a lossless medium and u = velocity of the wave in the homogeneous-material medium with electrical parameters μ, ϵ, and σ.

Traveling-field theory assumes that the force is produced in a conductor placed in a traveling magnetic field by absorbing the momentum of the field. Thus, if the power absorbed by the conductor is P and the velocity of the field (moving tangentially over the surface of the conductor) is u_o, the force F on the conductor is given by

$$F = \frac{P}{u_o}$$

Now, if the conductor is allowed to move with a velocity u (where $u < u_o$), a quantity called *slip*, s, is defined as

$$s = \frac{u_o - u}{u_o} \tag{4.9}$$

Evidently, slip is a fractional measure of the relative velocity between the conductor and the field. In this case the force on the moving conductor is

$$F = \frac{P}{su_o} \tag{4.10}$$

where P is the power absorbed from the traveling field.

An inspection of Eq. (4.10) indicates that to find the force on a conductor moving in a traveling field the power dissipated in the conductor has to be evaluated. The steps involved are illustrated by the following example.

EXAMPLE 4–1

A conducting plate of conductivity σ, thickness h, and length l, moves with a velocity u in a traveling field given by

$$B = B_m \cos(\omega_o t - \beta x)$$

as shown in Fig. 4E-1. Using traveling-field theory, we shall find the mechanical force on the plate arising from the absorption of power from the traveling field. We assume that the flux penetrates the full thickness of the plate without any substantial decay.

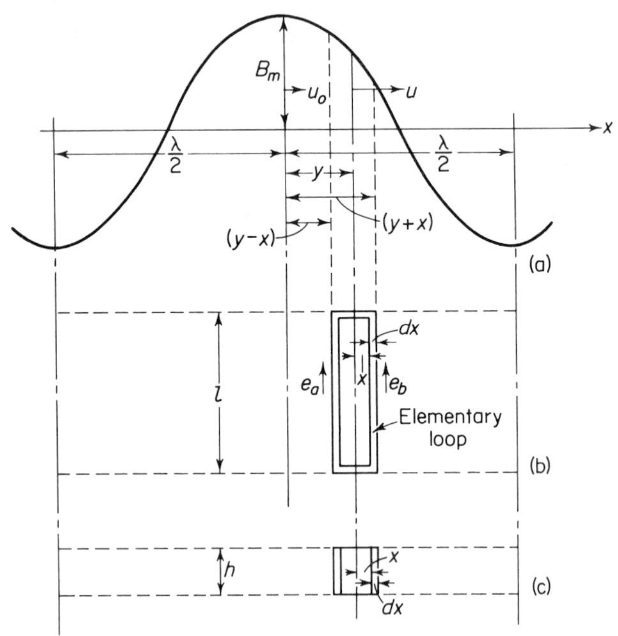

Fig. 4E-1 (a) Traveling B field; (b) plan; (c) cross-section. (Taken from Reference 1, with permission.)

An elementary loop of length l and width $2x$ is considered within the plate as shown in Fig. 4E-1. The resistance of the loop, R, is given by

$$R = \frac{2l}{\sigma h\, dx}$$

and its inductance is (from Reference 2)

$$L = \frac{\mu l}{\pi}\left(\log_e \frac{2x}{h} + 1.5\right)$$

From Fig. 4E-1, the velocity of the traveling field is also given by

$$u_o = \lambda f_o$$

where f_o is the frequency and λ is the wavelength of the traveling wave. Due to this traveling field, according to Faraday's law, an EMF e is induced in the loop. In terms of the dimensions shown in Fig. 4E-1, the EMF e is given by

$$e = e_a - e_b = B_m l(u_o - u)\left[\cos\frac{2\pi}{\lambda}(y-x) - \cos\frac{2\pi}{\lambda}(y+x)\right]$$

In terms of slip s, Eq. (4.9),

$$y = (u_o - u)t = su_o t$$

and

$$\frac{2\pi}{\lambda} y = s\omega_o t$$

Hence the expression for the EMF becomes

$$e = 2B_m lsu_o \left(\sin s\omega_o t \sin \frac{2\pi}{\lambda} x \right)$$

and the corresponding eddy current i induced in the plate is such that

$$e = Ri + L\frac{di}{dt}$$

Or,

$$i = \frac{1}{Z} 2B_m lsu_o \sin \frac{2\pi x}{\lambda} \sin (s\omega_o t - \psi)$$

where

$$Z = \sqrt{R^2 + (s\omega_o L)^2} \quad \text{and} \quad \tan \psi = \frac{s\omega_o L}{R}$$

The time-average power dissipated in the elementary loop under consideration is then

$$\langle dP \rangle = \langle i^2 R \rangle = \frac{B_m^2 ls^2 u_o^2 \sigma h}{1 + \tan^2 \psi} \sin^2 \left(\frac{2\pi x}{\lambda} \right) dx$$

At low frequencies (of the order of 60 cycles) and at small slips, $\tan^2\psi \ll 1$. This term can thus be neglected in the above expression. Therefore, the total power over a pole pitch, is

$$P = \int_0^{\lambda/2} \langle dP \rangle = \tfrac{1}{4} B_m^2 ls^2 u_o^2 \sigma h \lambda$$

And finally, the total mechanical force on the plate is, from Eq. (4.10),

$$F = \tfrac{1}{4} B_m^2 l \sigma h \lambda s u_o$$

4.5 The Concept of Wave Impedance

The preceding example illustrates the application of traveling-field theory in determining the force developed in a conducting plate moving in a traveling field. In Chapter 2 (Examples 2-1 and 2-2), we used the Lorentz force equation, Eq. (2.23), to determine the force exerted on a conducting sheet placed in a time-varying magnetic field. For energy-conversion devices with continuous media, the Lorentz force equation is still applicable. However, in contrast with conducting sheets of negligible thickness, the current-density distribution (of the induced currents) through conducting-material media of finite thickness is nonuniform. (*Note:* The reader familiar with *skin effect* will recall that the current-density distribution is frequency-dependent.) In this and the next section, we shall consider two methods of obtaining the force developed by a conducting plate moving in a traveling field. In using both methods, the problem reduces to a determination of the current-density distribution through the thickness of the conducting plate.

To illustrate the procedure, assume there exists a traveling field B_y given by

$$B_y = B_m \cos(\omega_o t - \beta x) = \text{Re}\,(B_m e^{j\omega_o t}\, e^{-j\beta x}) \tag{4.11a}$$

at the surface of the conducting plate ($y = 0$) shown in Fig. 4-3. In Eq.

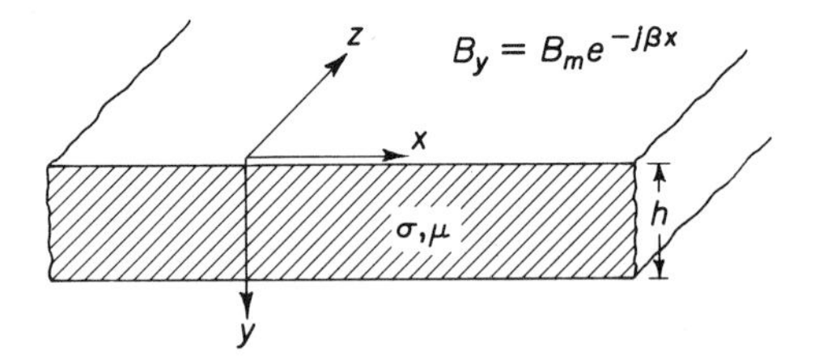

Fig. 4-3 A conducting plate in a traveling field.

(4.11a), Re denotes the real part of the quantity in parentheses. If the plate moves in this field with a slip s, Eq. (4.9), the field given by Eq. (4.11a) can be transformed to a set of coordinate axes attached to the moving plate by replacing ω_o by ω where $\omega = s\omega_o$. In other words, it may be said that the traveling field induces slip-frequency currents in the plate. Thus, the field referred to the plate-coordinate system is expressed as

$$B_y = \text{Re}\,(B_m e^{j\omega t} e^{-j\beta x}) \tag{4.11b}$$

where Re denotes the real part of the quantity in parentheses. For convenience, Eq. (4.11b) is sometimes written as

$$B_y = B_m e^{-j\beta x} \tag{4.11c}$$

where Re and $e^{j\omega t}$ are assumed to be understood.

As mentioned in Sec. 4.2, currents are assumed to flow in the z direction only. For a point (x, y, z) within the conducting plate, Maxwell's equation

$$\nabla \times \mathbf{E} = -\frac{\partial \mathbf{B}}{\partial t}$$

yields

$$-\frac{\partial E_z}{\partial y} = j\omega\mu H_x \tag{4.12}$$

Similarly,

$$\nabla \times \mathbf{H} = \mathbf{J}$$

together with

$$\mathbf{J} = \sigma \mathbf{E}$$

yields (see Problem 4-16)

$$-\frac{\partial H_x}{\partial y} = \left(\sigma + \frac{\beta^2}{j\omega\mu}\right) E_z \tag{4.13}$$

In Eqs. (4.12) and (4.13) sinusoidal space and time variations imply that $-\partial/\partial x = j\beta$ and that $\partial/\partial t = j\omega$. Note that Eqs. (4.12) and (4.13) are, respectively, similar to the uniform transmission-line equations

$$-\frac{\partial V}{\partial y} = ZI \tag{4.14a}$$

and

$$-\frac{\partial I}{\partial y} = YV \tag{4.14b}$$

where the symbols have their usual meanings. Therefore, a characteristic wave impedance Z_o and a propagation constant, γ, can be defined as

$$Z_o = \frac{j\omega\mu}{\gamma} \tag{4.15a}$$

and

$$\gamma = \sqrt{j\omega\mu\left(\sigma + \frac{\beta^2}{j\omega\mu}\right)} \tag{4.15b}$$

It is further seen from Eqs. (4.12–4.15) that E_z is analogous to the voltage V and that H_x corresponds to the current I. Hence, transmission-line theory is applicable to the determination of the electric-field distribution within the conducting plate. From the field E_z, the current density J_z is obtained (using Ohm's law) and an application of the Lorentz force equation gives the force density at the point under consideration. The following example should clarify the various steps involved in applying the theory developed in this section.

EXAMPLE 4–2

There exists a type of induction motor known as the *sleeve-rotor motor*. As the name suggests, the rotor of such a motor consists of an iron core with a "sleeve" of conducting material, such as copper, mounted on it. The airgap flux-density distribution is given by

$$B_y = B_m \cos(\omega_o t - \beta x)$$

This flux is produced by stator polyphase excitation of frequency ω_o. The rotor runs in this field with a slip s. If the thickness of the rotor sleeve is h, we wish to obtain the force per unit area of the rotor surface.

The idealized model of the rotor is shown in Fig. 4E-2. The conducting sleeve

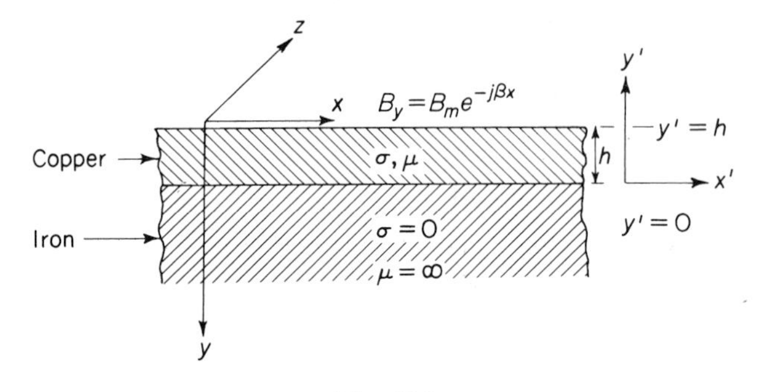

Fig. 4E-2

is backed by an iron boundary having infinite permeability. The airgap field can be transferred to the rotor coordinates by replacing ω_o by ω, where $\omega = s\omega_o$. Now, using the concept of wave impedance, the electric field E_z for a point within the conducting sleeve can be obtained from Eqs. (4.12) and (4.13). The boundary condition is that at the interface of the conducting sleeve and the iron core $H_x = 0$, since the permeability of iron is infinity. This corresponds to the case of an open-circuited transmission line. For convenience, therefore, a new origin is chosen and the new set of axes are labeled as x' and y', as shown in Fig. 4E-2. Recall from the analysis of the last section that an electric field is analogous to a voltage on a transmission line. We know from the theory of transmission lines that for an open-circuited line of length h and propagation constant γ, having a sending-end voltage V_1, the voltage distribution along the line is given by

$$V(y) = V_1 \frac{\cosh \gamma y}{\cosh \gamma h} \tag{a}$$

Correspondingly, the electric field within the sleeve is then given by

$$E_z = E_{z1} \frac{\cosh \gamma y'}{\cosh \gamma h} \tag{b}$$

where $E_z = E_{z1}$ at $y' = h$ corresponds to the sending-end voltage of the transmission line. The propagation constant γ is given by Eq. (4.15b).

The force per unit volume of the rotor sleeve is

$$f_x = -J_z B_y \tag{c}$$

and, from

$$\nabla \times \mathbf{E} = -\frac{\partial \mathbf{B}}{\partial t} \tag{d}$$

we have, since $\partial/\partial x = -j\beta$ and $\partial/\partial t = j\omega$,

$$E_{z1} = -\frac{\omega}{\beta} B_y \tag{e}$$

so that the time-average value of the force, $\langle f_x \rangle$ is (for sinusoidal time-variations)

$$\langle f_x \rangle = -\tfrac{1}{2}\,\mathrm{Re}(J_z B_y^*) \tag{f}$$

$$= \frac{\sigma\beta}{2\omega} E_z E_z^*, \quad \text{from Eq. (e)} \tag{g}$$

$$= \frac{\sigma\omega}{2\beta} B_m^2 \frac{\cosh \gamma y' \cosh \gamma^* y'}{\cosh \gamma h \cosh \gamma^* h}, \text{ from Eq. (b)} \tag{h}$$

where the superscript * denotes the complex conjugate.

Or, the force density per unit area of the rotor F_x is

$$F_x = \int_0^h \langle f_x \rangle \, dy' \tag{i}$$

Substituting the above expression for $\langle f_x \rangle$ in this integral finally yields (with $\gamma = \alpha + j\alpha$)

$$F_x = \frac{\omega\sigma}{8\alpha\beta} B_m^2 \frac{\sinh 2\alpha h + \sin 2\alpha h}{\cosh \gamma h \cosh \gamma^* h} \tag{j}$$

where

$$\alpha = \sqrt{\frac{s\omega_o \mu\sigma}{2}} \tag{k}$$

If the following approximations are made (Problem 4-17)

$$\gamma = \alpha(1+j) \tag{l}$$

$$\cosh \gamma h = 1 \quad \text{and} \quad \sinh \gamma h = \gamma h \tag{m}$$

the approximate value of the force per unit area of the rotor surface is given by

$$F_x|_{\text{approx}} = \frac{1}{2}\frac{\sigma\omega}{\beta} B_m^2 h$$

It can be readily verified that this expression is identical to that derived in Example 4-1.

The slip-force characteristics are given by the expression derived above if ω is replaced by $s\omega_o$.

4.6 An Alternative Approach

So far we have discussed two methods of force calculation in eddy-current devices. We have also seen that to obtain the force using the Lorentz force equation, the electric-field distribution within the conducting medium must be found. One way of doing this is by using the concept of wave impedance. An alternate, but not entirely independent, method is by using Maxwell's equations and deriving the electric-field equation as follows.

Differentiating Eq. (4.12) with respect to y and substituting for $-(\partial H_x/\partial y)$ from Eq. (4.13) leads to the equation

$$\frac{d^2E_z}{dy'^2} = \gamma^2 E_z \tag{4.16}$$

The solution to this equation is of the form

$$E_z = C_1 e^{\gamma y'} + C_2 e^{-\gamma y'} \tag{4.17}$$

where C_1 and C_2 are constants to be evaluated from boundary conditions. From Eq. (4.17), the electric field E_z is known for any point within the conducting sheet. Having determined the electric-field intensity, the current density J_z and the force density f_x can be determined in exactly the same manner as in the last example.

From Eqs. (4.12) and (4.17) we have, at $y' = 0$,

$$\left.\frac{\partial E_z}{\partial y'}\right|_{y'=0} = C_1\gamma - C_2\gamma = j\omega\mu H_x|_{y'=0} = 0$$

since

$$H_x|_{y'=0} = 0$$

Hence,

$$C_1 = C_2$$

At $y' = h$ (Fig. 4-3), $E_z = E_{z1}$. Consequently,

$$E_{z1} = C_1 e^{\gamma h} + C_1 e^{-\gamma h} = 2C_1 \cosh \gamma h$$

Or,

$$C_1 = C_2 = \frac{E_{z1}}{2 \cosh \gamma h} \tag{4.18}$$

From Eqs. (4.17) and (4.18) we have

$$E_z = \frac{E_{z1}}{2 \cosh \gamma h}(e^{\gamma y'} + e^{-\gamma y'})$$

Or,

$$E_z = E_{z1}\frac{\cosh \gamma y'}{\cosh \gamma h}$$

which is identical to the expression for the electric field distribution derived in Example 4-2 using the concept of wave impedance and the transmission-line analogy. The subsequent steps for determining the force density are, therefore, identical to those of Example 4-2.

4.7 The Airgap-Field Equation[8]

The three methods—traveling-field theory, the concept of wave impedance, and the alternative approach—presented in the preceding sections begin with the assumption that the magnetic fields acting on the conducting material are known. We had indicated in Sec. 4.3 that these fields are determined by solving the appropriate field equation. We shall now formulate the field equation, the solution of which determines the fields.

We know that an electrical machine consists of a stationary member, the *stator*, and the moving member, the *rotor*. Stator and rotor both carry currents, and we take J^s and J^r as the stator and rotor linear-current densities, respectively. We choose an idealized model and a rectangular coordinate system as shown in Fig. 4-4. We assume the currents to flow in the

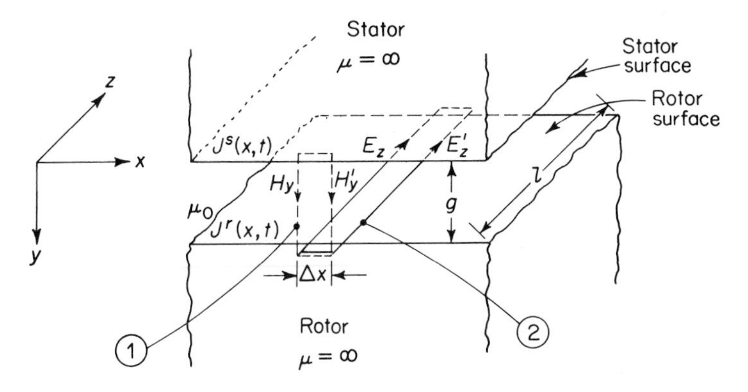

Fig. 4-4 An idealized model and paths of integration.

z direction only and the permeability of iron to be infinity. The stator and rotor surfaces support the respective current sheets, and the dimensions are as shown in Fig. 4-4. Let H_y and H_y' be the magnetic field intensities, and E_z and E_z' the electric-field intensities as shown. Then

$$H_y' = \frac{\partial H_y}{\partial x}\Delta x + H_y \tag{4.19a}$$

and

$$E_z' = \frac{\partial E_z}{\partial x}\Delta x + E_z \tag{4.19b}$$

Recall Maxwell's equations in the integral form as

$$\oint \mathbf{E}\cdot d\mathbf{l} = -\frac{\partial}{\partial t}\int_s \mathbf{B}\cdot d\mathbf{s} \tag{4.20a}$$

and

$$\oint \mathbf{H}\cdot d\mathbf{l} = \int_s \mathbf{J}\cdot d\mathbf{s} \tag{4.20b}$$

Integrating along path 1, from Eqs. (4.19a) and (4.20b) we have

$$g \frac{\partial H_y}{\partial x} \Delta x = (J^s + J^r)\,\Delta x$$

Or,

$$g \frac{\partial H_y}{\partial x} = J^s + J^r \tag{4.21a}$$

Similarly, integrating along path 2, from Eqs. (4.19b) and (4.20a) we have

$$-\frac{\partial E_z}{\partial x} \Delta x\cdot l = -\frac{\partial}{\partial t}(B_y\cdot \Delta x\cdot l)$$

Or,

$$-\frac{\partial E_z}{\partial x} = -\frac{\partial B_y}{\partial t} = -\mu_o \frac{\partial H_y}{\partial t} \tag{4.21b}$$

But, from Ohm's law,

$$J^r = \sigma E_z \tag{4.22}$$

Therefore, from Eqs. (4.21a–4.22) we have the desired airgap-field equation

$$g \frac{\partial^2 H_y}{\partial x^2} = \mu_o \sigma \frac{\partial H_y}{\partial t} + \frac{\partial J^s}{\partial x} \tag{4.23a}$$

Recalling that $B_y = \mu_o H_y$, we have the field equation in the final form as

$$\frac{\partial^2 B_y}{\partial x^2} - \frac{\mu_o \sigma}{g}\frac{\partial B_y}{\partial t} = \frac{\mu_o}{g}\frac{\partial J^s}{\partial x} \tag{4.23b}$$

Thus, if the stator current-sheet distribution is known we can obtain the airgap field, and proceed from there to carry out the force calculations by one of the methods discussed in the last three sections.

In the above derivation, we have assumed no relative motion between stator and rotor. If the rotor moves with a velocity u_x with respect to the stator, the electric field induced by this motion can be taken into account by using the equation

$$\nabla \times \mathbf{E} = -\frac{\partial \mathbf{B}}{\partial t} + \nabla \times \mathbf{u} \times \mathbf{B}$$

The modified airgap-field equation then becomes

$$\frac{\partial^2 B_y}{\partial x^2} - \frac{u_x \mu_o \sigma}{g}\frac{\partial B_y}{\partial x} - \frac{\mu_o \sigma}{g}\frac{\partial B_y}{\partial t} = \frac{\mu_o}{g}\frac{\partial J^s}{\partial x} \tag{4.23c}$$

The details of this derivation are left as an exercise (Problem 4-8).

4.8 Liquid-Metal Pumps

Liquid metals are of interest in nuclear reactors because of the heat-transfer properties of certain liquid metals, such as sodium, sodium-potassium, etc. The high conductivity of these liquid metals permits the use of electromagnetic pumps in circulating liquid coolants. Electromagnetic pumps utilize the pressure developed within the liquid metal when carrying current in the presence of magnetic fields. In this respect, an application of the Lorentz force equation leads to the force and pressure calculations for the pump.

Depending on how the current flow is imparted to the circulating liquid in the pump, a liquid-metal pump may be a *conduction pump* or an *induction pump*. Although the method developed in the preceding sections is applicable to the analysis of induction pumps, a slightly different approach is taken here. The induction pump is first considered and the conduction pump is discussed subsequently.

4.8.1 The Induction Pump

A schematic arrangement of an ac induction pump is shown in Fig. 4-5. Stator blocks carry the exciting windings embedded in a slotted structure (not shown). The exciting windings produce a traveling flux wave between the two stator blocks. Let the flux ϕ be given by

$$\phi = \phi_m \cos(\omega t - \beta x) \tag{4.24}$$

where $\beta = 2\pi/\lambda$, λ = wavelength or twice *pole pitch*, ω = supply frequency, and ϕ_m = maximum value of the flux. The flux given by the above equation is in the y direction. Since

$$\phi = \phi(x, t)$$

$$d\phi = \frac{\partial \phi}{\partial x} dx + \frac{\partial \phi}{\partial t} dt$$

or,

$$\frac{d\phi}{dt} = \frac{\partial \phi}{\partial x}\frac{dx}{dt} + \frac{\partial \phi}{\partial t} \tag{4.25a}$$

But since $dx/dt = u_x$, the velocity of the conductor in the x direction, Eq. (4.25a) can also be written as

$$\frac{d\phi}{dt} = \frac{\partial \phi}{\partial t} + u_x \frac{\partial \phi}{\partial x} \tag{4.25b}$$

Thus, the voltage induced in the moving fluid is given by Eq. (4.25b), in which the first term corresponds to a *transformer voltage* and the second to a *speed voltage*. If ρ_f is the resistivity of the fluid and J_f the current density in the fluid, then, from Ohm's law and Eq. (4.25b),

$$J_f \rho_f b = -\left(\frac{\partial \phi}{\partial t} + u_x \frac{\partial \phi}{\partial x}\right) \tag{4.26}$$

where b is the width of the channel, as shown in Fig. 4-5. It should be recog-

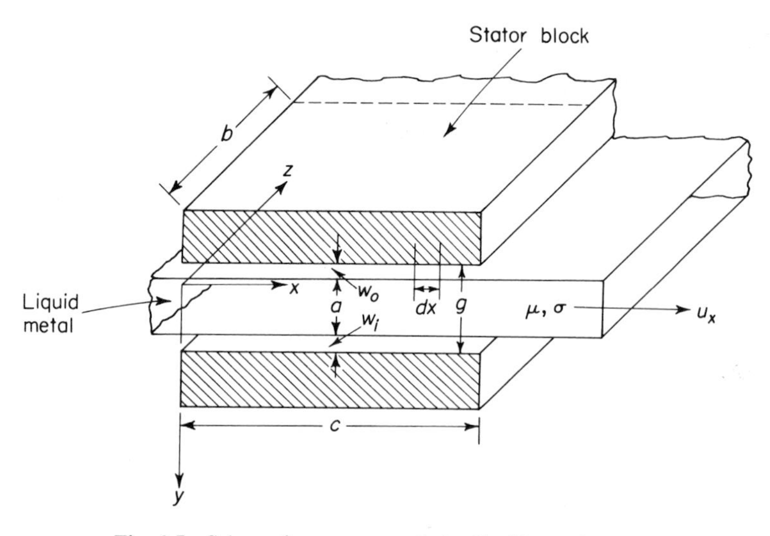

Fig. 4-5 Schematic arrangement of a liquid-metal pump.

nized here that edge effects are not taken into account in this simplified analysis, so that $J_f = 0$ for $c \geqslant x \geqslant 0$. Because the channel is stationary with respect to the moving fluid and the traveling flux wave, only the transformer voltage is induced in the channel walls. Proceeding as above, if ρ_w is the wall resistivity, the wall current density J_w is given by

$$J_w \rho_w b = -\frac{\partial \phi}{\partial t} \tag{4.27}$$

If B is the flux density in an elementary area $dA = b\,dx$, the flux through the area dA is

$$d\phi = B\,dA = Bb\,dx = \mu b H\,dx \tag{4.28a}$$

since $B = \mu H$. From Eq. (4.28a), therefore,

$$H = \frac{1}{\mu b}\frac{d\phi}{dx} \tag{4.28b}$$

Equations (4.24) and (4.28b) yield

$$H = H_m \sin(\omega t - \beta x) \tag{4.29a}$$

where

$$H_m = \frac{2\pi\phi_m}{\lambda\mu b} \tag{4.29b}$$

Now, if $\tau = \lambda/2 =$ *pole pitch*, the velocity of the traveling field u_o can be related to the supply frequency by

$$u_o = \lambda f = \frac{\lambda\omega}{2\pi} \tag{4.30a}$$

Furthermore, the slip s can be defined as before, Eq. (4.9), as

$$s = \frac{u_o - u}{u_o} \tag{4.30b}$$

For the case under consideration, $u = u_x$, so that from Eqs. (4.24), (4.26), (4.29a–4.30b), the fluid current density can be expressed as

$$J_f = \frac{su_o}{\rho_f} \mu H_m \sin(\omega t - \beta x) \tag{4.31}$$

Similarly from Eqs. (4.24) and (4.27), the wall current density is

$$J_w = \frac{u_o}{\rho_w} \mu H_m \sin(\omega t - \beta x) \tag{4.32}$$

The gross pressure p_x developed by the pump can be obtained from Eq. (4.29a) in conjunction with the Lorentz force equation. Thus, by equating the forces acting on an element dx,

$$\frac{\partial p_x}{\partial x} = \mu J_f H \tag{4.33a}$$

and

$$p_x = \mu \int_0^\lambda J_f H \, dx \tag{4.33b}$$

so that from Eqs. (4.29), (4.31), and (4.33b), we have

$$p_x = \frac{1}{2\rho_f} \mu^2 s u_o \lambda H_m^2 \tag{4.33c}$$

If q is the flow through the pump, then

$$q = u_x ab \tag{4.33d}$$

and the gross output power P_o of the pump becomes, from Eqs. (4.33c) and (4.33d),

$$P_o = p_x q = P_p\,(1-s)s \tag{4.34}$$

where

$$P_p = \frac{1}{2\rho_f}\,\mu^2 ab\lambda u_o^2 H_m^2 \tag{4.35}$$

is sometimes called the *induction-pump parameter.*

The ohmic loss in the fluid P_f is given by

$$P_f = \rho_f ab \int_0^\lambda J_f^2\,dx = P_p s^2 \tag{4.36}$$

and the ideal efficiency of the pump is

$$(\text{efficiency})_{\text{ideal}} = (1-s) \tag{4.37}$$

Other losses in the pump are tube-wall loss, hydraulic-friction loss, winding loss, and losses due to edge effects. These are beyond the scope of the present work and are not considered here.

EXAMPLE 4-3

We shall now use the preceding analysis in designing an ac induction pump. The pump is required to pump liquid sodium-potassium eutectic alloy at the rate of about 475 gal/min (gallons per minute) at a pressure of 15 lbs/in^2. The resistivity of the liquid is 45.5 $\mu\Omega$-cm. The pump is to operate at 60-cycle input frequency. We wish to find the dimensions of a rectangular cross-section pump and the required strength of the magnetic field. We are free to choose data not specified above.

In an electromagnetic pump the highest permissible fluid velocity is chosen, the upper limit of the velocity being limited by the hydraulic loss. The general dimensions of the pump are shown in Fig. 4-5. We begin our design by choosing $u_o = 12$ m/sec, so that from Eq. (4.30a) we have

$$\lambda = 20 \quad \text{cm}$$

Next we assume that the pump runs at slip of 0.35. Therefore, from Eq. (4.30b) we have

$$u = u_x = 7.8 \quad \text{m/s}$$

Now, 475 gal/min $= 0.03$ m^3/s $= q$. Thus, from Eq. (4.33d),

$$ab = \frac{q}{u_x} = \frac{0.03}{7.8} = 38.4 \quad \text{cm}^2$$

Because a (Fig. 4-5) should be small to reduce the effective airgap, let $a = 2$ cm, so that $b = 19.2$ cm.

Turning now to the pressure requirement, a pressure of 15 lbs/in^2 approximately corresponds to 10×10^4 N/m^2 in MKS units. Substituting this value and the above data in Eq. (4.33c) we have

$$10 \times 10^4 = \frac{(4\pi \times 10^{-7})^2 \times 0.35 \times 12 \times 0.2}{2 \times 45.5 \times 10^{-8}}\,H_m^2$$

from which

$$H_m = 2.60 \times 10^5 \quad \text{A/m}$$

The maximum flux ϕ_m is obtained from Eq. (4.29b), from which

$$\phi_m = 0.2 \text{ Wb}$$

The output power is, from Eq. (4.34),

$$P_o = p_x q = 3000 \quad \text{W}$$

The ohmic loss in the fluid is, from Eq. (4.36),

$$P_f = 1620 \quad \text{W}$$

and the ideal efficiency of the pump is, from Eq. (4.37),

$$\text{efficiency} = 65\%$$

The actual efficiency is lower than the ideal efficiency, since in the above calculations we have not taken into account various other losses, such as the armature-copper loss, iron loss, hydraulic loss, tube-wall loss, etc. When all these are taken into account, the overall efficiency of the pump may be about 40%

4.8.2 The Hydromagnetic dc Converter

Instead of considering the dc conduction pump alone, we shall now consider a system in which the pumping and generating modes are both studied. Such a system, sometimes called a *hydromagnetic converter*, is a liquid-metal device with the terminal characteristic of an ideal transformer. It is analogous to the motor-generator set in conventional rotating machinery in that a liquid-metal conduction pump serves as a motor and the electromagnetic fluid generator as a generator, with the liquid metal providing the coupling. The interaction of a conducting fluid and magnetic field results in the pumping and generating modes of operation. Such a device might find application to meet the need for a dc converter capable of handling several kilowatts at an input level of 1 volt or less and then raising the voltage to levels at which it can be handled by conventional methods.

The force of electromagnetic origin acting on the fluid is the usual $l\mathbf{I} \times \mathbf{B}$ force of the fluid current. In the electromagnetic fluid generator the moving fluid generates $l\mathbf{u} \times \mathbf{B}$ voltage, and supplies current to the external load. The generator and the pump are identical except that the direction of flow of current in the generator is opposite that in the pump. The schematic hydromagnetic dc converter is shown in Fig. 4-6. In addition to the pump and the generator, also shown are the interconnecting pipes, the fluid reservoir, and the manometers for measuring pump and generator pressure. The channels for the pump and generator are identical, and the pertinent dimensions are shown in Fig. 4-7.

The equations which yield the characteristics of the converter relate to

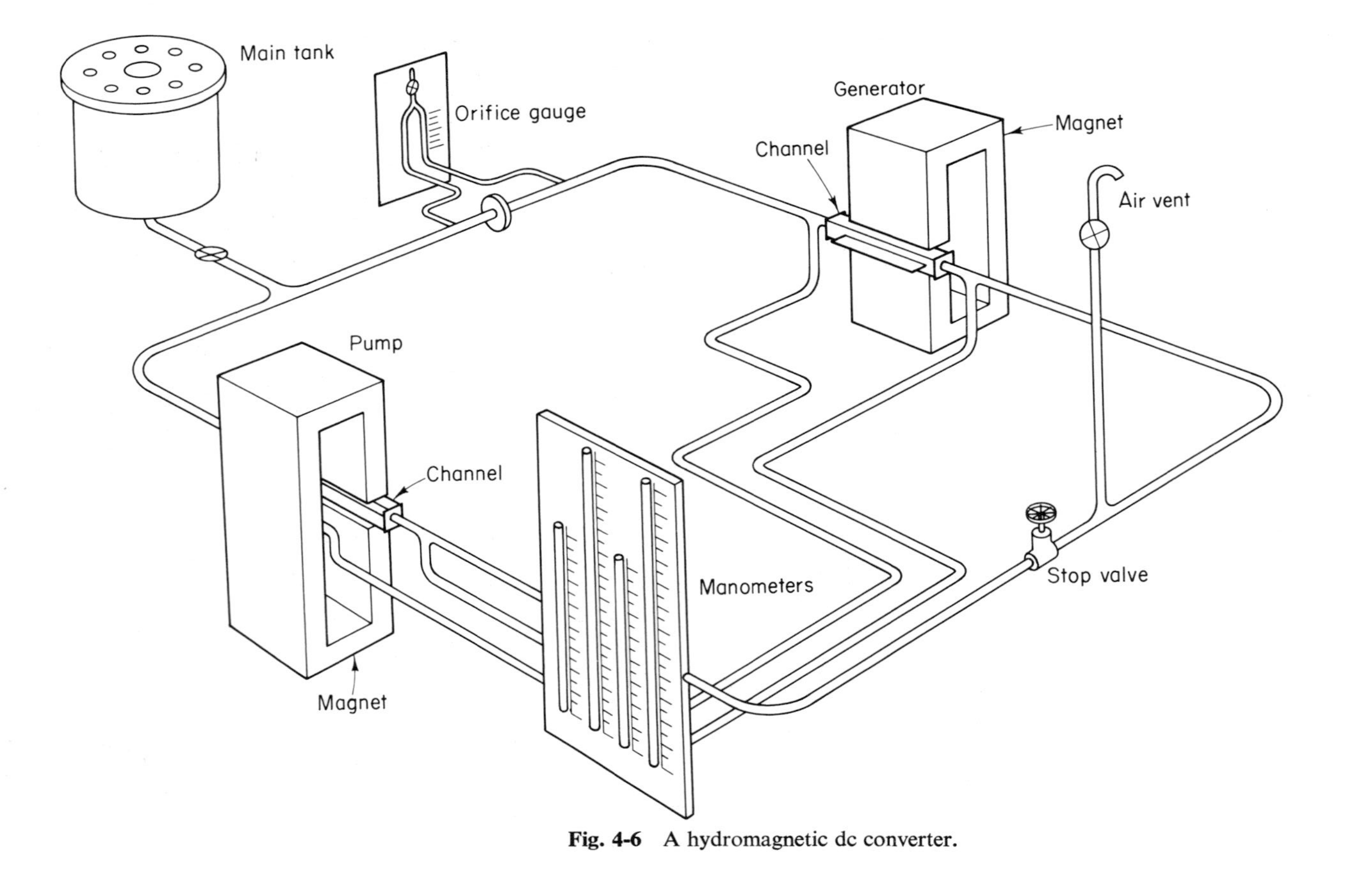

Fig. 4-6 A hydromagnetic dc converter.

electromagnetism and hydrodynamics. If we assume: (1) no time variations (or dc steady state); (2) an incompressible, nonpolarizable, and Newtonian (linear-stress/rate-of-strain relationship) fluid; (3) no fluid sources or sinks; and (4) negligible gravitational effects, the basic equations for a point within the channel become:

Maxwell's equations: $$\nabla \times \mathbf{E} = 0 \tag{4.38}$$

$$\nabla \times \mathbf{H} = \mathbf{J} \tag{4.39}$$

Ohm's law for fluid: $$\mathbf{J} = \sigma(\mathbf{E}+\mathbf{u} \times \mathbf{B}) \tag{4.40}$$

Conservation of mass
(or continuity) equation: $$\nabla \cdot \mathbf{u} = 0 \tag{4.41}$$

For our analysis, we shall only consider the idealized conditions and neglect the fluid-viscosity force. In this simplified case,

$$\nabla p = \mathbf{J} \times \mathbf{B} \tag{4.42}$$

and we notice that the pressure gradient ∇p arises from the Lorentz force $\mathbf{J} \times \mathbf{B}$.

The boundary conditions are: (1) the tangential components of **E** and **H** and the normal component of **B** are continuous; (2) the discontinuity in the normal component of **D** is equal to the surface-charge density; and (3) there is no motion relative to the boundary surface of the fluid in immediate contact with it.

From Eq. (4.41), assuming unidirectional flow,

$$u_x ab = u'_x a'b' \tag{4.43}$$

where the dimensions a, b, and c relate to the channel as shown in Fig. 4-7.

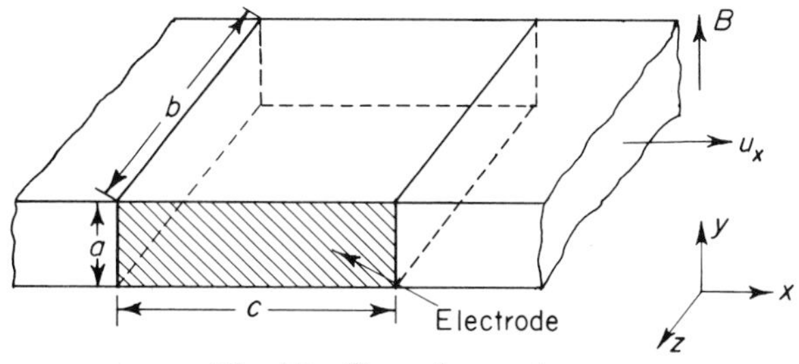

Fig. 4-7 Channel geometry.

The pressure rise p in the pump and the pressure drop p' in the generator are respectively given by

$$p = \frac{IB}{a} \tag{4.44}$$

and

$$p' = \frac{I'B'}{a'} \tag{4.45}$$

where the primed quantities relate to the generator. From the above equations, the hydraulic efficiency η_h and the voltage and current conversion ratios K_v and K_i are respectively given by

$$\eta_h = \frac{p'}{p} \tag{4.46}$$

$$K_v = \frac{V'}{V} = \frac{aB'}{a'B} \tag{4.47}$$

and

$$K_i = \frac{I'}{I} = \frac{\eta_h}{K_v} \tag{4.48}$$

For the pump, the output power P_o is given by

$$P_o = pq = \frac{IB}{a} u_x ab = Iu_x Bb = VI \tag{4.49}$$

and, finally, the fluid ohmic loss is

$$P_f = I^2 R = \rho J^2 abc = \frac{\rho I^2 b}{ac} \tag{4.50}$$

One aspect of the operation of the liquid electromagnetic dc device (pump or generator) that has not been considered so far is the *armature reaction*, by which is meant the interaction of the armature (or fluid) current with the externally applied magnetic field. For the pumping mode, the effects of armature reaction are shown in Fig. 4-8(a). Note that the flux-density and current-density distributions both become distorted. This leads to low pump pressure and low efficiency. Armature reaction can, however, be compensated for by returning the main current (which flows through the fluid) back through the magnet gap by means of a pole-face winding. A simple arrangement for compensating the armature reaction is shown in Fig. 4-8(b), from which it is seen that the net current flow under the pole face is essentially zero, and consequently the armature reaction is reduced to a minimum.

EXAMPLE 4-4

The cross-section of a hydromagnetic channel is to be obtained from the following data. The channel is of the form shown in Fig. 4-7. The flow through the channel is $\frac{1}{8}$ m^3/s at a velocity of 2 m/s and at a pressure of 8×10^4 N/m^2. If the dc magnetic flux density is 1 Wb/m^2, the voltage induced, the current through the fluid, and the power output are also to be obtained.

From Eq. (4.43),

$$q = uab$$

Or,

$$\frac{1}{8} = 2ab$$

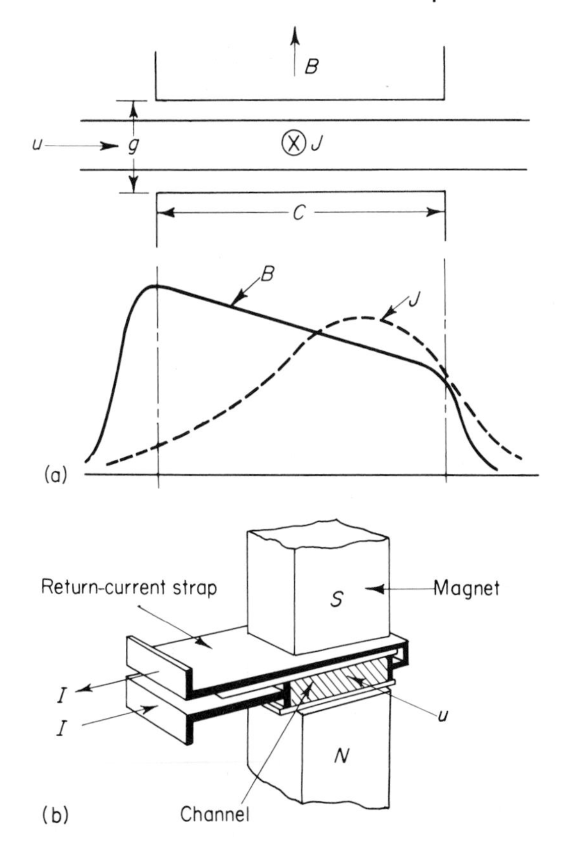

Fig. 4-8 (a) Distortion of B-field and J due to armature reaction; (b) compensation of armature reaction.

If the ratio a/b is taken as $\frac{1}{4}$, the required dimensions are

$$a = \tfrac{1}{8} \quad \text{m}$$

and

$$b = \tfrac{1}{2} \quad \text{m}$$

The induced voltage is simply obtained from the equation

$$e = uBb = 1 \quad \text{V}$$

The current can be calculated using Eq. (4.44), from which it follows that

$$I = \frac{ap}{B} = 10{,}000 \quad \text{A}$$

Finally, the power output is given by Eq. (4.49):

$$P_o = VI = 10 \quad \text{kW}$$

If the conductivity of the fluid is 6×10^6 mho/m, the ohmic loss in the fluid per unit length of the channel is, from Eq. (4.50),

$$P_f = 66.66 \quad \text{W}$$

4.8.3 Comparison of Pump Types[10]

In the last two sections we considered induction and conduction pumps with channels of rectangular cross-sections only. Such pumps are also called flat pumps, such as the flat linear induction pump (abbreviated *FLIP*). On the other hand, a tubular induction pump is called an annular linear induction pump (abbreviated *ALIP*). A third kind of induction pump also exists known as the spiral induction pump (abbreviated *SIP*). The stator of a SIP is similar to that of a conventional induction motor. But the core remains stationary and the liquid is deflected in the forward direction by vanes. The vanes impart a spiral motion to the liquid, whereas the circumferential motion is caused by the traveling (or rotating) field of the stator.

Typical performance data comparing various kinds of electromagnetic pumps are summarized in Table 4-1.

4.9 Acyclic or Homopolar Machines[15–18]

In Sec. 4.8 we discussed liquid-metal pumps. Induction pumps are quite convenient to operate, but are not suitable for such high-density, high-viscosity, and high-resistivity liquid metals as mercury and lead. For such metals, the dc pump is more suitable. But with these pumps the supply problem becomes quite severe. For example, to pump liquid bismuth at a rate of 2000 gal/min at 75 lbs/in^2 pressure requires 100,000 A at 2.5 V. About the only efficient supply at such low voltages and high currents is the *homopolar* or *acyclic* generator.

The acyclic generator is perhaps the oldest electrical machine. In the *Faraday disk* (Fig. 4-9) proposed in 1831, which can be considered the first basic acyclic generator, a copper disk rotates between the poles of a magnet. The voltage induced in the disk is available at the sliding brushes on the rim of the disk and on the shaft, as shown in Fig. 4-9. In order to obtain any appreciable voltage between the brushes the angular velocity of the disk and the total flux crossing the airgap must be as large as possible. However, mechanical considerations limit the speed of the disk, while the saturation of the magnetic circuit limits the total flux. As a result, this elementary disk-type acyclic generator is of no practical interest.

Table 4.1*

PERFORMANCE DATA OF CONDUCTION AND INDUCTION PUMPS

Conduction Pumps

Type	*Fluid*	*Flow, gal/min*	*Pressure, lb/in²*	*Output, hp*	*Efficiency, per cent*	*P.F.*	*Supply*		*Power*	
							Voltage	*Current, kA*	*hp/ton*	*hp/ft³*
ac	mercury	6	15	0.067	~4	0.26	—	—	1.7	0.2
ac	NaK 400°C	20	10	0.14	—	—	—	—	1.3	0.08
spiral dc	bismuth 500°C	0.66	60	0.027	~1	—	1.2	1.4	1.9	0.4
dc	bismuth 200°C	10	60	0.42	12	—	0.6	4.4	5	0.7
dc	NaK 250°C	300	40	8.4	44	—	0.75	19	—	—
dc	bismuth 550°C	2000	75	105	30	—	2.6	200	40	6
dc	sodium 410°C	8300	75	435	~50	—	2.5	100	—	~2
					~45					~0.5

Induction Pumps

Type	*Fluid*	*Flow, gal/min*	*Pressure, lb/in²*	*Output, hp*	*Efficiency, per cent*	*P.F.*	*Supply frequency, Hz*	*Power*		*Cooling required*
								hp/ton	*hp/ft³*	
SIP	sodium 400°C	25	60	1.05	22	0.56	50	4	0.3	3–4 kW
SIP	sodium 400°C	312	40	8.7	18.5	0.8	25	13	—	10^3 ft³/min of air
ALIP	NaK 175°C	420	14	4.1	36	0.22	50	11	0.9	3.5 kW
ALIP	sodium 500°C	400	50	14	36	0.30	50			10 kW
FLIP	sodium 370°C	1200	40	34	36	0.45	60	12	—	2000 ft³/min of air
ALIP	sodium 400°C	8300	75	435	45	0.48	50	64	4.5	

* Reprinted from L. E. Blake, "Conduction and Induction Pumps for Liquid Metals," *Proc. IEE* (London), Vol. 104(a), 1957, pp. 49–63, with permission.

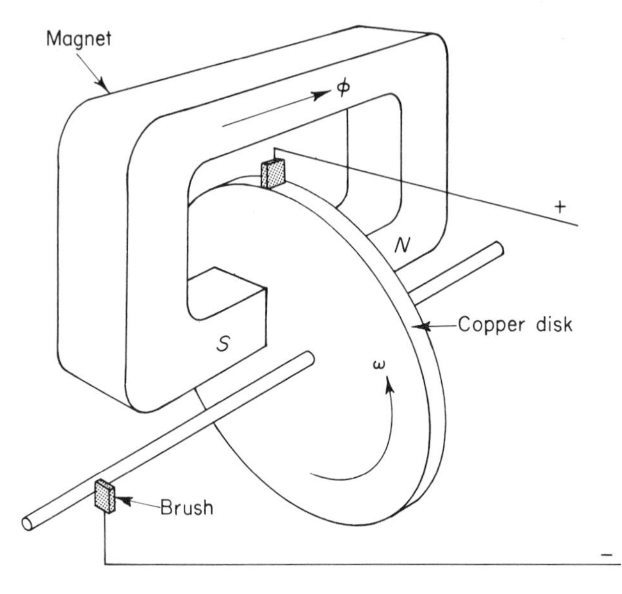

Fig. 4-9 A Faraday-disk acyclic generator.

4.9.1 The Drum-Type Acyclic Generator

To circumvent some of the difficulties of the disk-type generator, a conducting drum is used in place of the disk. A simplified form of such a machine is shown in Fig. 4-10. The generator consists of a conducting drum rotating in a dc magnetic field produced by the exciting coils. The voltage induced in the conducting drum is available at the sliding brushes, as in the disk-type machine. According to Faraday's law, the magnitude of the induced voltage is proportional to the total flux and angular velocity of the

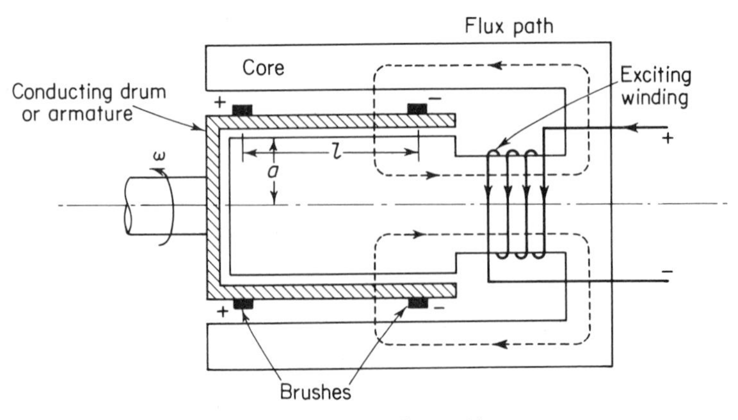

Fig. 4-10 An acyclic machine.

drum. A better magnetic circuit permits more flux in this machine than in a disk-type machine.

Let a be the mean radius of the conducting drum, c be its thickness, and l be the distance between the two brushes as shown in Fig. 4-10. Assuming uniform flux density B in the airgap (in the radial direction) and uniform current density J in the conducting drum (in the axial direction), the magnitude of the terminal voltage V is given by

$$V = \left(uB - \frac{J}{\sigma}\right) l = \text{(induced voltage} - \text{voltage drop)} \tag{4.51}$$

where σ is the conductivity and u is the linear velocity of the cylinder. But

$$u = a\omega \tag{4.52}$$

the total flux ϕ is

$$\phi = \int_S \mathbf{B} \cdot d\mathbf{S} = B(2\pi a l) \tag{4.53}$$

the total current I is

$$I = \int_S \mathbf{J} \cdot d\mathbf{S} = J(2\pi a c) \tag{4.54}$$

and the resistance R is

$$R = \frac{b}{\sigma} 2\pi a c \tag{4.55}$$

so that, from Eqs. (4.51–4.55), the voltage equation for the homopolar generator becomes

$$V = \frac{\omega}{2\pi} \phi - IR \tag{4.56}$$

However, as with most dc machines, the terminal voltage changes appreciably with the load current. This is mainly due to armature reaction (discussed in Sec. 4.8.2), which results in saturating the iron, distorting the flux-density distribution, and eventually causing a voltage drop. Armature reaction can be compensated for to a certain extent by using a compensating shield as shown in Fig. 4-11. The resultant field produced by the armature current and shield current is essentially zero. Because the shield is connected in series with the armature, the two carry the same current but in opposite directions.

4.9.2 *The Acyclic Generator with Liquid-Metal Brushes*

We have considered the development of the acyclic generator from the disk type to the drum type. For a long time there were no further significant developments in acyclic generators. The drum-type generator, compensated for armature reaction, but with sliding brushes, had the following inherent

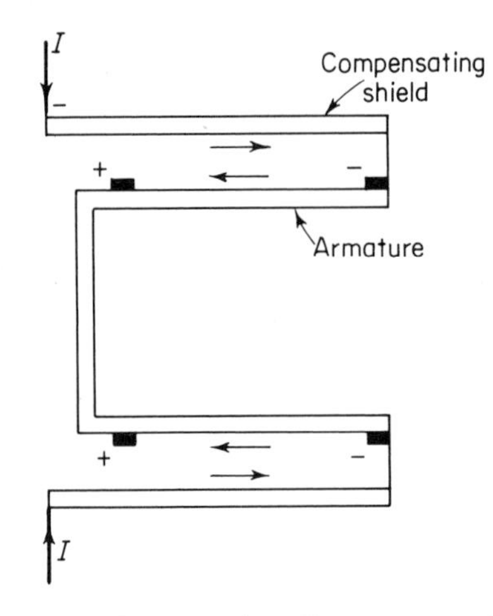

Fig. 4-11 Compensation of armature reaction.

undesirable features: high friction losses, high brush-contact voltage drop, awkward current-collector mechanism, difficult brush maintenance, etc. All these factors limited the overall output and applications of acyclic generators. A typical rating of this type generator is 2000 kW, 7700 A, 260 V, 1200 r/min. The unique feature of the modern acyclic generator is a liquid metal, such as a eutectic alloy of sodium and potassium (called *NaK*), between the rotating and stationary parts of the collector. This liquid metal facilitates the current collection without the use of sliding brushes, and has low electrical and viscous losses.

The liquid-metal collector system offers the following advantages: low collector losses, because the frictional and resistive losses are very low in a liquid metal such as NaK; high machine efficiency—efficiency of about 98% has been reported; dependability and low maintenance; low excitation—the excitation power is approximately 0.1% the output rating; high peak-current capability; high-speed operation, which permits the use of economical prime movers and purifies the output of voltage ripples.

All the above mentioned advantages in an acyclic generator are due to NaK collector system. Some of the properties of NaK are: low melting point—about 10°F; high boiling point—about 1400°F; low viscosity and light weight—about 85% of that of water; and desirable wetting properties. Because of excellent heat-transfer properties, losses in the fluid are removed by circulating NaK through the collector by means of an external pump.

4.9.3 Some Applications

The development of the liquid-collector system in acyclic generators has considerably increased their output-power capacities. Some typical ratings are:

1. 2000 kW, 35 V, 57,000 A, 7200 r/min
2. 36,000 kW, 73 V, 495,000 A, 3600 r/min
3. 210,000 kW, 450 V, 470,000 A, 1800 r/min

These high ratings have opened up possibilities of applications of acyclic generators in many industries. A typical example is in the chemical industry concerned with the chlorine-cell line. Such electrochemical loads require a few thousand amperes current per volt. An aluminum-pot line rated at, say, 150,000 A, 400 V can be efficiently served by an acyclic generator. Power supply for particle-accelerator magnets for pulse duty is yet another application. Other applications for not-so-large acyclic generators include power supplies for electromagnetic pumps. Motor–generator combinations of acyclic machines have been proposed for use in ship propulsion and in electric-car transmission systems.

EXAMPLE 4-5[18]

In this example, the operation of a homopolar machine is considered as a motor. Furthermore, a disk-type configuration, as shown in Fig. 4E-5, is considered. The conducting disk is mounted on a shaft which rotates in a magnetic field. The performance equations are to be derived and numerical results are to be calculated from the given data.

An annular ring of thickness dx is considered. The voltage induced in this ring is

$$dV = Bu\ dx = \omega Bx\ dx$$

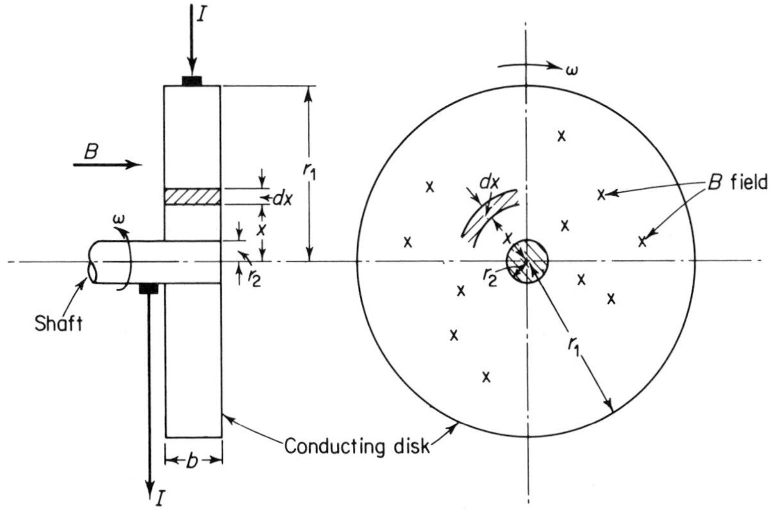

Fig. 4E-5

so that the total voltage induced in the disk is

$$V = \omega B \int_{r_2}^{r_1} x\,dx = \tfrac{1}{2}\,\omega B(r_1^2 - r_2^2)$$

where ω is the angular velocity of the disk. We note that the induced voltage is also the open-circuit voltage and follows from Eq. (4.56), in which the voltage drop term IR is zero.

If ϕ is the total flux, the above equation can be expressed as

$$V = \frac{\omega}{2\pi}\,\phi$$

Since the disk is fed with a current I, the torque dT on the annular ring is

$$dT = BIx\,dx$$

Or, the total torque T becomes

$$T = BI \int_{r_2}^{r_1} x\,dx = \frac{\phi I}{2\pi}$$

Finally, if ρ is the resistivity of the disk and b is its thickness, the total resistance of the disk is given by

$$R = \int_{r_2}^{r_1} \rho \frac{dx}{2\pi x b} = \frac{\rho}{2\pi b} \ln \frac{r_1}{r_2}$$

Now, for the sake of illustration, if we let $b = 0.31$ cm, $r_1 = 5.36$ cm, $r_2 = 0.795$ cm, $\rho = 17.4$ μmho-cm, $\phi = 1.6$ megalines, $I = 100$ A, and $n = 6000$ r/min, then

$$V = 1.6 \quad \text{V}$$
$$T = 36 \quad \text{oz-in.}$$
$$R = 1.71 \times 10^{-6} \quad \Omega$$

From the above example we conclude, first, that the resistance of the disk is much too small compared with the resistance of the brushes and contact resistance. In a good machine these should be reduced to a minimum. Secondly, the induced voltage V is a *back EMF* and acts in a direction opposite to the applied voltage. Therefore, in order that the machine may develop any torque that can be delivered to a load running at 6000 r/min, the applied voltage must be greater than 1.6 V at 100 A.

4.10 MHD Power Conversion[19]

One of the most important examples of energy conversion in continuous media is magnetohydrodynamic (MHD) energy conversion. It is not practicable to include here a detailed treatment of this subject and allied problems. Therefore, we shall only consider underlying principles, derive the basic equations, study some elementary configurations, and discuss certain aspects of dc and ac power conversion in MHD generators.

4.10.1 dc MHD Power Conversion[20]

The principle of operation of a dc MHD generator is the same as that of a conventional dc generator. The major difference between the two, however, arises from the fact that in an MHD generator an ionized gas, or plasma, serves the purpose of the conductor. The presence of an ionized gas in an electromagnetic field leads to all sorts of complicated analysis. For instance, even the electrical conductivity of an ionized gas depends upon a number of variables, and in fact the conductivity is not a scalar quantity. Another factor that considerably affects the performance of an MHD generator is the *Hall effect* (discussed in the next section). However, concentrating our attention only on the most pertinent and simplified governing equations, we shall consider the elementary MHD generator shown in Fig. 4-12. Through the channel an ionized gas, of electrical conductivity σ, flows with a velocity u in a crossed electrical field E and magnetic field B. The net electric field is then $(E-uB)$, so that from Ohm's law, the current density J is given by

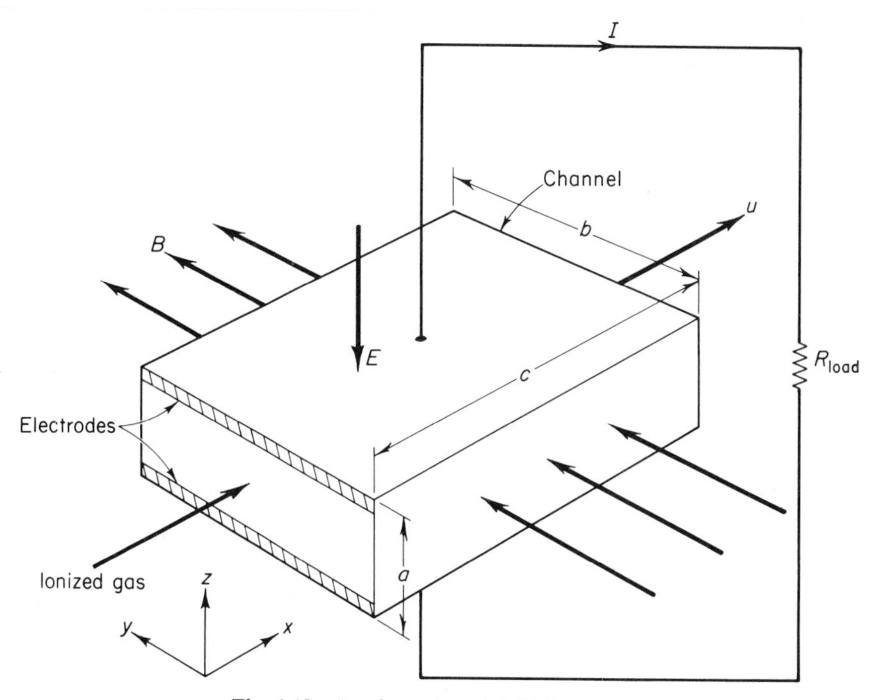

Fig. 4-12 An elementary dc MHD generator.

$$J = \sigma(E - uB) \tag{4.57}$$

If $K = E/uB$, then

$$J = -\sigma uB(1-K) \tag{4.58}$$

where K is called the *loading factor*.

The output power density P_o per unit volume of the gas (noting that power is the product of current and voltage) is

$$P_o = -JE \tag{4.59}$$

Or, from Eqs. (4.58) and (4.59),

$$P_o = \sigma u^2 B^2 K(1-K) \tag{4.60}$$

For maximum output-power density, it can be readily verified that $K = \frac{1}{2}$. This corresponds to the matched-resistance condition for which the load resistance equals the internal resistance of the generator for maximum power transfer. The constant K can thus be interpreted as the ratio of the load resistance to the total resistance. Since the electric field $E = KuB$, the voltage V across the load is given by

$$V = KuBa \tag{4.61}$$

where a is the separation, where the gas flows, between the channel electrodes. If the electrode area is bc, the load current I is, from Eq. (4.58),

$$I = -Jbc = \sigma(1-K)uBbc \tag{4.62}$$

The load resistance R_L is

$$R_L = \frac{V}{I} = \frac{Ka}{\sigma(1-K)bc} \tag{4.63}$$

But the internal resistance of the generator is $R_G = a/\sigma bc$. Thus, Eq. (4.63) can also be expressed as

$$R_L = \left(\frac{K}{1-K}\right)R_G \tag{4.64}$$

Or,

$$K = \frac{R_L}{R_T} \tag{4.65}$$

where $R_T = R_L + R_G$ = total resistance, which further validates the statement that K is the ratio of load resistance to total resistance.

Further analysis of this generator can be made by considering the fact that the output power density P_o, as given by Eqs., (4.59) and (4.60), is obtained by work done by ionized gas as it flows against the body force per unit volume. Consequently, the force density JB and the work done $-JBu$ are respectively given by

$$JB = -(1-K)\sigma uB^2 \tag{4.66}$$

$$-JBu = (1-K)\sigma u^2 B^2 \tag{4.67}$$

The ohmic dissipation in the gas is, from Eqs. (4.60) and (4.67), the difference between the work done by the gas and the output power. Or,

$$P_{\text{ohmic}} = (1-K)^2 \sigma u^2 B^2 \tag{4.68}$$

In the above analysis, fluid-flow equations have not been considered. Nevertheless, certain important features of the dc MHD generator emerge from the equations derived above. For example, from Eq. (4.60) we see that for maximum-output power density, $K = \frac{1}{2}$. But this is not the only consideration for the operation of an MHD generator. Equations (4.60) and (4.67) show that only a fraction K of the work done by the gas appears as electrical power, and the remaining $(1-K)$ dissipates as ohmic heating in the gas. This ohmic heating does not represent a loss of energy from the system. Rather it denotes an increase in entropy, a degradation of energy and a reduction in its availability for conversion. The factor K is analogous to the *isentropic efficiency* factor in a steam or gas turbine. Therefore, the balance between maximum-output power density and adequate isentropic efficiency leads to values of K between 0.7 and 0.8 for MHD generators.[10]

Apart from the loading factor K, there are a number of other thermodynamic, hydraulic, and electrical factors which affect the performance of MHD generators. The first two factors are beyond the scope of this book and only a few of the electrical parameters are considered here.

†4.10.2 *Electrical Conductivity and the Hall Effect*

The preceding analysis clearly indicates that for a successful operation of an MHD generator electrical conductivity of the gas must be large. However, the electrical conductivity of an ionized gas does not obey any linear laws. Furthermore, it depends on the degree of seeding also. (*Note:* The conductivity of a gas is increased by the injection of seed material; for example, a conductivity of approximately 10 mho/m at 2000°K can be obtained by seeding air with 1% potassium vapor.) The electrical conductivity of an ionized gas is approximately given by[20]

$$\sigma \cong 0.5 \frac{n_e e^2}{\sqrt{m_e k T}} \frac{1}{\sum_i N_i Q_i} \tag{4.69}$$

where n_e = number of electrons per unit volume; Q_i = collision cross-section (defined as an area about the center of a particle such that a collision occurs as soon as another particle comes within the area Q) of the ith particle with the electron; N_i = number of such particles per unit volume in the gas; e = electronic charge, which is 1.6×10^{-19} C (coulomb); m_e = mass of electron (9.1×10^{-31} kg); k = Boltzmann constant (1.38×10^{-23} J/°C); and T = temperature in °K.

Another factor that has not been considered so far is the *Hall effect*, which causes a voltage to appear in a direction perpendicular to the current flow and applied magnetic field. This causes the electrical conductivity to be a tensor rather than a scalar quantity. The magnitude of the Hall effect depends upon the product $\omega\tau$, sometimes called the *Hall number*, where ω = cyclotron frequency of the electron and τ = mean time between collision for electrons. (*Note:* The *cyclotron frequency* $\omega = eB/m_e$ is the angular velocity with which an electron moves in a uniform transverse magnetic field B.) Taking into account the Hall effect, Ohm's law for fluid, Eq. (4.40), modifies to [20]

$$\mathbf{J} = \sigma(\mathbf{E}+\mathbf{u} \times \mathbf{B}) - \frac{\omega\tau}{B}(\mathbf{J} \times \mathbf{B}) \tag{4.70}$$

In terms of components this equation can be written as

$$J_x = \frac{\sigma}{1+(\omega\tau)^2}[\omega\tau(uB-E_y)+E_x] \tag{4.71a}$$

$$J_y = \frac{\sigma}{1+(\omega\tau)^2}(-uB+E_y+\omega\tau E_x) \tag{4.71b}$$

where it is assumed that $\mathbf{u} = \mathbf{a}_x u$, $\mathbf{B} = \mathbf{a}_z B$, and $\mathbf{E} = \mathbf{a}_x E_x + \mathbf{a}_y E_y$. The electric field component E_x is called the *Hall field* and J_x the *Hall current*.

In a generator such as the one shown in Fig. 4-13, which has a single pair of electrodes in the direction of flow of the gas, $E_x = 0$ (since the electrodes act as a short circuit), so that Eq. (4.71b) reduces to

$$J_y = \frac{\sigma}{1+(\omega\tau)^2}(-uB+E_y) \tag{4.72}$$

A comparison of Eqs. (4.57) and (4.72) indicates that the effective conductivity of the gas is reduced from σ to $\sigma/(1+\omega^2\tau^2)$ and, consequently, output-power density, from Eq. (4.60), reduces to

$$P_o = \frac{\sigma u^2 B^2 K(1-K)}{1+(\omega\tau)^2} \tag{4.73}$$

We may thus conclude that the Hall effect reduces output power, and that the simple (*Faraday-type*) MHD generator considered in the preceding section is not satisfactory for cases where $\omega\tau > 1$.

Modifications are made to overcome this difficulty in MHD generators by segmenting the electrodes and making the load connections as shown in Fig. 4-13. In the segmented-electrode generator the axial current is prevented from flowing ($J_x = 0$) and an axial electric field is generated as given by

$$E_x = -\omega\tau(uB-E_y) \tag{4.74}$$

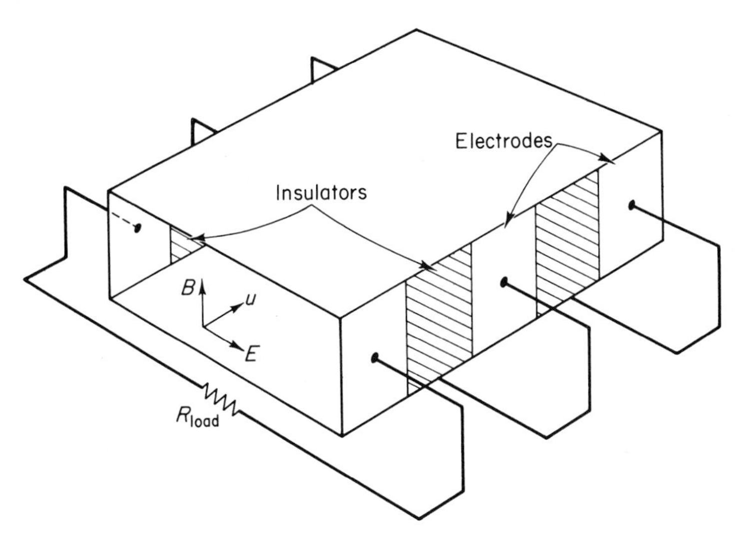

Fig. 4-13 Segmented electrode MHD generator.

which when substituted in Eq. (4.71b) yields

$$J_y = \sigma(-uB+E_y) \tag{4.75}$$

Notice that Eqs. (4.57) and (4.75) are identical. The effect of segmenting the electrodes is thus to restore the effective electrical conductivity to the original Hall-field-free value.

Another method of minimizing the Hall effect in an MHD generator is shown in Fig. 4-14, in which the opposite pairs of segmented electrodes are

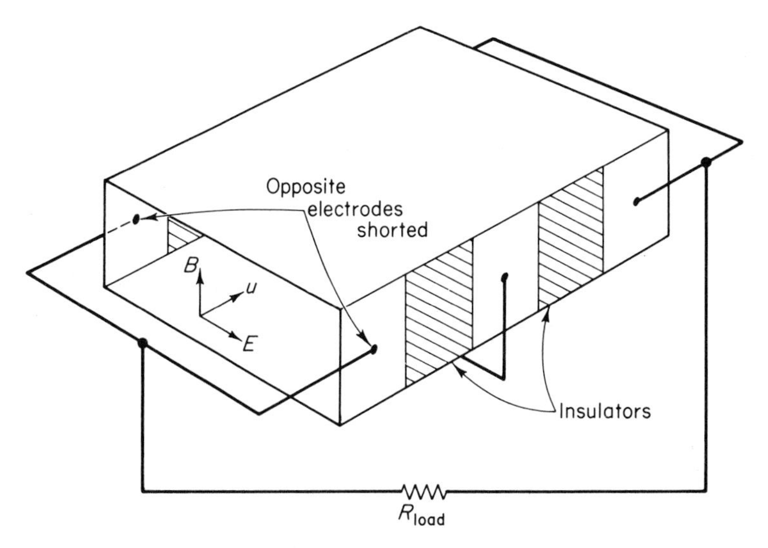

Fig. 4-14 A Hall generator to reduce Hall effect.

short-circuited and an axial current is fed through a single load. In this case, $E_y = 0$ and the output-power density is

$$P_o = \frac{(\omega\tau)^2}{1+(\omega\tau)^2}\sigma u^2 B^2 L(1-L) \tag{4.76}$$

where

$$L = \frac{E_x}{\omega\tau u B} \tag{4.77}$$

For large values of the Hall number ($\omega\tau \gg 1$), the output-power density is the same as in the elementary Faraday-type generator considered in Sec. 4.10.1.

EXAMPLE 4–6

The channel dimensions of the MHD generator shown in Fig. 4-12 are a = 0.2 m, b = 0.2 m, c = 0.5 m. An ionized gas of conductivity 10 mho/m traverses the channel at a velocity of 1000 m/s at 2500°K. Given $B = 2$ Wb/m^2, for maximum-power conditions, the performance characteristics of this generator are to be obtained if (a) the Hall effect is negligible, and (b) the Hall number is 10.

(a) Recall that for maximum-power conditions, the loading factor $K = 0.5$. Therefore, from Eqs. (4.58) and (4.60),

$$J = \sigma u B(1-K) = 10{,}000 \quad \text{A/m}^2$$
$$P_o = \sigma u^2 B^2 K(1-K) = 10{,}000 \quad \text{kW/m}^3$$

Since the volume of the channel is 0.02 m^3, the total output power is

$$P_{oT} = 200 \quad \text{kW}$$

The voltage across the load is, from Eq. (4.61),

$$V = KuBa = 200 \quad \text{V}$$

and the load current is, from Eq. (4.62),

$$I = Jbc = 1000 \quad \text{A}$$

It can be readily verified that

$$P_{oT} = VI = 200 \quad \text{kW}$$

The load resistance is, from Eq. (4.63),

$$R_L = \frac{V}{I} = 0.2 \quad \Omega$$

The generator's internal resistance is, from Eq. (4.64),

$$R_G = (1-K)\frac{R_L}{K} = 0.2 \quad \Omega$$

Finally, the ohmic dissipation in the gas is, from Eq. (4.68),

$$P_{\text{ohmic}} = (1-K)^2\sigma u^2 B^2 = 10{,}000 \quad \text{kW}$$

(b) The Hall effect can be included in the above solution simply by replacing the conductivity σ by the effective conductivity $\sigma/(1+\omega^2\tau^2) = \frac{10}{101} \simeq \frac{1}{10}$.

The simplifying assumptions in the above analysis are: (1) the flux density is uniform and edge effects and fringing are negligible; (2) the gas is frictionless, compressible, and perfect; (3) the electrical conductivity is a scalar quantity; and (4) the heat transfer from the channel is neglected.

4.10.3 ac MHD Power Conversion[21–24]

In the preceding sections the possibility of dc power generation by means of interaction between a moving ionized gas and an electromagnetic field was demonstrated. The feasibility of ac MHD generators was also explored. The analysis of ac MHD generators is more complicated than that of dc generators. Nevertheless, in this section some of the underlying principles are discussed. Broadly speaking, the ac MHD generator may be either (a) a conduction-type generator, or (b) an induction-type generator. These are now considered in some detail.

(a) *ac MHD Conduction Generators.*[24] The novel conduction-type ac generator proposed by Woodson is shown in Fig. 4-16. For purposes of analysis, it is assumed that the fluid is incompressible, flows uniformly in a constant-area channel, and has a constant scalar conductivity. Edge effects, fringing, and armature reaction are neglected. Furthermore, the skin depth is large as compared to the width of the channel, and the B field is uniform and at right angles to the direction of flow of the gas.

The ac generation is accomplished by cross-coupling two self-excited transverse-current MHD conduction generators. In the nomenclature shown in Fig. 4-15 the flux densities in the generators are

$$B_1 = \frac{\mu_o N i_1}{a} - \frac{\mu_o N_m i_2}{a} \tag{4.78a}$$

and

$$B_2 = \frac{\mu_o N i_2}{a} + \frac{\mu_o N_m i_1}{a} \tag{4.78b}$$

For the two circuits, the voltage-balance equations are

$$ubB_1 = Ri_1 + Ncb\frac{dB_1}{dt} + N_m cb\frac{dB_2}{dt} \tag{4.79a}$$

$$ubB_2 = Ri_2 + Ncb\frac{dB_2}{dt} - N_m cb\frac{dB_1}{dt} \tag{4.79b}$$

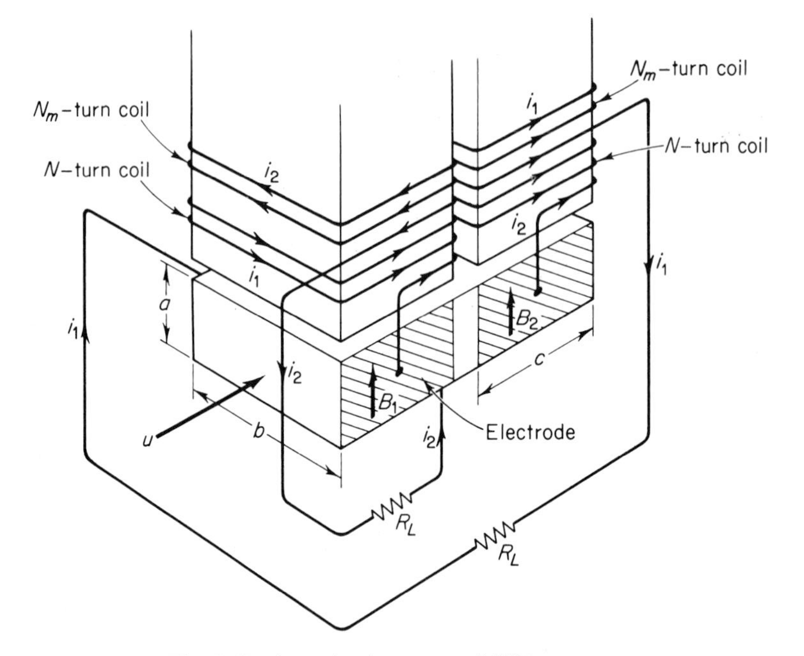

Fig. 4-15 A conduction-type ac MHD generator. (Adapted from Ref. 15 with permission.)

where $R = R_L + R_i + R_c$ = total resistance of the circuit; R_L = load resistance; $R_i = b/ac\sigma$, the internal resistance of each pair of electrodes; and R_c = resistance of the exciting windings of the magnet. We define the parameters

$$G = \frac{\mu_o ubN}{a}$$

$$G_m = \frac{\mu_o ubN_m}{a} \tag{4.80}$$

and

$$L = \frac{\mu_o cb}{a}(N^2 + N_m^2)$$

Equations (4.78a–4.80) then yield

$$Gi_1 - G_m i_2 = Ri_1 + L\frac{di_1}{dt} \tag{4.81a}$$

and

$$Gi_2 + G_m i_1 = Ri_2 + L\frac{di_2}{dt} \tag{4.81b}$$

Eliminating i_2 from these equations,

$$\frac{d^2 i_1}{dt^2}+\frac{2(R-G)}{L}\frac{di_1}{dt}+\left[\frac{G_m^2+(R-G)^2}{L^2}\right]i_1 = 0 \tag{4.82}$$

which has the solution of the form

$$i_1 = I_o \exp\left[-\frac{(R-G)}{L}t\right]\cos\left[\left(\frac{G_m}{L}\right)t+\theta\right] \tag{4.83}$$

where the constants I_o and θ are obtained from the initial conditions. Clearly, from Eq. (4.83), when $R = G$ the current i_1 is a sinusoid of constant amplitude. With this condition fulfilled, Eqs. (4.81a) and (4.81b) reduce to

$$-G_m i_2 = L\frac{di_1}{dt} \tag{4.84a}$$

$$G_m i_1 = L\frac{di_2}{dt} \tag{4.84b}$$

and the solutions to these equations are

$$i_1 = I_o \cos(\omega t+\theta) \tag{4.85a}$$

and

$$i_2 = I_o \sin(\omega t+\theta) \tag{4.85b}$$

where

$$\omega = \frac{G_m}{L} = \frac{uN_m}{c(N^2+N_m^2)} \tag{4.86}$$

Under the above conditions, the device is capable of working as a balanced 2-phase ac generator. From Eq. (4.86) it is evident that for maximum frequency $N = N_m$, in which case the maximum frequency ω_m is

$$\omega_m = \frac{\mu_o b u^2}{2acR} \tag{4.87}$$

The presence of R in the above expression indicates that the maximum frequency depends on gas properties. This analysis then shows that it is possible to generate ac directly by using cross-coupled MHD self-excited generators.

Yet another conduction-type ac MHD generator can be developed by replacing the dc magnetic field of a dc MHD generator by an ac field. However, it has been found[23] that such generators are uneconomical and not very promising.

†(b) *ac MHD Induction Generator.*[22,23] An ac electrodeless generator has been proposed which is analogous to the induction generator, or to a transmission line, or to a traveling-wave tube. In such a generator, the plasma flows in a traveling field, as shown in Fig. 4-16, and the output power is due

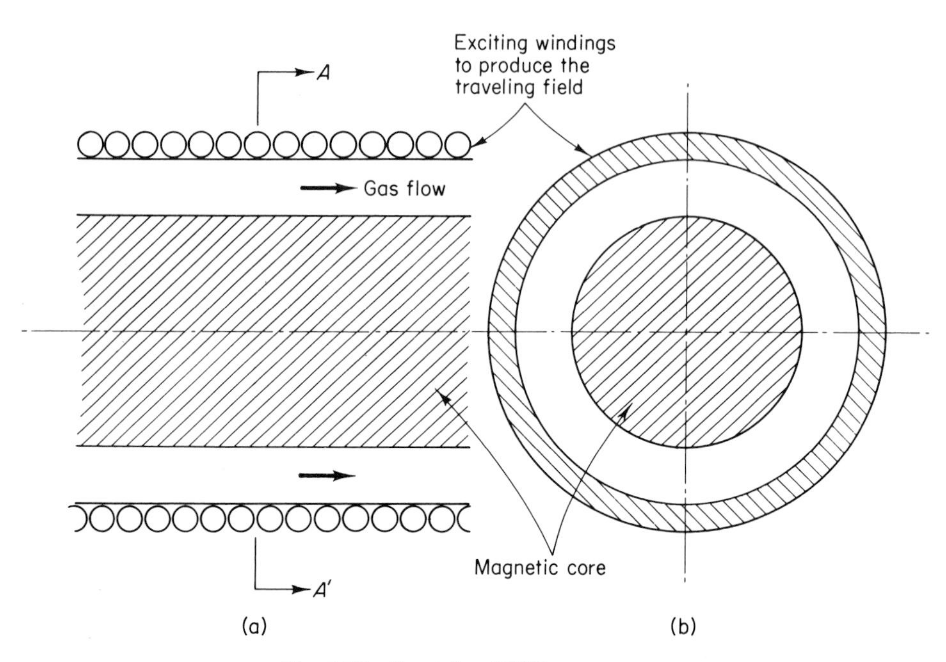

Fig. 4-16 Coaxial ac MHD generator.

to the electromagnetic coupling between the traveling wave and the plasma.

Using the previously stated simplifying assumptions, within the plasma the electric and magnetic fields, **E** and **B** respectively, are related to the current density **J** by the following equations (see Problem 4-14):

$$\Box^2\mathbf{E} = \mu\mathbf{J} \tag{4.88}$$

$$\Box^2\mathbf{B} = -\mu\nabla \times \mathbf{J} \tag{4.89}$$

and

$$\mathbf{J} = \sigma(\mathbf{E}+\mathbf{u} \times \mathbf{B}) \tag{4.90}$$

where $\Box$ is the *D'Alembertian operator* defined by

$$\Box^2 = \nabla^2 - \mu\epsilon\frac{\partial^2}{\partial t^2} \tag{4.91}$$

and **u** is the velocity of the plasma.

If

$$B = B_m \operatorname{Re}[e^{j(\omega t - \beta x)}] \tag{4.92}$$

it can be readily verified (see Problem 4-14) that

$$J = \sigma u B\left(\frac{u}{u_o} - 1\right) \tag{4.93}$$

where $u_o = \omega/\beta$. Therefore, proceeding as in Sec. 4.10.1, the time-average output-power density (per unit volume) $\langle P_o \rangle$ is

$$\langle P_o \rangle = -\frac{1}{2}JE^* = \sigma u^2 B^2\left(1-\frac{u_o}{u}\right)\frac{u_o}{u} \tag{4.94}$$

and the maximum output-power density is

$$\langle P_{om} \rangle = \tfrac{1}{4}\sigma u^2 B^2 \tag{4.95}$$

which occurs at $u_o/u = \frac{1}{2}$.

Comparison with the dc MHD generator, Eqs. (4.60) and (4.94), shows that the ratio u_o/u plays the same role in an ac generator as does K in a dc generator. Moreover, the above analysis shows that the output power is zero if $u_o = 0$, that is, if the magnetic field does not travel in space. Again, the output power is zero if $u_o = u$, that is, if the plasma travels with the same velocity as the velocity of the traveling field, called the *synchronous velocity*. For positive power output, the velocity u must be greater than the velocity u_o, and the circuit can be connected so that power can be extracted as it is generated at each point along the channel.

4.11 Summary

In this chapter we studied the energy-conversion process for continuous media. For ac induction-type or eddy-current devices three methods of analysis were given. Certain ac and dc hydromagnetic pumps and converters were discussed and a simplified treatment of the homopolar, or acyclic, machine was given. Performance equations for these devices were derived from the field theory developed in Chapter 2. Finally, an outline of dc and ac MHD power conversion was presented.

PROBLEMS

4–1. Two B fields are given:

$$B_1 = B' \cos\left(\omega t - \frac{\phi}{2}\right) \cos\left(\theta - \frac{\delta}{2}\right)$$

and

$$B_2 = B'' \cos\left(\omega t - \frac{\pi}{2} + \frac{\phi}{2} + \epsilon\right) \cos\left(\theta - \frac{\pi}{2} + \frac{\delta}{2}\right)$$

Show that the resultant field can be expressed as the sum total of a forward traveling field, a backward traveling field, and a pulsating field. Assume ϵ is small, that $B' = B_o + b$, and that $B'' = B_o - b$.

4–2. Derive compact and simplified expressions for the three field components in Problem 4-1.

4–3. Derive Eq. (4.10) from the principle of conservation of power.

4–4. Using traveling-field theory, calculate the driving torque on the disc of an energy meter having the following specifications: $\lambda = 5.6 \times 10^{-2}$ m; $h = 10^{-3}$ m; $l = 1.2 \times 10^{-2}$ m; $\sigma = 3.5 \times 10^7$ mho/m; $f = 50$ c/s; $u_o = 2.8$ m/s; and $u = 0.126$ m/s. The flux densities produced by the current and voltage coils are $B_1 = B_2 = 0.05$ Wb/m^2 at 10 A load and unity power factor.

4–5. Complete the problem of Example 4-2 assuming the motor has q pairs of poles and the axial length of the sleeve is l.

4–6. Verify Eq. (4.18), assuming that the conducting sleeve is backed by a perfectly permeable iron boundary as in Example 4-2. Clearly identify the boundaries and the boundary conditions.

4–7. (a) A conducting plate of thickness $2h$ (shown in Fig. 4P-7) has a traveling B field on each side of its surface. The y components of the fields are given by

$$B'_y = B_m e^{-j\beta x} e^{j\omega_o t} \qquad \text{at} \quad y = +h$$

and

$$B''_y = B'_y e^{-j\psi} \qquad \text{at} \quad y = -h$$

If the plate moves with a slip s in the above fields, obtain an expression for the time-average force density on the plate.

(b) What values of ψ yield the maximum and minimum force densities? Determine the maximum and minimum force densities.

(c) For a slip of 10%, calculate the force density on the plate from the following data: $h = 1$ cm; $B_m = 0.1$ Wb/m^2; $\sigma = 5.8 \times 10^7$ mho/m; supply frequency $= 60$ c/s; and $\psi = 30°$.

4–8. Derive Eq. (4.23c). How can this equation be solved for sinusoidal space and time variations of B_y and J^s?

4–9. Electromagnetic flowmeters are designed using MHD principles. Derive an expression relating to the potential difference between two selected points in the fluid and the mean velocity of the fluid. Assume a configuration of a rectangular channel with a dc magnetic field at right angles to the direction of flow of the fluid. Neglect edge effects and displacement current.

4–10. Design an elementary dc electromagnetic pump of rectangular cross-section to pump mercury at a rate of 1 m^3/s at a pressure of 1000 N/m^2. Explicitly, determine the dimensions of the channel, the flux density to be obtained from the electromagnet, and the voltage and current ratings of the power supply. Make reasonable assumptions. For mercury, $\sigma = 1.016 \times 10^6$ mho/m and mass density $= 1.355$ kg/m^3. Calculate the ohmic loss in the mercury per unit length of the channel.

4–11. Define: (a) laminar flow; (b) turbulent flow; (c) Reynolds' number; and (d) Mach number.

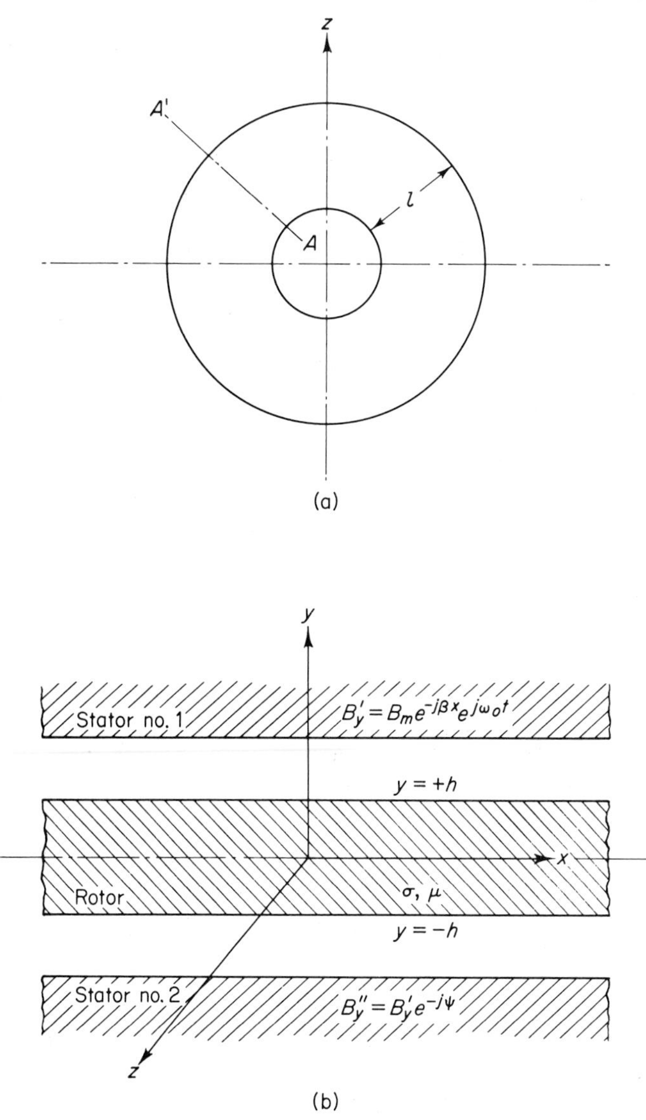

Fig. 4P-7 (a) The disk rotor; (b) motor "cut" along AA' (a) and developed.

4–12. It is desired to obtain a load current of 1000 A at 150 V from a dc MHD generator. The gas is seeded CO_2 at 3000°K and has a conductivity of 80 mho/m. The gas velocity is 2500 m/s and the flux density is 0.5 Wb/m^2. For loading factors of (a) 0.5, and (b) 0.7, obtain approximate channel dimensions neglecting Hall effect.

4–13. Repeat Problem 4-12 if Hall number is 5.

4–14. From Maxwell's equations, derive Eqs. (4.88) and (4.89).

4–15. Based on the discussions of Sec. 4.10.3(b), derive an analogy between an induction machine and a traveling-wave tube, and hence show that it is possible to develop a traveling-wave electromagnetic machine.

4–16. Derive Eq. (4.13) from Maxwell's equations, assuming sinusoidal space and time variations.

4–17. Justify the approximations given by Eqs. (l) and (m) of Example 4-2.

4–18. Show that Eq. (4.26) is dimensionally correct.

4–19. Verify that Eq. (4.56) can be derived from Eqs. (4.51–4.55).

4–20. Derive Eqs. (4.93) and (4.94).

REFERENCES

1. Hague, B., *Electromagnetic Problems in Electrical Engineering*. London: Oxford University Press, 1929; New York: Dover Publications, 1962.

2. Barlow, H. E. M., "Travelling-Field Theory of Induction Instruments and Motors," *Proc. IEE* (London), Vol. 112 (6), 1955, pp. 1208–1214.

3. Barlow, H. E. M., "Simplified Treatment of Mechanical Force on Materials in an Electromagnetic Field," *Proc. IEE* (London), Vol. 113 (2), 1966, pp. 373–377.

4. Javid, M., and P. M. Brown, *Field Analysis and Electromagnetics*, pp. 129–133. New York: McGraw-Hill Book Company, 1963.

5. Fano, R. M., L. J. Chu, and R. B. Adler, *Electromagnetic Fields, Energy, and Forces*, Chapters 7–10. New York: John Wiley & Sons, Inc., 1960.

6. Cullen, A. L., and T. H. Barton, "A Simplified Electromagnetic Theory of the Induction Motor Using the Concept of Wave Impedance," *Proc. IEE*, Vol. 105(c), 1958, p. 331.

7. Ramo, S., J. R. Whinnery, and T. Van Duzer, *Fields and Waves in Communication Electronics*, p. 46. New York: John Wiley & Sons, Inc., 1965.

8. Hesmondhalgh, D. E., and E. R. Laithwaite, "Method of Analysing the Properties of 2-Phase Servo-motors and AC Tachometers," *Proc. IEE* (London), Vol. 110 (11), 1963, pp. 2039–2054.

9. Nasar, S. A., "An Axial-Airgap Variable-Speed Eddy-Current Motor," *IEEE Trans.*, Vol. PAS-87 ("Power Apparatus and Systems"), 1968, p. 1599.

10. Blake, L. R., "Conduction and Induction Pumps for Liquid Metals," *Proc. IEE* (London), Vol. 104(a), 1957, pp. 49–63.

11. Okhremenko, N. M., "An Investigation of the Spatial Distribution of the Magnetic Fields and Electromagnetic Effects in Induction Pumps," *Magnetohydrodynamics* (New York: The Faraday Press, Inc.), Vol. 1, No. 1, 1965, pp. 72–80. (*Note:* This publication is a translation from Russian *Magnitnaya Gidrodinka*, and subsequent issues contain many interesting, but advanced, research papers on MHD generators, liquid-metal pumps, and allied topics.)

12. Panholzer, R., "Electromagnetic Pumps," *Electrical Engineering*, Vol. 82, 1963, pp. 128–135.

13. Pierson, E. S., "A Hydromagnetic Converter," *MIT ASD Technical Report* No. 61–102.

14. Harris, L., *Hydromagnetic Channel Flows*. Cambridge, Mass.: MIT Press, 1960.

15. Watt, D. A., "The Development and Operation of a 10kW Homopolar Generator with Mercury Brushes," *Proc. IEE* (London), Vol. 105(a), 1958, p. 233.

16. Zeisler, F. L., "A High Power Density Electric Machine Element," *IEEE Trans.*, Vol. PAS-86 ("Power Apparatus and Systems"), 1967, pp. 811–818.

17. Levi, E., and M. Panzer, *Electromechanical Power Conversion*, Chapter 5. New York: McGraw-Hill Book Company, 1966.

18. Ku, Y. H., and A. Kamal, "A New Homopolar Motor," *Jour. Franklin Inst.*, Vol. 258, 1954, p. 7.

19. Jeffery, A., *Magnetohydrodynamics*. New York: Interscience Publishers, Inc., 1966.

20. Dunn, P. D., and J. K. Wright, "Unconventional Methods of Electricity Generation," *Proc. IEE* (London), Vol. 110 (10), 1963, p. 1837.

21. Haus, H. A., "Alternating-current Generation with Moving Conducting Fluids," *Jour. Appl. Phys.*, Vol. 33, No. 7, 1962, p. 2161.

22. Carter, R. L., and R. A. Laubenstein, "A Non-equilibrium Alternating Current Magnetogasdynamic Linear Induction Motor," *Third Symposium on Engineering Aspects of Magnetohydrodynamics*, p. 291. New York: Gordon and Breach, Publishers, 1962.

23. Clark, R. B., D. T. Swift-Hook, and J. K. Wright, "Prospects of ac MHD Generation," *Brit. Jour. Appl. Phys.*, Vol. 14, 1963, p. 10.

24. Woodson, H. H., "AC Power Generation with Transverse-Current Magnetohydrodynamic Conduction Machines," *IEEE Trans.*, Vol. PAS-84 ("Power Apparatus and Systems"), 1965, p. 1066.

25. Koch, W. H., "Equivalent Circuits with Transformer Elements for Eddy-Current Rotor Induction Motors Derived from Field Equations, Parts I and II," *IEEE Trans.*, Vol. PAS-84 ("Power Apparatus and Systems"), 1964, pp. 567–583.

5

Energy Conversion in Lumped-parameter Rotary Devices

5.1 Introduction

In the last chapter energy conversion in continuous media was studied. Explicitly, expressions for forces and voltages were derived for distributed-parameter devices. It was seen that for such devices the basic equations, derived from Maxwell's equations, are partial-differential equations. On the other hand, for lumped-parameter systems, discussed in Chapter 3, the basic equations are total-differential equations. However, in Chapter 3 only incremental-motion devices were considered. The equations describing the energy-conversion process in lumped-parameter rotary devices, or rotating electrical machines, are the same as those for incremental-motion transducers, but for the former there are certain constraints that must be observed. Moreover, the resulting differential equations are solved by techniques other than the linearization method discussed previously. In this chapter, rotary devices are considered from a "lumped-circuits-in-relative-motion" point of view. The study will include parameter determination, formulation of equations of motion, and solutions of these equations for specified constraints.

Before embarking upon the analysis of rotating machines, we should recall some of their physical features and the constraints to be observed for their successful operation. A hypothetical rotating energy-converting device is shown in Fig. 5-1. If the device is to convert electrical energy into mechani-

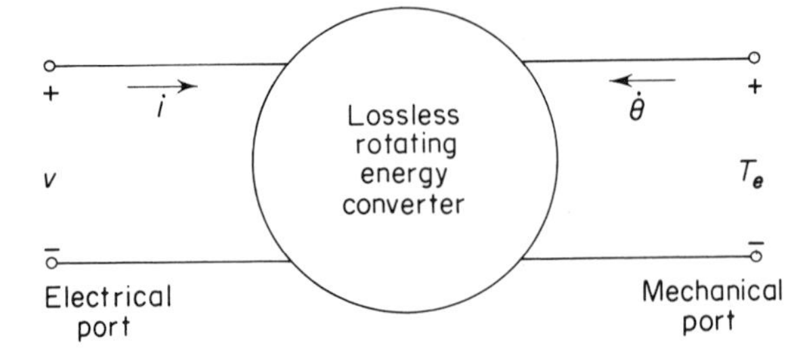

Fig. 5-1 A hypothetical rotating energy converter.

cal form, first of all the electromagnetic torque T_e must be nonzero for any position of the rotor (or of the moving member of the device) for starting. This constraint can be expressed in the form of the following inequality:

For starting

$$T_e > 0 \qquad \text{for} \quad 0 \leqslant \theta \leqslant 2\pi \tag{5.1}$$

where θ is the mechanical position of the rotor.

Now, since mechanical power is the product of torque and angular velocity, energy conversion requires that the velocity be nonzero. This implies that the time-average torque $\langle T_e \rangle$ be positive over a complete revolution. Expressed mathematically:

For running

$$\langle T_e \rangle > 0 \qquad \text{for} \quad \dot{\theta} > 0 \tag{5.2a}$$

where

$$\langle T_e \rangle = \frac{1}{2\pi} \int_0^{2\pi} T_e \, d\theta \tag{5.2b}$$

For conversion of mechanical energy into electrical form, the required condition is

$$\langle vi \rangle > 0 \tag{5.3}$$

And, for a lossless electromechanical energy converter, the following equation holds:

$$T_e \dot{\theta} = vi \tag{5.4}$$

where the symbols are as shown in Fig. 5-1 and the quantities are instantaneous values. The constraints for the conversion of electrical energy to mechanical form are shown in Fig. 5-2. The device having the characteristic shown by curve a will start but will not run, because the average torque is zero; the one having the torque variation shown in curve b will not start, but if started by some means it will continue to run; and the one with the torque variation shown in curve c is self-starting and will run continuously. In each of these cases it has been assumed that the rotor has inertia.

From Chapters 3 and 4 it may be seen that electromagnetic torque, as in a motor, is produced by the interaction between current-carrying con-

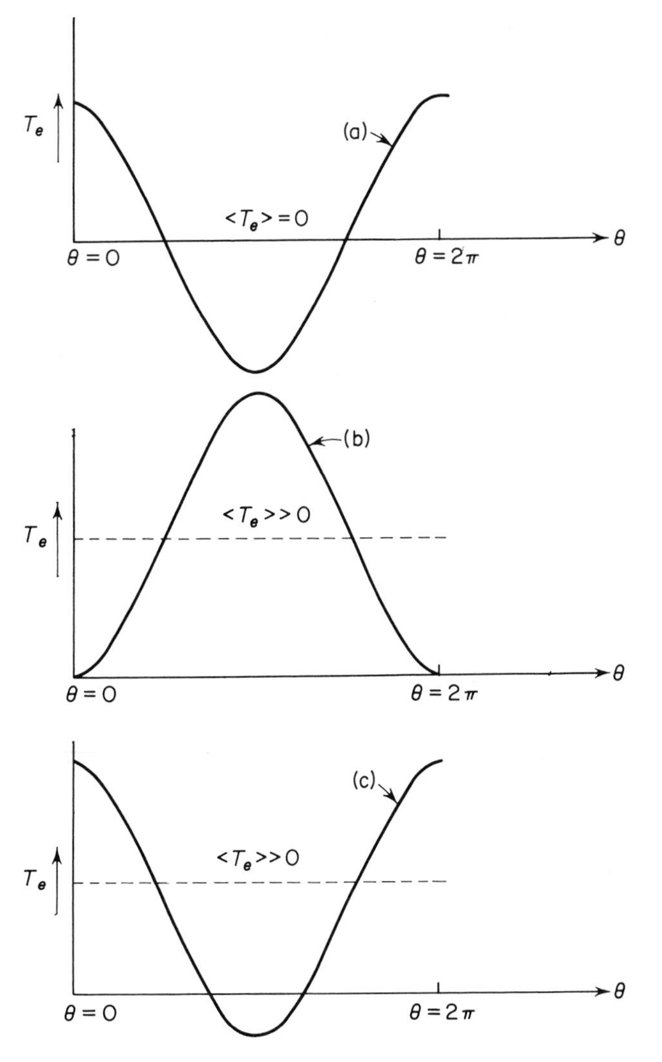

Fig. 5-2 T_e/θ variations for running and/or starting.

ductors and magnetic fields. The magnitude of the torque or force is obtainable from Ampere's law or from the energy relations. For generator action, according to Faraday's law a voltage is induced in a conductor if the flux linking the conductor changes or if the conductor cuts the magnetic lines of force. All rotating machines must, therefore, have (1) conductors, with arrangements for carrying current, mounted on suitable structures for rotation, and (2) an appropriate arrangement for the production of the magnetic field. A machine may be a dc machine or an ac machine depending upon how the requirements (1) and (2) just mentioned are satisfied. Further classification of machines can also be made on the basis of the fulfillment of the constraints. To illustrate this point some simplified machine configurations are now considered.

5.1.1 Synchronous Machines

Consider an N-turn coil of radius r, rotating with an angular velocity ω in a uniform magnetic field $\mathbf{B}$, as shown in Fig. 5-3(a). The voltage v induced in the coil is, according to Faraday's law,

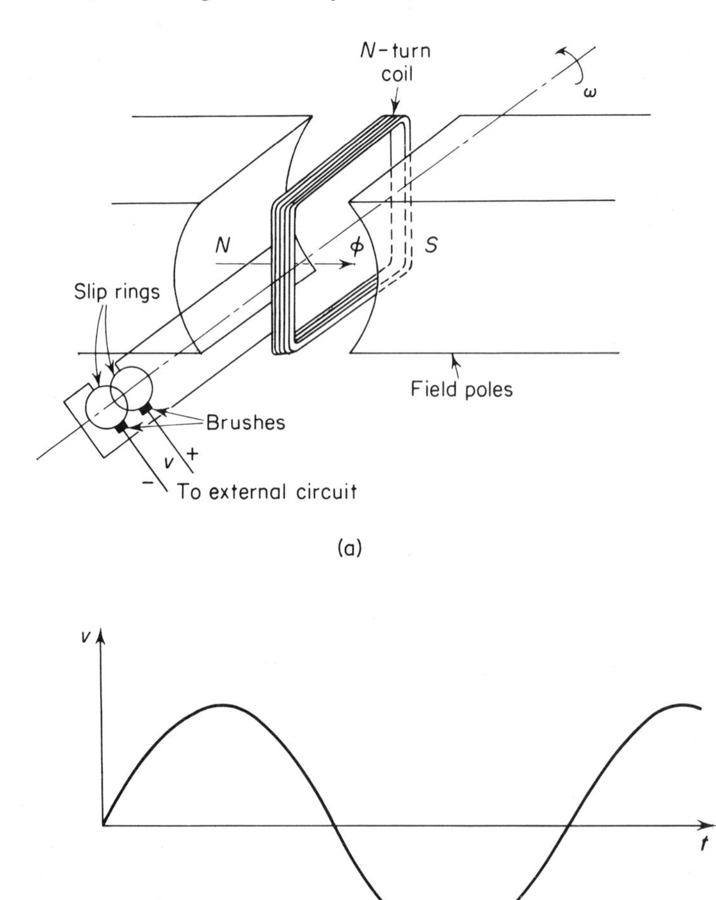

Fig. 5-3 (a) An elementary ac generator; (b) voltage variation with time.

$$v = l\mathbf{u} \times \mathbf{B} = \omega N\phi \sin \omega t \tag{5.5}$$

where

$$\phi = 2Brl \tag{5.6}$$

is the total flux per pole. The voltage thus induced in the coil can be measured at the *slip rings* to which the ends of the coil are connected for current collection by means of *brushes.* From Eq. (5.5) it follows that the voltage varies sinusoidally in time as shown in Fig. 5-3(b). This arrangement can therefore be used to generate ac, and the machine may be called an *ac generator*, an *alternator*, or a *synchronous generator*. It is called a synchronous generator because the frequency, $f = \omega/2\pi$, of the generated voltage is synchronized with the mechanical speed of rotation n_s, as given by the equation

$$n_s p = 120f \tag{5.7}$$

where p = number of poles.

In the above machine, the flux is produced by means of the *field windings,* which carry dc excitation. The windings which supply the load are called *armature windings.* Because only a relative motion between the field and the armature windings is required, for practical considerations it is more suitable to keep the armature windings stationary and place the field windings on the rotor. Two such arrangements are shown in Figs. 5-4(a) and (b), where current to the field winding is fed through slip rings. If the rotor is cylindrical in shape [Fig. 5-4(a)], the machine is called a *round-rotor* machine; but if the rotor has projecting poles [Fig. 5-4(b)], the machine is a *salient-pole* machine. In practice the armature windings are *distributed* over the entire periphery of the stator, and the machine may be a polyphase machine if it has more than one such independent armature winding wound symmetrically on the stator.

In the preceding paragraph, the production of voltage in an elementary synchronous generator was described. The machine is equally capable of operating as a *synchronous motor*, in which case the electrical energy supplied at the armature terminals of the machine is converted into mechanical form. The polyphase armature windings, which carry ac excitation, produce a rotating magnetic field in the *airgap* of the machine (as discussed in Section 4.3). The rotor, carrying dc excitation, *locks in* or *pulls into synchronism* with the magnetic field produced by the stator when the speed of the rotor is very close to synchronous speed. The synchronous motor is a constant-speed motor, but is not self-starting as such. It is started by some auxiliary means, such as by *damper windings* (which are copper or aluminum bars short-circuited at both ends and mounted on the rotor) or by an auxiliary motor. The angle between the axis of the stator magnetic field and the axis of the rotor magnetic field, denoted by δ, is a measure of the power developed by the machine and is called the *power angle* or *torque angle* (as in a reluctance machine).

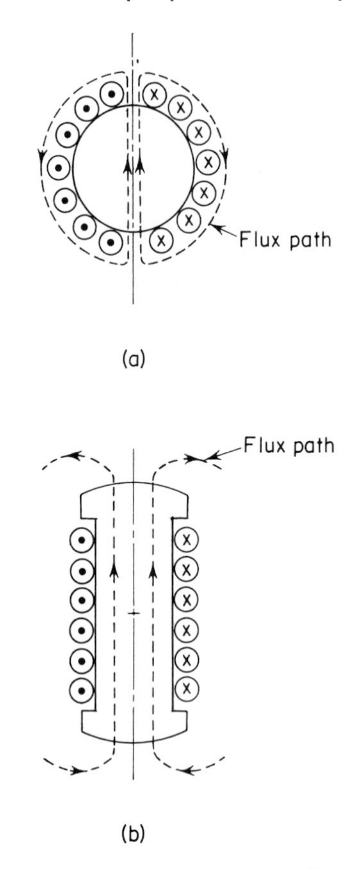

Fig. 5-4 Field windings: (a) round rotor; (b) salient rotor, 2-pole.

To summarize, the constraints on a synchronous machine are that it is a constant-speed machine, that the rotor and stator are both excited by external sources, and that one of these (generally the rotor) is fed by dc and the other (generally the stator) by an ac source (which is usually polyphase).

5.1.2 *Induction Machines*

The principle of operation of the induction motor is discussed at a number of places in Chapter 4. The stator of the polyphase induction machine is identical to that of a synchronous machine in that it carries a polyphase ac excitation, which results in the production of a rotating magnetic field in the airgap of the machine. The rotor of the induction machine is cylindrical and carries either (1) conducting bars short-circuited at both ends, as in a *cage-type* machine [Fig. 5-5(a)], or (2) a polyphase winding with terminals brought out to slip rings, as in a *wound-rotor* machine [Fig. 5-5(b)]. Sometimes the cage-type machine is also called a *brushless* machine and the wound-rotor machine termed a *slip-ring machine*. The stator rotating field induces a voltage, and hence a current, in the rotor circuit. Thus, according to Ampere's law, a torque is developed by the machine. The rotor may be driven by an external source and the speed of the rotor may be made to exceed the synchronous speed, in which case the machine could operate as an *induction generator*. On the other hand, if the rotor is allowed to run under the torque developed by the interaction of the stator rotating field and the rotor-induced currents, the speed of the rotor will always be less than the synchronous speed. The difference in the synchronous speed and the actual rotor speed is measured by *slip* as defined by Eq. (4.9). The rotor currents are of *slip frequency*.

The constraint of a definite relationship between the frequencies of the stator and rotor currents must be fulfilled for the successful operation of an induction machine. Although the machine is capable of functioning as a generator, it is most commonly used as a motor.

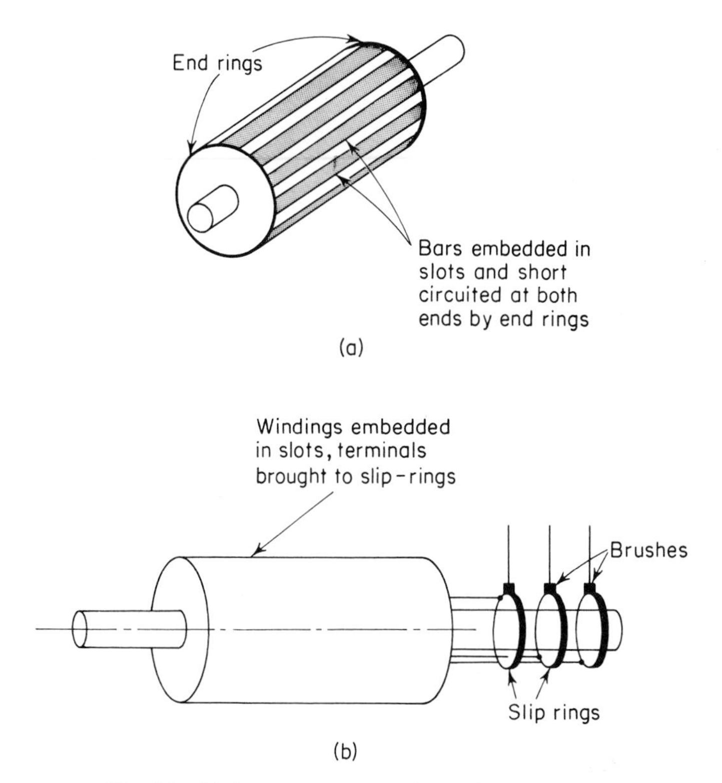

Fig. 5-5 (a) A cage-type rotor; (b) a slip-ring rotor.

5.1.3 dc Machines

In Section 5.1.1 we described an elementary form of an ac generator. If the slip rings shown in Fig. 5-3(a) are replaced by a pair of segments, called the *commutator*, as shown in Fig. 5-6(a), the voltage available at the terminals is unidirectional and of the form shown in Fig. 5-6(b). (This device is an oversimplified version of a dc generator.)

If direct current is fed into the armature windings, with the field excited by dc as in the generator described above, the machine operates as a motor. Because of the presence of the commutator the torque developed by the motor is unidirectional. For both modes of operation Eq. (5.4) is satisfied. It should

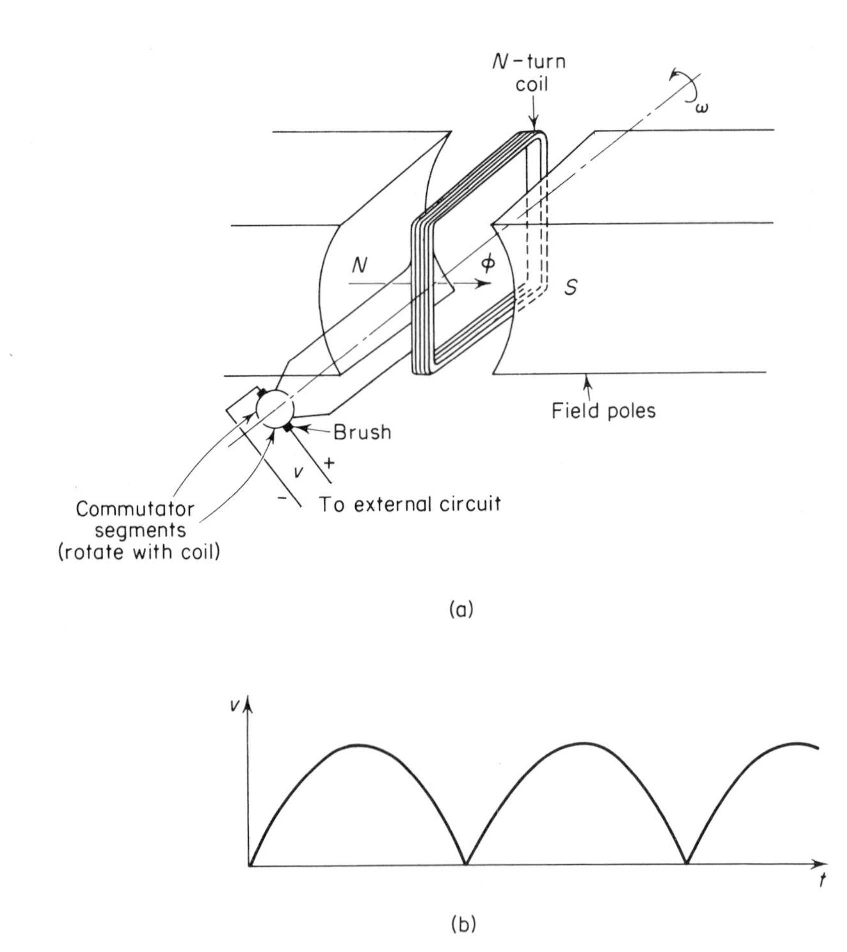

Fig. 5-6 (a) An elementary dc generator; (b) voltage waveform at the terminals.

be recognized that a dc machine while operating as a generator has an inherently developed electromagnetic torque opposing the prime-mover torque; and while operating as a motor it has a motional EMF induced in the armature windings opposing the applied terminal voltage. This induced voltage in a dc motor is called the *back EMF.*

The constraints in a dc machine are introduced by the presence of the commutator, which results in imparting definite polarities to the brushes and the corresponding terminals.

The above machines and other different machines are discussed in detail in later chapters. At present we shall attempt to derive different types of machines by imposing constraints on an idealized rotating machine, which is discussed in the next section. The parameters of this machine will be determined and its equations of motion will be derived and solved for different constraints.

5.2 An Idealized Rotating Machine

In the preceding section elementary forms of three most common types of rotating machines were described. We indicated how different constraints led to different types of machines, but we did not attempt to evaluate quantitatively the performance of the machine.

The performance characteristics of a number of important types of machines can be derived from the equations of motion of an idealized machine by imposing appropriate constraints. Although it is more general to choose a salient-pole model, a double-cylindrical type idealized machine is considered here. The effects of saliency are incorporated later, when necessary. The machine under consideration here consists of a slotted stator and slotted rotor each carrying a set of balanced 2-phase windings as shown in Fig. 5-7 (only phase a of a 2-phase, 6-pole stator winding is shown). This machine is idealized by replacing the slotted structures by smooth structures, and the actual current-carrying windings by infinitely thin current elements on the airgap side of the stator and rotor surfaces. These fictitious infinitely thin current elements are called *current sheets,* and have linear current densities. The element width is chosen as the original slot opening, and the linear current densities are such that they give the same fields in the airgap of the idealized model with smooth surfaces as the original windings produced in the slotted machine. An arbitrary current-sheet equivalent of an actual winding is shown in Fig. 5-8.

The current-sheet pulses of the form shown in Fig. 5-8 can be expressed in a Fourier series. In the idealized machine, however, it is assumed that the current sheet has a perfectly sinusoidal space distribution. This amounts

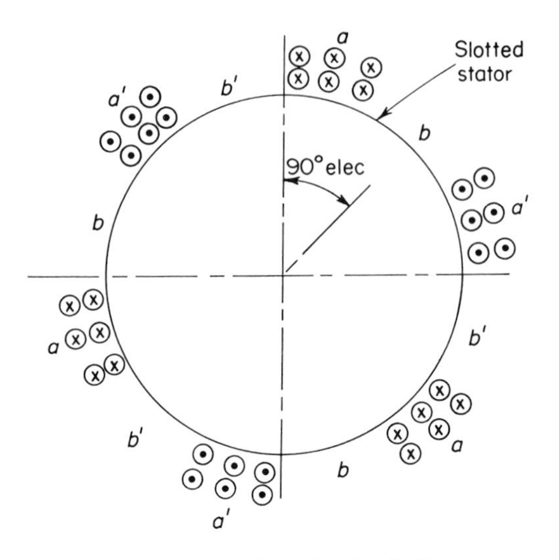

Fig. 5-7 A 2-phase, 6-pole winding.

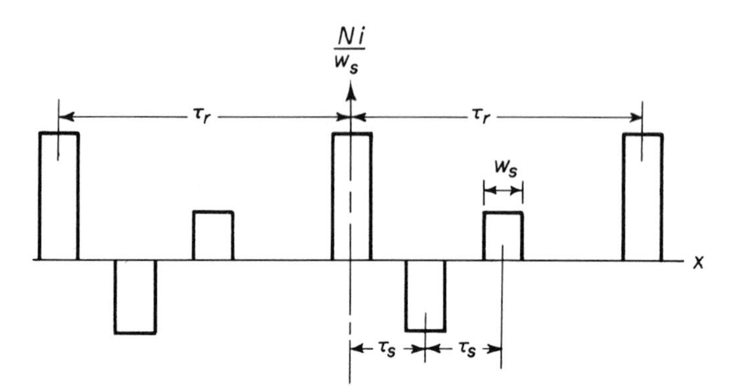

Fig. 5-8 Current-sheet representation of an actual winding.

to saying that only the fundamental component of the Fourier series representing the current-sheet pulses is considered. If the airgap is small as compared with the pole pitch or the length of the repeatable section τ_r (Fig. 5-8), a rectangular coordinate system can be used. For the idealized machine, two further assumptions are made: first, the currents are all z-directed; and secondly, the iron backing the current sheets is infinitely permeable.

To summarize, the idealized machine is a balanced 2-phase, double-cylindrical structure having sinusoidally distributed current sheets on the airgap side of the stationary and rotating members. The current sheets are backed by iron boundaries of infinite permeability.

The angular positions of the current sheets are measured in terms of *electrical degrees* θ_e, related to the mechanical degree θ by the equation

$$\theta_e = \tfrac{1}{2} p\theta = \nu\theta \tag{5.8}$$

where p = number of *poles*, defined as the region in which the radial flux is unidirectional. It is also the same as a half-cycle of a sinusoidally distributed current sheet, ν being the number of cycles in the current sheet; or ν can also be considered as the order of the space harmonics in the airgap fields.

EXAMPLE 5–1

In Sec. 5.1.1 we derived the EMF equation of an elementary generator using the "flux-cutting" rule. We would like to show by means of this example that identical results can be obtained using the "change-in-the-flux-linkage" rule.

Consider an N-turn coil, of axial length l and radius r, rotating at constant velocity ω in a sinusoidally distributed magnetic field as shown in Fig. 5E-1.

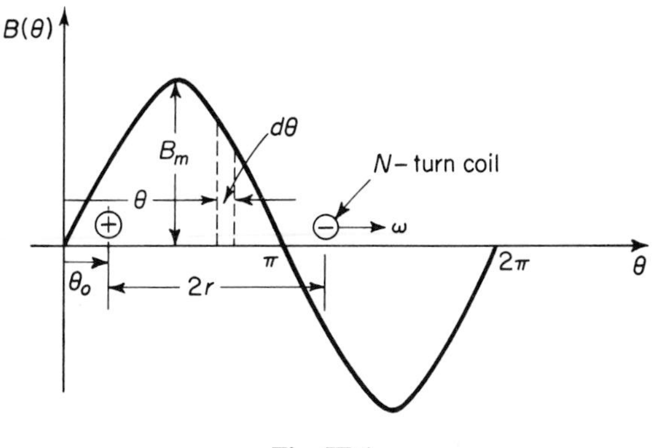

Fig. 5E-1

Let the flux-density distribution be given by

$$B(\theta) = B_m \sin \theta$$

Then the elemental flux $d\lambda$ linking with the coil within an angle $d\theta$ is

$$d\lambda = Nlrd\theta B_m \sin \theta$$

If we consider a full-pitch coil—that is, one in which the distance between the coil sides is the same as the pole pitch (which is π according to Fig. 5E-1)—the total flux linking the coil at a particular instant is

$$\lambda = \int_{\theta_o}^{\pi+\theta_o} d\lambda = \int_{\theta_o}^{\pi+\theta_o} NlB_m r \sin \theta \, d\theta$$

Or,

$$\lambda = 2NlrB_m \cos \theta_o \tag{a}$$

where θ_o is the displacement of the coil at the given instant. But the angular velocity ω and displacement θ_o are related by

$$\omega = \frac{d\theta_o}{dt}$$

or,

$$\theta_o = \omega t \tag{b}$$

so that, from Eqs. (a) and (b), we have

$$\lambda = 2NlrB_m \cos \omega t$$

and the voltage induced is, therefore,

$$v = \frac{d\lambda}{dt} = 2Nlr\omega B_m \sin \omega t$$

Or,

$$v = \omega N\phi \sin \omega t \tag{c}$$

where $\phi = 2B_m rl$. Notice that Eq. (c) above and Eq. (5.5) are identical.

5.3 Current Sheets[1]

In the preceding discussions we referred to the concept of current sheets at a number of places. We shall now consider this in some detail.

The field in the airgap of a machine is established by current sheets on the stator and rotor surfaces. To appreciate the physical picture, we consider the layout of a typical ac machine stator as shown in Figure 5-7. The stator in this case has 36 slots equally distributed around the periphery of the surface. The slots are cut in silicon steel and the conductors are embedded in these slots. Thus current-carrying conductors are distributed around the machine.

The slot-embedded conductors are replaced by a current sheet of infinitesimal thickness having the same distribution of current density as the slot-embedded conductor configurations. While it may be difficult to formulate the exact equivalent current sheet, an equivalent current sheet is a necessary adjunct to an analytic solution of the fields in the airgap of a machine. Figure 5-9 shows rotor and stator current sheets together with the coordinate system.

Each machine type has its own current-sheet type. For example, most ac machines attempt to achieve a sinusoidal or uniform pulse-current distribution, while dc machines use only uniform pulse distributions. In Figure 5-9, the current density varies in the x direction, but will be assumed constant for all z's for any value of x. This is not always the case, since conductors are often twisted or *skewed* from a parallel position with respect to the axis of

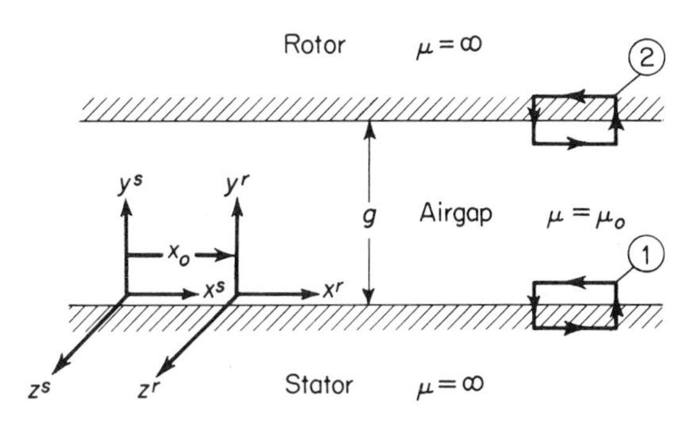

Fig. 5-9 A model for the solution of the field problem.

the machine. In such cases, the current will vary in both the x and z directions.

For the assumption that the current density varies only in the x direction, the cross-section appearing in Figure 5-9 applies at any z. In the y direction the sheets are assumed to be of negligible thickness. In effect, this negligible thickness places the steel at the actual airgap surface.

Even though the current flows in the z direction, the assumptions that the current density does not vary in the z direction, and that the sheet is of negligible thickness, force the current density h to be a function only of x and time; thus

$$\mathbf{h} = h_z(x, t)\mathbf{a}_z \tag{5.9a}$$

Now, this current sheet is produced by a current flowing through the terminals of the machine; thus it is necessary to relate the current density along the sheet with the terminal currents. Currents are related to the current density by the relationship

$$i' = \int_s \mathbf{h} \cdot ds \tag{5.9b}$$

where i' is a fictitious net current to be defined below. Because of Eq. (5.9a), Eq. (5.9b) becomes

$$i' = \int h_z(x, t)\, dx \quad \text{or} \quad \frac{di'}{dx} = h_z(x, t) \tag{5.9c}$$

According to Eq. (5.9c) a current density may be produced by a change in net current per unit length in the x direction. Some reflection will show that a change in net current may be effected by changing the number of conductors as a function of x and holding the actual current constant, by varying the current holding the number of conductors constant, or by varying both. In most machines a series circuit is used between any two terminals so that a single current flows; this requires that the conductors vary with x in some appropriate manner to yield the current density desired. Thus, as a slot with embedded conductors is traversed, there is a change in i' with x and a pulse of current density is produced. Each slot will generate its own pulse of current density. These in turn may be analyzed as a Fourier series producing a continuous current sheet of a fundamental plus harmonic expansion in x.

†5.3.1 *Fourier Analysis of Current Sheets*[1,3]

Distributed windings have the characteristic of having two coil sides per slot, thus filling all the slots as shown in Fig. 5-7. Each coil (one of whose sides is in the upper half of the slot and the other in the lower half) spans a certain fixed angle known as the coil pitch. Normally there is a group of coils in adjacent slots; this group is termed a *phase belt*. Coil grouping will sometimes repeat around the machine; each repetition constitutes a *repeatable section*. Fourier analysis need only be made for one repeatable section since the field conditions in the airgap are then repeated as the observer passes from one repeating section to the next.

An arbitrary current-sheet representation of an actual winding is shown in Fig. 5-8. Notice that the "strength" of current-sheet pulse is Ni/w_s where w_s is the width of the slot opening containing N conductors which carry a current i. If there are m slots in a repeatable section, with the origin chosen as shown, the Fourier expansion of the pulses of the current sheet shown in Fig. 5-8 is

$$h(x) = \sum_{k=1}^{m} h_m(x) = \sum_{k=1}^{m} \frac{N_k i_k}{w_s} \sum_{\nu=1}^{\infty} (a_\nu \cos \alpha_\nu x + b_\nu \sin \alpha_\nu x) \tag{5.10a}$$

where $\alpha_\nu = 2\pi\nu/\tau_r$, ν = order of the harmonic, and a_ν and b_ν are the Fourier coefficients (to be evaluated). Other symbols are as shown in Fig. 5-8.

The coefficients a_ν and b_ν are determined from orthogonality conditions. For example, for a periodic function $f(x)$ (see also Appendix D),

$$\begin{aligned} a_\nu &= \frac{2}{\tau_r}\int_{-\tau_r/2}^{\tau_r/2} f(x)\cos\frac{2\pi\nu}{\tau_r}\,dx \\ &= \frac{2}{\tau_r w_s \alpha_\nu}\sum_{k=1}^{m} N_k\left\{\sin\left[\alpha_\nu(k-1)\tau_s+\frac{\alpha_\nu w_s}{2}\right]-\sin\left[\alpha_\nu(k-1)\tau_s-\frac{\alpha_\nu w_s}{2}\right]\right\} \\ &= \frac{2K_{s\nu}}{\tau_r}\sum_{k=1}^{m} N_k\cos\left[\alpha_\nu(k-1)\tau_s\right] \end{aligned} \tag{5.10b}$$

where

$$K_{s\nu} = \left(\sin\frac{\alpha_\nu w_s}{2}\right)\Big/\frac{\alpha_\nu w_s}{2} \tag{5.10c}$$

Similarly,

$$b_\nu = \frac{2K_{s\nu}}{\tau_r} \sum_{k=1}^{m} N_k \sin\left[\alpha_\nu (k-1)\tau_s\right] \tag{5.10d}$$

and the series given by Eq. (5.10a) can be finally expressed as a cosine function and a phase angle as follows

$$h(x) = \sum_{k=1}^{m} \sum_{\nu=1}^{\infty} i_k c_{k\nu} \cos\left(\alpha_\nu x - \phi_{k\nu}\right) \tag{5.10e}$$

where

$$c_{k\nu}^2 = a_{k\nu}^2 + b_{k\nu}^2 \tag{5.10f}$$

and

$$\phi_{k\nu} = \tan^{-1} \frac{b_{k\nu}}{a_{k\nu}} \tag{5.10g}$$

Although we have assumed sinusoidal current-sheet distribution for the idealized machine of Section 5.2, the above analysis indicates that harmonics, if present, can be taken into account. We shall use this analysis in a later chapter. Now we shall consider an example of current-sheet analysis to get some idea of the magnitudes of harmonics present in an actual machine.

EXAMPLE 5–2

Consider a 36-slot, 3-phase stator with a double-layer distributed winding. The slots are evenly spaced so that there are ten mechanical degrees between adjacent slots. Because of the manner in which this winding is laid out, the most prominent component is the fifth, being larger than the fundamental. Thus the machine will run on the fifth space harmonic. Much of the literature terms the fifth space harmonic as the fundamental; in this context the machine will have a total of 1800 electrical degrees in all. Note that the electrical and mechanical degrees are related by Eq. (5.8).

The slot layout for this winding is shown in Fig. 5E-2(a). In this layout, the coils of any one phase lie in a 60° phase belt.

We wish to determine the magnitudes of the current-sheet harmonics of this winding.

This layout yields what is known as a *fractional-slot winding* as opposed to an *integral-slot winding* since the slots per pole per phase (SPP) is not an integer:

$$\text{SPP} = \frac{36}{3 \times 10} = 1\tfrac{1}{5} \text{ slots/pole/phase}$$

Each coil has a pitch of three slots in this case. The coil whose one side is the upper side of slot 1 will have the other side in slot 4. Coil in upper 2 is in lower 5, and so on.

The current sheet for phase a is pictured below the slot layout. In some slots there are two coil sides, giving rise to twice as dense a current sheet for that slot. The current pulses have an amplitude of

$$h = \frac{Ni}{w_s}$$

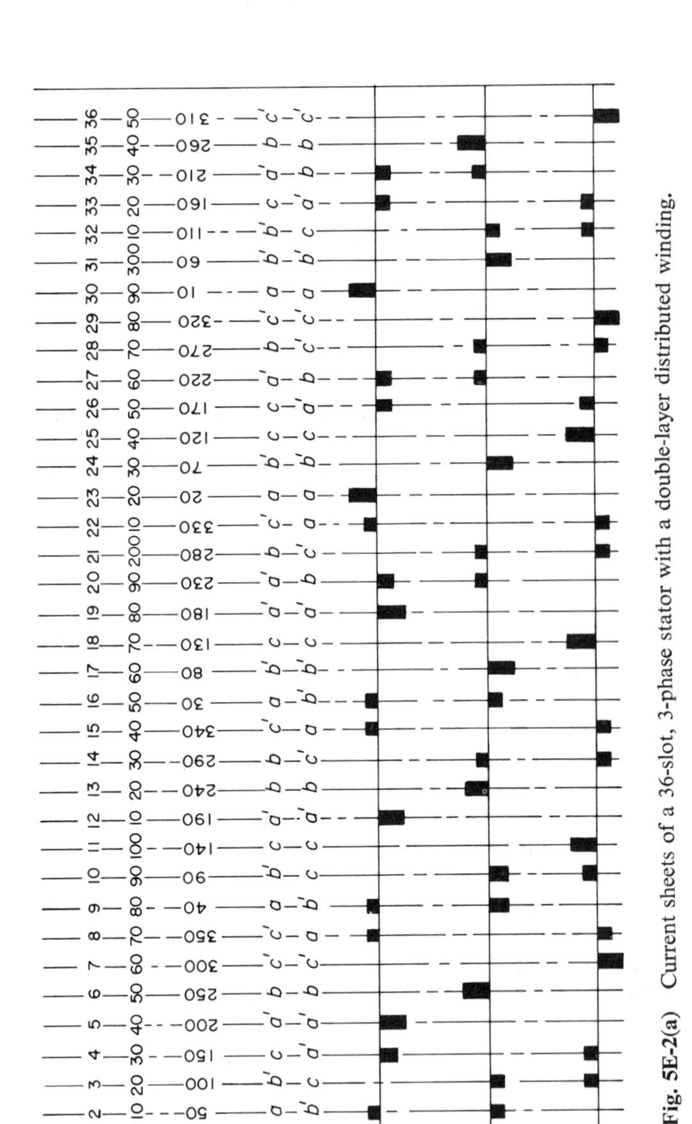

Fig. 5E-2(a) Current sheets of a 36-slot, 3-phase stator with a double-layer distributed winding.

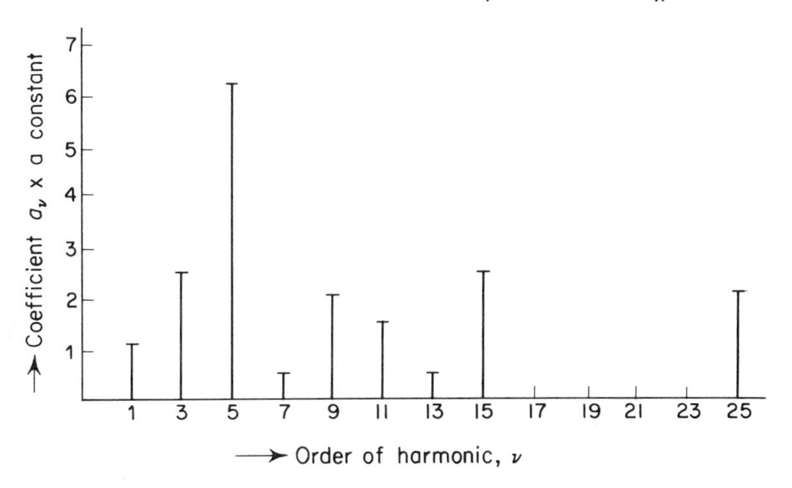

Fig. 5E-2(b) Fourier series coefficients of the current sheets of Fig. 5E-2(a).

To evaluate the coefficients in a Fourier series we find the axis of symmetry. One such axis is about slot 12; we call this $x = 0$ for the purposes of analysis. Since $f(-x) = f(x)$, the function is even and the coefficients b_ν are all zero. Thus, only the constants a_ν are to be evaluated. We recall that these constants are given by Eq. (5.10b), with the constant defined by Eq. (5.10c). If we assume $\tau_s = 2w_s$ and substitute in Eqs. (5.10b,c), the relative magnitudes of the current-sheet harmonics are obtained, as shown in Fig. 5E-2(b).

It is quite convenient to calculate the numerical results on a digital computer. A suitable program in Fortran is given in Appendix C.

From Fig. 5E-2(b) we notice that the amplitude of the fifth harmonic is the largest and that the machine will run on this predominant harmonic. This machine is, therefore, a 10-pole machine.

This example illustrates the concept of the current sheet and its resolution into space harmonics. However, we now return to the idealized machine having a sinusoidal current sheet, with no harmonics present.

5.4 Determination of Inductance Coefficients[2–4]

It was pointed out in Chapter 2 that the performance of an electromagnetic device depends very much on the various associated inductances. In Sec. 2.1.5 a method of evaluating the inductances of lumped-parameter electromechanical systems was given. For rotating machines, the situation is slightly complicated due to the distribution of the windings. In general, however, the inductance coefficients can be determined by one of the following methods.

1. The energy-storage method, using the relationship

$$\frac{1}{2}\int_v \mathbf{H}\cdot\mathbf{B}\,dv = \frac{1}{2}\sum_k\sum_m L_{km}i_ki_m \tag{5.11a}$$

where the symbols have their usual meanings defined earlier—see Eqs. (2.31) and (2.60).

2. The permeance-function method, using the relationship

$$L = N^2\mathscr{P} \tag{5.11b}$$

where N is the number of turns and $\mathscr{P}$ is the permeance function.

3. The flux-linkage method, using the defining equation

$$L = \frac{\lambda}{i} \tag{5.11c}$$

Only method 1 is considered in detail here. Methods 2 and 3 are given only in outline. Details are available in Reference 4. Note that these methods are closely related to each other in that the fields in the airgap of the machine have to be determined in each case.

†5.4.1 *The Energy-Storage Method*

An idealized machine is described in Section 5.2. The model and the coordinate system for this machine for the evaluation of inductances are shown in Fig. 5-9 (p. 159). Let the stator and rotor current sheets, $\mathbf{h}^s(x)$ and $\mathbf{h}^r(x)$, be

at stator surface, $y = 0$:

$$\mathbf{h}_a^s(x) = \mathbf{a}_z i_a^s N_a^s \cos \alpha_\nu x \tag{5.12a}$$

$$\mathbf{h}_b^s(x) = \mathbf{a}_z i_b^s N_b^s \cos\left(\alpha_\nu x - \frac{\pi}{2}\right) \tag{5.12b}$$

at rotor surface, $y = g$:

$$\mathbf{h}_a^r(x) = \mathbf{a}_z i_a^r N_a^r \cos \alpha_\nu (x - x_o) \tag{5.13a}$$

$$\mathbf{h}_b^r(x) = \mathbf{a}_z i_b^r N_b^r \cos\left[\alpha_\nu (x - x_o) - \frac{\pi}{2}\right] \tag{5.13b}$$

In the above expressions, $\alpha_\nu = 2\pi\nu/\tau_r$ is a winding constant, ν = number of pole pairs, τ_r = length of repeatable section, or twice the pole pitch, and N = conductor density (per unit length in the x direction). The superscripts s and r denote the stator and rotor quantities respectively, and the subscripts a and b correspond to phases a and b respectively, of the windings. Since the idealized machine is assumed to be balanced, the nomenclature can be simplified by putting $N_a^s = N_b^s = N^s$; $N_a^r = N_b^r = N^r$; $i_a^s = i_b^s = i^s$; and $i_a^r = i_b^r = i^r$ in the above expressions. The unit vector $\mathbf{a}_z$ can also be dropped because all currents are z-directed. Knowing the current sheets, the $\mathbf{B}$ fields can be found as follows.

In the airgap of the machine, since there are no free charges, $\nabla \cdot \mathbf{B} = 0$

and $\nabla \times \mathbf{H} = 0$, so that the tangential and radial components of the **B** field satisfy Laplace's equation, that is,

$$\nabla^2 B_x = 0 \tag{5.14a}$$

and

$$\nabla^2 B_y = 0 \tag{5.14b}$$

Since the current sheets on the stator and rotor are both periodic in x, the solution to Eq. (5.14a) takes the form (see Appendix D)

$$B_x(x, y) = \sum_q (K_{1q} \cosh \alpha_v y + K_{2q} \sinh \alpha_v y) \cos (\alpha_v x - \phi_q) \tag{5.15}$$

Now, let the stator alone be excited. The boundary condition at the stator surface ($y = 0$, line integral 1, Fig. 5-9) gives

$$K_{1q}^s = -\mu_o i_q^s N_q^s \tag{5.16a}$$

At the rotor surface ($y = g$, line integral 2, Fig. 5-9), $B_x^s = 0$ yields

$$K_{2q}^s = -K_{1q}^s \coth \alpha_v g \tag{5.16b}$$

Having evaluated the constants, the tangential component of the **B** field is known. The divergence equation, $\partial B_x^s/\partial x = -\partial B_y^s/\partial y$, gives the radial component B_y from Eqs. (5.15) and (5.16).

Next, only the rotor is considered to be excited. The general solution is still of the form of Eq. (5.15). Proceeding as for the case with stator excitation, $B_x^r = 0$ at the stator surface gives

$$K_{1q}^r = 0 \tag{5.17a}$$

At the rotor surface, the line integral 2 yields

$$K_{2q}^r = \frac{\mu_o i_q^r N_q^r}{\sinh \alpha_v g} \tag{5.17b}$$

The radial component of the field is determined from the divergence equation as before.

Having evaluated the constants, the solutions to Eqs. (5.14a,b) for stator and rotor current sheets become, respectively,

$$B_x^s(x, y) = -\mu_o \sum_q N_q^s i_q^s (\cosh \alpha_v y - \coth \alpha_v g \sinh \alpha_v y) \cos (\alpha_v x - \phi_q^s) \tag{5.18a}$$

$$B_y^s(x, y) = -\mu_o \sum_q N_q^s i_q^s (\sinh \alpha_v y - \coth \alpha_v g \cosh \alpha_v y) \sin (\alpha_v x - \phi_q^s) \tag{5.18b}$$

$$B_x^r(x, y) = \frac{\mu_o}{\sinh \alpha_v g} \sum_p N_p^r i_p^r \sinh \alpha_v y \cos [\alpha_v (x - x_o) - \phi_p^r] \tag{5.18c}$$

$$B_y^r(x, y) = \frac{\mu_o}{\sinh \alpha_v g} \sum_p N_p^r i_p^r \cosh \alpha_v y \sin [\alpha_v (x - x_o) - \phi_p^r] \tag{5.18d}$$

Now, the stored magnetic energy W_m per unit volume per repeatable section is

$$W_m = \tfrac{1}{2}\,\mu_o(B_x^2+B_y^2) \tag{5.19a}$$

and the total energy stored is

$$(W_m)_T = \tfrac{1}{2}\,\mu_o\zeta \int_v (\,B_x^2+B_y^2)\,dv \tag{5.19b}$$

where ζ = number of repeatable sections and the integral is over the entire volume of the airgap; $B_x = (B_x^s+B_x^r)$; and $B_y = (B_y^s+B_y^r)$. When Eqs. (5.18a–d) are substituted in Eqs. (5.19a,b), the final form for the total stored energy is

$$\begin{aligned}(W_m)_T = {} & \frac{1}{4}\,\zeta\mu_o l\tau_r \sum_k \sum_m i_k^s i_m^s N_k^s N_m^s \cos\,(\phi_k^s-\phi_m^s)\,\frac{\coth \alpha_\nu g}{\alpha_\nu} \\ & +\frac{1}{2}\,\zeta\mu_o l\tau_r \sum_k \sum_q i_k^s i_q^r N_k^s N_q^r \cos\,(\alpha_\nu x_o+\phi_q^r-\phi_k^s)\,\frac{\text{cosech}\, \alpha_\nu g}{\alpha_\nu} \\ & +\frac{1}{4}\,\zeta\mu_o l\tau_r \sum_p \sum_q i_p^r i_q^r N_p^r N_q^r \cos\,(\phi_p^r-\phi_q^r)\,\frac{\coth \alpha_\nu g}{\alpha_\nu}\end{aligned} \tag{5.20a}$$

In Eqs. (5.18) and (5.20) the summations are over the phases a and b. In terms of self- and mutual inductances, the same magnetic stored energy can be expressed as

$$(W_m)_T = \frac{1}{2}\Big(\sum_k \sum_m i_k^s i_m^s L_{km}^{ss} + \sum_k \sum_q i_k^s i_q^r L_{kq}^{sr} + \sum_p \sum_q i_p^r i_q^r L_{pq}^{rr}\Big) \tag{5.20b}$$

A comparison of Eqs. (5.20a) and (5.20b) gives the elements of the machine inductance matrix as follows:

$$L_{km}^{ss} = \frac{1}{2}\,\zeta\mu_o l\tau_r N_k^s N_m^s \cos\,(\phi_k^s-\phi_m^s)\,\frac{\coth \alpha_\nu g}{\alpha_\nu} \tag{5.21a}$$

$$L_{pq}^{rr} = \frac{1}{2}\,\zeta\mu_o l\tau_r N_p^r N_q^r \cos\,(\phi_p^r-\phi_q^r)\,\frac{\coth \alpha_\nu g}{\alpha_\nu} \tag{5.21b}$$

$$L_{kq}^{sr} = \zeta\mu_o l\tau_r N_k^s N_q^r \cos\,(\alpha_\nu x_o+\phi_q^r-\phi_k^s)\,\frac{\text{cosech}\, \alpha_\nu g}{\alpha_\nu} \tag{5.21c}$$

For the 2-phase machine under consideration, the subscripts $k, m, p,$ and q range over the phases a and b. Thus for this machine there are eight inductances in all. These can be expressed in matrix notation as follows:

$$L = \left[\begin{array}{c|c} L^{ss} & L^{sr} \\ \hline L^{rs} & L^{rr} \end{array}\right] = \left[\begin{array}{cc|cc} L^s & 0 & L^{sr}\cos\nu\theta & -L^{sr}\sin\nu\theta \\ 0 & L^s & L^{sr}\sin\nu\theta & L^{sr}\cos\nu\theta \\ \hline L^{sr}\cos\nu\theta & L^{sr}\sin\nu\theta & L^r & 0 \\ -L^{sr}\sin\nu\theta & L^{sr}\cos\nu\theta & 0 & L^r \end{array}\right] \tag{5.22}$$

We note that the zeros in Eq. (5.22) indicate that the phase windings are in space quadrature and that there is no coupling between them. The quantities L^s, L^r, L^{sr}, and θ can be expressed in terms of the constraints μ_o, ζ, l, N's, g, etc., of Eqs. (5.21a–c) (see Problem 5-1).

5.4.2 *The Permeance-Function and Flux-Linkage Methods*[4]

Expressions for machine inductances can also be derived using the concept of permeance function and the defining equation for inductance. The permeance function $\mathscr{P}(\theta)$ is related to the MMF $F(\theta)$ and the flux density $B(\theta)$ by the equation [see Eq. (2.47)]

$$B(\theta) = F(\theta)\frac{\mathscr{P}(\theta)}{l} \tag{5.23}$$

where l = length in the axial (or z) direction. A knowledge of current-sheet distribution gives $F(\theta)$, and if $\mathscr{P}(\theta)$ is known, $B(\theta)$ can be determined. The permeance function for a double-cylindrical structure is[4]

$$\mathscr{P}(\theta) = \frac{\mu_o/g}{g\nu\theta/\sinh g\nu\theta} \tag{5.24}$$

Expressions for $\mathscr{P}(\theta)$ for other forms of magnetic structures can be derived by field theory methods.[5] The flux density distribution $B(\theta)$ is thus known in terms of the machine dimensions and current sheets. If $N(\theta)$ is the winding distribution, the flux linkage λ is given by

$$\lambda = \int_0^{2\tau_r} N(\theta)B(\theta)lr\, d\theta \tag{5.25}$$

This expression, if evaluated on a basis of unit current excitation, gives the value of the inductance.

5.4.3 *Effects of Saliency*

There are certain machines, such as dc machines and salient-pole synchronous machines, which do not fall in the category of double-cylindrical structures. A salient-pole rotor is shown in Fig. 5-4(b). In these machines one member of the magnetic structure (the stator in dc machines, the rotor in synchronous machines) has saliency. Evidently the model shown in Fig. 5-9 cannot be used to solve the field problem and thereby obtain the machine inductances. There are at least three methods available which take into account the effect of saliency on machine inductances. These are outlined below.

1. Graphical field mapping:[4,5] The permeance function and the corresponding flux density are determined by freehand flux plotting. Knowing the permeance function, the flux linkages and the inductances can be determined. The accuracy of the results depends on the accuracy of the field plots.

2. Space-dependent permeability function:[2] This method assumes that the radial permeability μ_r is a function of the position of the nonsalient member with respect to the salient member. In particular it is assumed that

$$\mu_r = \mu_1 - \mu_o \cos \nu\theta \tag{5.26}$$

in a double-cylindrical structure. The field problem is then solved as in Sec. 5.4.1, and the inductances are thus determined. The objection to this method is that permeability does not change with position, and determination of equivalent permeability is difficult. To obtain the equivalent permeability, such as the one given by Eq. (5.26), the field equations have to be solved.

3. Conformal transformation:[6] A semigraphical method leading to the conformal transformation of salient-pole structures is also applicable to the determination of inductances. The method has several advantages over methods 1 and 2, but is rather complicated in analytical details.

The discussions of Sec. 5.4 indicated that inductance determination is a somewhat involved procedure and that numerical evaluation of inductance coefficients, as given by Eqs. (5.21), requires the use of a digital computer. For the present discussions, we shall consider two simple examples which illustrate the procedures of obtaining the inductances by (1) the energy-storage method and (2) the use of the permeance function and some physical arguments.

EXAMPLE 5–3

The fundamental components of the stator and rotor MMF's of a smooth-airgap machine are respectively given by

$$F^s = k^s N^s i^s \sin \theta$$
$$F^r = k^r N^r i^r \sin (\theta + \theta_o)$$

as shown in Fig. 5E-3, where k^s and k^r are winding constants, and N^s and N^r are the number of turns. The machine dimensions are as shown in Fig. 5E-3(a). Assume $\mu_{\text{iron}} \gg \mu_o$, $r \gg g$, and $H_y \gg H_x$. Determine the self- and mutual inductances of the machine by the energy-storage method.

In Sec. 5.4.1 we obtained the flux density by solving Laplace's equation. For the example under consideration, we use a simplified approach and recall Eqs. (2.4) and (2.22), from which

$$B_y = \mu_o H_y = \frac{\phi}{dS} \tag{a}$$

where ϕ is the total flux coming out of a small area dS. But flux, MMF, and

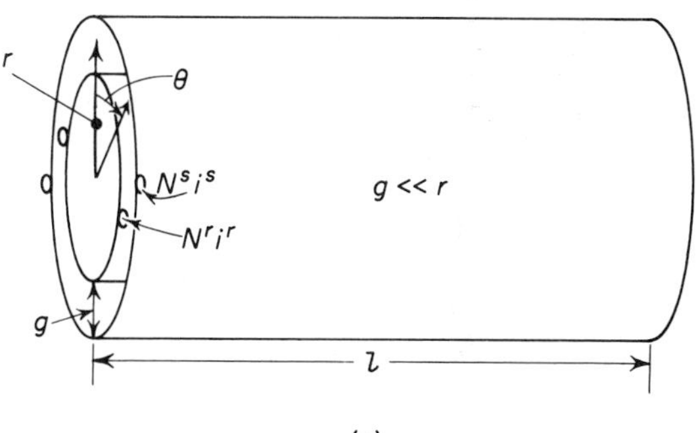

(a)

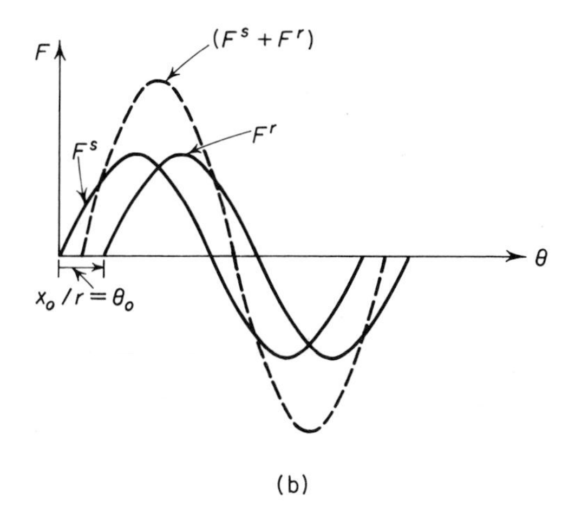

(b)

Fig. 5E-3

reluctance are related by Eq. (2.47). For the case under consideration, the reluctance $\mathscr{R}$ is given by

$$\mathscr{R} = \frac{g}{\mu_o \, dS} \tag{b}$$

Consequently, from Eqs. (a), (b), and (2.47),

$$H_y = \frac{\phi}{\mu_o dS} = \frac{Ni}{\mu_o g} \frac{\mu_o \, dS}{dS} = \frac{Ni}{g} \tag{c}$$

where Ni is the resultant maximum MMF. For the case under consideration, the resultant MMF, Ni, is found from the phasor addition of the stator and rotor MMF's; that is,

$$(Ni)^2 = (k^s N^s i^s)^2 + (k^r N^r i^r)^2 + 2k^s k^r N^s N^r i^s i^r \sin \theta_o \tag{d}$$

We now recall, from Eq. (2.31), that in terms of the magnetic field intensity H_y, the density of the magnetic energy stored in the airgap is $\frac{1}{2}(\mu_o H_y^2)$.

Substituting $x = r\,\theta$ in Eq. (d), the total magnetic energy is

$$W_m = \frac{1}{2}\mu_o \int_{z=0}^{l} \int_{y=0}^{g} \int_{x=0}^{2\pi r} \left(H_y^2 \sin^2 \frac{x}{r}\, dz\, dy\, dx \right) \tag{e}$$

Substituting Eq. (d) in Eq. (c) and the resulting expression for H_y in Eq. (e) we obtain

$$W_m = \mu_o \frac{\pi r l}{2g} [(k^s N^s)^2 (i^s)^2 + (k^r N^r)^2 (i^r)^2 + 2(k^s k^r N^s N^r i^s i^r \sin\theta_o)] \tag{f}$$

In terms of the inductances L^s, L^r, and L^{sr}, the same magnetic energy can also be expressed as

$$W_m = \tfrac{1}{2} L^s (i^s)^2 + \tfrac{1}{2} L^r (i^r)^2 + L^{sr} i^s i^r \tag{g}$$

From Eqs. (f) and (g), therefore,

$$L^s = \frac{1}{2}\mu_o \pi \frac{rl}{g} (k^s N^s)^2$$

$$L^r = \frac{1}{2}\mu_o \pi \frac{rl}{g} (k^r N^r)^2$$

$$L^{sr} = \mu_o \pi \frac{rl}{g} k^s k^r N^s N^r \sin\theta_o$$

To illustrate the use of permeance function we consider the following example.

EXAMPLE 5–4

A 2-phase salient-pole machine is shown in Fig. 5E-4. It has two windings on the stator and one on the rotor. We wish to determine the machine inductances

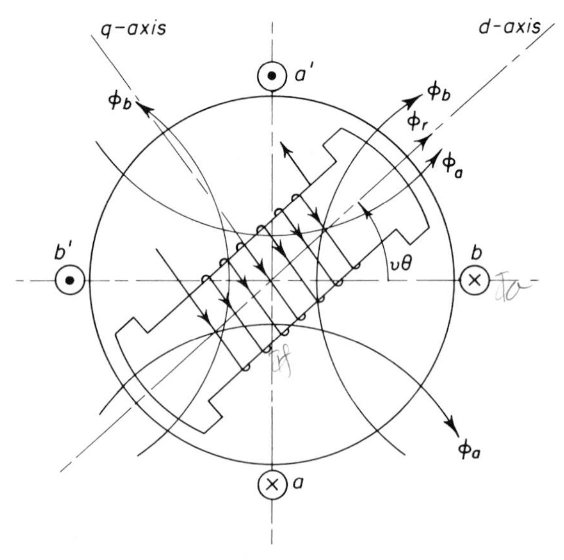

Fig. 5E-4

and express the result in matrix form, assuming sinusoidal MMF and flux distributions.

First of all, the inductance matrix for the machine can be written as

$$L = \left[\begin{array}{cc|c} L_{aa}^{ss} & L_{ab}^{ss} & L_{ar}^{sr} \\ L_{ba}^{ss} & L_{bb}^{ss} & L_{br}^{sr} \\ \hline L_{ra}^{rs} & L_{rb}^{rs} & L_{rr}^{rr} \end{array}\right] \tag{a}$$

where the nomenclature is the same as given in Eq. (5.21).

From Fig. 5E-4, evidently the rotor self-inductance is independent of the rotor position, so that

$$L_{rr}^{rr} = L_f = \text{a constant} \tag{b}$$

It is also clear from Fig. 5E-4 that the mutual inductance between the stator and rotor windings is periodic in θ. If the stator winding has ν pairs of poles then, for phase a,

$$L_{ar}^{sr} = L_{ra}^{rs} = L^{sr} \cos \nu\theta \tag{c}$$

where L^{sr} is a constant.

To find the stator self- and mutual inductances, consider the equivalent MMF, $F_a^s = N_a^s i_a^s$, due to phase a, and resolve it along the d and q axes (Fig. 5E-4) to obtain

$$\left.\begin{aligned} F_{da}^s &= F_a^s \cos \nu\theta = N_a^s i_a^s \cos \nu\theta \\ F_{qa}^s &= -F_a^s \sin \nu\theta = -N_a^s i_a^s \sin \nu\theta \end{aligned}\right\} \tag{d}$$

Defining P_d and P_q as permeances along the d and q axes, the airgap fluxes can be expressed (see also Example 2-6) as

$$\left.\begin{aligned} \phi_{da}^s &= P_d F_{da}^s = P_d F_a^s \cos \nu\theta \\ \phi_{qa}^s &= P_q F_{qa}^s = -P_q F_a^s \sin \nu\theta \end{aligned}\right\} \tag{e}$$

Or, the flux-linking phase a, due to the current i_a^s, is therefore

$$\lambda_{aa}^{ss} = N_a^s \phi_a^s = N_a^s(\phi_{da}^s \cos \nu\theta - \phi_{qa}^s \sin \nu\theta) \tag{f}$$

Equations (d), (e), and (f) finally yield

$$\lambda_{aa}^{ss} = \tfrac{1}{2} (N_a^s)^2 i_a^s [(P_d + P_q) + (P_d - P_q) \cos 2\nu\theta] \tag{g}$$

Or,

$$L_{aa}^{ss} = \frac{\lambda_{aa}^{ss}}{i_a^s} = L^s + L_o^s \cos 2\nu\theta \tag{h}$$

where

$$L^s = \tfrac{1}{2} (N_a^s)^2 (P_d + P_q)$$

and

$$L_o^s = \tfrac{1}{2} (N_a^s)^2 (P_d - P_q)$$

Substituting $\theta = \theta + \pi/2$ in Eq. (h) yields

$$L_{bb}^{ss} = L^s - L_o^s \cos 2\nu\theta \tag{i}$$

To determine the mutual inductances we recall that

$$L_{ab}^{ss} = L_{ba}^{ss} = \frac{\lambda_{ab}^{ss}}{i_a^s} \tag{j}$$

The flux-linking phase b, λ_{ab}^{ss}, is obtained from Eqs. (e) and (f) by substituting $\theta = \theta + \pi/2$ so that

$$\lambda_{ab}^{ss} = N_b^s \left[\phi_{da}^s \cos \nu \left(\theta + \frac{\pi}{2} \right) - \phi_{qa}^s \sin \nu \left(\theta + \frac{\pi}{2} \right) \right] \tag{k}$$

Substituting Eq. (d) in Eq. (k) yields

$$\lambda_{ab}^{ss} = N_b^s N_a^s i_a^s \left[P_d \cos \nu\theta \cos \nu \left(\theta + \frac{\pi}{2} \right) + P_q \sin \nu\theta \sin \nu \left(\theta + \frac{\pi}{2} \right) \right] \tag{l}$$

Finally, from Eqs. (j) and (l)

$$L_{ab}^{ss} = L_{ba}^{ss} = -L_o^s \sin 2\nu\theta \tag{m}$$

where

$$L_o^s = \tfrac{1}{2} (N_a^s)^2 (P_d - P_q)$$

with $N_a^s = N_b^s$ (the windings being symmetrical and balanced).

Consequently, the required L matrix is

$$\left[\begin{array}{c|c} L^{ss} & L^{sr} \\ \hline L^{rs} & L^{rr} \end{array} \right] = \left[\begin{array}{cc|c} L^s + L_o^s \cos 2\nu\theta & -L_o^s \sin 2\nu\theta & L^{sr} \cos \nu\theta \\ -L_o^s \sin 2\nu\theta & L^s - L_o^s \cos 2\nu\theta & -L^{sr} \sin \nu\theta \\ \hline L^{sr} \cos \nu\theta & -L^{sr} \sin \nu\theta & L_f \end{array} \right] \tag{5.27}$$

5.4.4 *General Remarks on Effects of Leakages, Slots, Saturation, and Harmonics*

The expressions for inductances derived so far are for an idealized machine. An actual machine differs considerably from an idealized model. Some of the factors which affect machine performance are

1. *Leakages.* In a coupled electromagnetic circuit, leakage fluxes are those which link with one circuit but not with the other. For example, in Fig. 2-5 a flux linking with N_1 turns alone, but not with any others, is a *leakage flux*. In rotating machines, there are a number of possible leakage paths for fluxes. With each of these is associated a *leakage reactance*, which affects all the important operating characteristics of a machine. Some of the leakages encountered in a rotating machine are the following: end-connection leakage, slot leakage, and tooth-top, zigzag, and belt leakages. These and some of the following factors are considered in Chapter 9.

2. *Saturation.* Because the fluxes depend upon the MMF's, the saturation of the magnetic circuit influences the performance of the machine. A typical saturation curve is shown in Fig. 2-4. When the excitation is such that the machine operates under saturated conditions, the inductance coefficients, derived earlier on the basis of a linear analysis, are no longer valid. Graphical

and semiempirical methods are available to account for saturation and such nonlinearities. These are discussed in Chapter 9.

3. *Harmonics*. In the preceding derivation of an idealized machine, sinusoidal current-sheet distributions were assumed. However, in actual machines the airgap fields can be resolved into a series of harmonic contents. These space harmonics are mainly due to nonsinusoidal winding distributions and to permeance variations due to slot openings, etc. Space harmonics are responsible for effects such as crawling, magnetic noise, vibration, locking, and harmonic voltage ripples. Methods for taking into account the effects of space harmonics are given in Chapter 9.

5.5 The Equations of Motion

Once inductance coefficients have been obtained, writing down the equations of motion is a rather routine matter. For the idealized machine, the volt-amp and torque equations are obtained from Eqs. (3.45) and (3.46). These are

$$\begin{bmatrix} v^s \\ \hline v^r \end{bmatrix} = \left[\begin{array}{c|c} R^s + pL^{ss} & pL^{sr} \\ \hline pL^{rs} & R^r + pL^{rr} \end{array}\right] \begin{bmatrix} i^s \\ \hline i^r \end{bmatrix} \tag{5.28a}$$

$$T_e = -\frac{1}{2}[i^s \quad i^r] \frac{\partial}{\partial \theta} \left[\begin{array}{c|c} L^{ss} & L^{sr} \\ \hline L^{rs} & L^{rr} \end{array}\right] \begin{bmatrix} i^s \\ \hline i^r \end{bmatrix} \tag{5.28b}$$

where $p = d/dt$.

In these equations, R^s, R^r, L^{ss}, L^{sr}, L^{rs}, L^{rr} are all 2×2 submatrices, and the voltages and currents are column submatrices. Inductances can be substituted in Eqs. (5.28a) and (5.28b) in terms of Eq. (5.22) for a uniform-airgap machine, or in terms of Eq. (5.27) for a salient-pole machine.

In order to obtain any quantitative information about the machine, the equations of motion, Eqs. (5.28a,b), have to be solved for specified conditions. However, it is immediately seen that these equations are simultaneous nonlinear differential equations with time-varying coefficients. As such it is not always practicable to solve these equations under all operating conditions unless some constraints are applied. These are now considered.

5.6 Constraints

In Sec. 5.1 criteria for torque production and energy conversion were discussed. It was also indicated there that conventional machines can be evolved from an idealized machine by imposing certain constraints. These constraints, first in a uniform-airgap machine and then in a salient-pole machine, are now considered.

5.6.1 *Uniform-Airgap Machines*

From Eqs. (5.22) and (5.28b), the expression for the electromagnetic torque becomes

$$T_e = \nu L^{sr}[(-i_a^r i_b^s + i_b^r i_a^s) \cos \nu\theta + (i_b^r i_b^s + i_a^r i_a^s) \sin \nu\theta] \tag{5.29}$$

Now, if it is assumed that the machine runs at a speed $\dot{\theta} = \omega_m$, then

$$\theta = \omega_m t + \delta \tag{5.30}$$

If it is further assumed that the stator and rotor currents are sinusoidal, then from Eqs. (5.29) and (5.30) it can be verified that the time-average value of the torque $\langle T_e \rangle$ is nonzero only if the stator and rotor frequencies, ω^s and ω^r respectively, are related to each other by the following equation (see Problem 5-11):

$$\omega^r = \pm\omega^s \pm \nu\omega_m \tag{5.31}$$

This equation can be considered a constraint for the operation of an induction machine where the given stator frequency appears to be different when viewed from a moving reference frame. In other words, there is a Doppler shift in frequency when measured in a reference frame moving with respect to the source.

The various constraints imposed on Eqs. (5.28a,b) leading to different types of machines are summarized in Table 5-1.

Table 5-1

Constraint	*Machine*
$i_a^r = i^r$; $i_b^s = i_b^r = v_b^s = v_b^r = 0$; $v_a^r = V^r$ (dc voltage); $\dot{\theta} = \omega_m =$ constant	1-phase synchronous generator
$i_a^r = i^r$; $i_b^r = v_b^r = 0$; $v_a^r = V^r$ (dc voltage); $\dot{\theta} = \omega_m$ = constant	2-phase synchronous generator
$i^s = I^s \sin \nu\omega_m t$; $i^r = I^r \sin 2\nu\omega_m t$; $\dot{\theta} = \omega_m =$ constant	synchronous-induction machine
$v_a^r = v_b^r = 0$; $i_a^s = I^s \sin \omega^s t$; $\omega^r = \omega^s - \nu\omega_m$; $i_b^s = I^s \cos \omega^s t$	2-phase induction motor

In Table 5-1 are summarized the end results applicable to a smooth-airgap machine having two windings on the stator and two on the rotor. If we accept these end results, we can see that various apparently different types of machine emerge from the idealized 2-phase machine. In later chapters (Chapters 6, 7, and 8) we shall use the constraints listed in Table 5-1 to obtain some of the performance characteristics of the machine. The fact we wish to emphasize at this stage is that, in a smooth-airgap machine, energy conversion is not possible unless a set of constraints specified for a particular type of machine is fulfilled.

For the salient-pole machine, similarly, a set of constraints must hold for energy conversion. These are considered in the next section.

5.6.2 Salient-Pole Machines

From Eqs. (5.27) and (5.28b), the expression for the electromagnetic torque for a salient-pole machine becomes

$$T_e = \nu\{L^{sr}(i_a^s \sin \nu\theta - i_b^s \cos \nu\theta)i^r + L_o^s[(i_a^s)^2 - (i_b^s)^2] \sin 2\nu\theta + 2L_o^s i_a^s i_b^s \cos 2\nu\theta\} \tag{5.32}$$

As for the smooth-airgap machine, Eq. (5.29)—and for the salient-pole machine too—it follows from Eq. (5.32) that for the average torque to be nonzero at a constant angular velocity $\dot{\theta} = \omega_m$,

$$\omega^r = \pm\omega^s \pm \nu\omega_m \tag{5.33a}$$

In addition to Eq. (5.33a), saliency leads to the last two terms in Eq. (5.32), which yield nonzero average torque, for a speed ω_m and $\omega^r = 0$, when (see Problem 5-12)

$$\omega_a^s = \pm\omega_b^s \pm 2\nu\omega_m \tag{5.33b}$$

$$\omega_a^s = \nu\omega_m \quad \text{or} \quad \omega_b^s = \nu\omega_m \tag{5.33c}$$

For the smooth-airgap machine as well as for the salient-pole machine, the constraints have both mathematical and physical significance. For example, the condition $\omega^r = \omega^s - \nu\omega_m$ in an induction motor implies that the frequency of the induced rotor currents depends on the stator frequency as well as on the mechanical speed of the rotor. Similarly, the condition $\omega^s = \nu\omega_m$ in a synchronous machine implies that the synchronous machine is a constant-speed machine.

There is one type of machine, namely the dc machine, that has not been included in the preceding discussions. However, the dc machine emerges from the concept of the idealized machine, the frequency constraint in this case being introduced mechanically by the commutator.

To summarize, we have indicated here a common link between conventional electric machines and the idealized machine. The performance of these machines can be calculated by solving the equations of motion. This is considered in the next section.

5.7 Solution of the Equations of Motion

As was shown in the last section, conventional machines evolve from the idealized machine when we apply appropriate constraints. The equations of motion can also be solved using these constraints. However, under transient or unbalanced operating conditions, it is difficult to solve the resulting equations simply by using frequency constraints. In such cases it has been found that the equations of motion can be considerably simplified by introducing a change of variables. Although these *transformations* (or change of variables) are well known, it seems more logical to derive the transformations from linear algebra before they are used to solve any problems. Such an approach provides a formal basis to the well-known transformations.

†5.7.1 Introduction to Linear Transformations[7]

Before getting into any details of linear transformations, it is best to review some matrix algebra. A resume is given in Appendix B. Some pertinent definitions and theorems are stated below.

Definition 1. A square matrix Z, of order n, is *cyclic-symmetric* if its elements satisfy the conditions

$$z_{mk} = z_{(m-k+1)1} = z_{(k-m+1)1}, \quad \text{indices modulo } n$$

For example, the following 3×3 matrix is cyclic-symmetric:

$$\begin{bmatrix} z_{aa} & z_{ab} & z_{ac} \\ z_{ac} & z_{aa} & z_{ab} \\ z_{ab} & z_{ac} & z_{aa} \end{bmatrix}$$

with $z_{ab} = z_{ac}$.

Definition 2. In an n-dimensional vector space, the *scalar product* of two vectors $\mathbf{X}$, with components $x_1, x_2, \ldots, x_n$, and $\mathbf{Y}$, with components $y_1, y_2, \ldots, y_n$, is defined as

$$\langle \mathbf{X}, \mathbf{Y} \rangle = \sum_{k=1}^{n} x_k y_k$$

Definition 3. An operator $\mathbf{S}$ is *linear* provided it has the property that if a vector $\mathbf{Z}_1$ corresponds to a vector $\mathbf{X}_1$, and $\mathbf{Z}_2$ corresponds to $\mathbf{X}_2$, then $a\mathbf{Z}_1 + b\mathbf{Z}_2$ corresponds to $a\mathbf{X}_1 + b\mathbf{X}_2$, where $\mathbf{Z}$ and $\mathbf{X}$ are related by the operator $\mathbf{S}$, such that $\mathbf{Z} = \mathbf{SX}$.

A transformation of this type is a *linear transformation*. The elements of

$\mathbf{S}$ are independent of $\mathbf{X}$ and $\mathbf{Z}$, and the elements of $\mathbf{S}$ completely characterize the transformation.

Definition 4. A linear operator $\mathbf{S}^+$ (represented by a matrix) is the *adjoint* of the operator $\mathbf{S}$ (also represented by a matrix) if

$$\langle \mathbf{Y}, \mathbf{SX} \rangle = \langle \mathbf{S}^+\mathbf{Y}, \mathbf{X} \rangle$$

Definition 5. A real matrix $\mathbf{S}$ which satisfies the relation $\mathbf{S}^+\mathbf{S} = \mathbf{I}$ (where $\mathbf{I}$ is the identity matrix) is called the *orthogonal matrix*, where $\mathbf{S}^+$ is the adjoint of $\mathbf{S}$.

Theorem 1. A real self-adjoint operator ($\mathbf{S}^+ = \mathbf{S}$) can be reduced to a diagonal form by an orthogonal transformation.

Theorem 2. An orthogonal transformation leaves the scalar product of two vectors invariant.

Theorem 3. If $S^{-1}ZS = \Delta_{mm}$, where Z is a real cyclic-symmetric self-adjoint matrix and Δ_{mm} is a diagonal matrix (Theorem 1), the elements of S are

$$s_{pk} = \frac{1}{\sqrt{n}} e^{(j2\pi/n)(p-1)(k-1)} = \frac{(s_n)^{(p-1)(k-1)}}{\sqrt{n}}$$

where $j = \sqrt{-1}$ and $s_n = n$th root of unity $= e^{j2\pi/n}$ $\quad (n>2)$

Theorem 4. If Z_{mn} is an $m \times n$ matrix with typical elements

$$z_{pq} = \cos \nu \left[\theta_o + (q-1)\frac{\pi}{n} - (p-1)\frac{\pi}{m} \right]$$

and S denotes an orthogonal transformation, then $S_m^{-1} Z_{mn} S_n$ contains only two nonzero terms, where S_m and S_n are orders of m and n respectively. (*Note*: By order m is meant an $m \times m$ matrix.)

Theorem 5. If Z_{mn} is of the form given in Theorem 4, then

$$S_n^{-1} Z_{nm} S_m = (S_m^{-1} Z_{mn} S_n)^*$$

where the superscript * denotes the complex conjugate of the matrix.

†5.7.2 *An Application to Electric Circuits*

As mentioned earlier, the purpose of using transformations is to reduce the degree of complexity in solving a set of n simultaneous equations. First, this is illustrated by the following simple example, and then its application to electric machines is considered.

A set of volt-ampere equations for an electric circuit can be written as

$$v = Zi \tag{5.34}$$

where Z is a square matrix representing the impedance, and v and i are column matrices denoting the voltages and currents respectively.

Now, a change of variables is introduced such that

$$v = Sv' \tag{5.35a}$$

and

$$i = Si' \tag{5.35b}$$

Substitution of Eqs. (5.35a,b) in Eq. (5.34) yields

$$v' = (S^{-1}ZS)i' = Z'i' \tag{5.36}$$

where

$$Z' = S^{-1}ZS \tag{5.37}$$

If Z is cyclic-symmetric (Definition 1), then by Theorems 1 and 3 a matrix S can be found such that $S^{-1}ZS$ is a diagonal matrix. The elements of S are obtained from Theorem 3.

To see the advantage gained by such a transformation, Eqs. (5.34) and (5.36) may be compared, as these equations describe the same physical system. However, whereas Eq. (5.34) is a set of simultaneous equations involving n unknowns, Eq. (5.36) is a set of n "decoupled" equations each involving one unknown at a time. Thus the degree of complexity of the problem is reduced. The advantages will become more obvious as the transformations are developed for machines (Chapters 7 and 8).

†5.7.3 *Invariance of Power*

We can easily verify that the orthogonal transformation, as given by Eqs. (5.35a,b), keeps the power of the system (or circuit) invariant; that is, the instantaneous power associated with the original circuit is the same as the power associated with the new circuit after introduction of the new variables. We know that instantaneous power p is the scalar product of the voltage and current. Or,

$$\begin{aligned} p &= \langle v, i \rangle \\ &= \langle Sv', Si' \rangle \quad \text{from Eq. (5.35a,b)} \\ &= \langle S^{+}Sv', i \rangle \quad \text{from Definition 4} \\ &= \langle v', i' \rangle, \quad \text{from Definition 5} = p' \end{aligned}$$

The above may also be considered to constitute the proof of Theorem 2.

5.7.4 Transformations for Electric Machines

From the remarks of Sec. 5.7.2, we see that certain transformations facilitate the solution of the equations of motion. In this section, using the previously stated definitions and theorems, certain transformations are derived which are suitable for the analysis of electric machines. The derivations are illustrated by the following examples.

EXAMPLE 5–5

Given the 2-phase salient-pole machine shown in Fig. 5E-5. It is desired to obtain a transformation such that the angular dependence of the coefficient matrix is eliminated and the equations of motion are linearized.

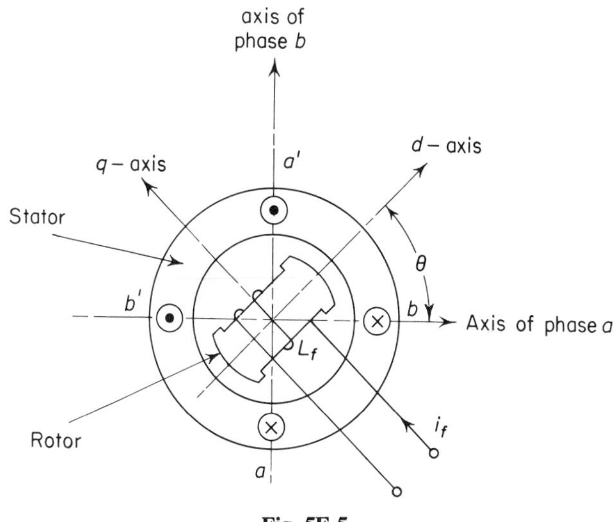

Fig. 5E-5

The inductance of an idealized salient-pole machine is given by Eq. (5.27). The rotor has only one winding and the stator has a 2-phase winding. The inductance matrix is

$$\left[\begin{array}{c|c} L^{ss} & L^{sr} \\ \hline L^{rs} & L^{rr} \end{array}\right] = \left[\begin{array}{cc|c} L^s + L_o^s \cos 2\nu\theta & -L_o^s \sin 2\nu\theta & L^{sr} \cos \nu\theta \\ -L_o^s \sin 2\nu\theta & L^s - L_o^s \cos 2\nu\theta & -L^{sr} \sin \nu\theta \\ \hline L^{sr} \cos \nu\theta & -L^{sr} \sin \nu\theta & L_f \end{array}\right] \tag{5.38}$$

The eigenvalues of the stator inductance submatrix L^{ss} (obtained by solving the determinantal equation det $|L^{ss} - \lambda I| = 0$ for the characteristic roots) are

$$\lambda_1 = L^s + L_o^s \quad \text{and} \quad \lambda_2 = L^s - L_o^s$$

The corresponding normalized eigenvectors are

$$\begin{bmatrix} \cos\nu\theta \\ -\sin\nu\theta \end{bmatrix} \quad \text{and} \quad \begin{bmatrix} \sin\nu\theta \\ \cos\nu\theta \end{bmatrix}$$

The transformation matrix S_{dq} constructed from these eigenvectors is

$$S_{dq} = \begin{bmatrix} \cos\nu\theta & \sin\nu\theta \\ -\sin\nu\theta & \cos\nu\theta \end{bmatrix} \tag{5.39}$$

It can be readily verified that

$$S_{dq}^{-1} = \tilde{S}_{dq} \tag{5.40}$$

and

$$S_{dq}^{-1} L^{ss} S_{dq} = \begin{bmatrix} L_d & 0 \\ 0 & L_q \end{bmatrix} \tag{5.41}$$

where the superscripts -1 and $\sim$ respectively denote the inverse and transpose of S_{dq} (see Appendix B), and $L_d = L^s + L_o^s$ and $L_q = L^s - L_o^s$ are respectively called the *direct-axis* and *quadrature-axis inductances.*

The angular dependence from the inductance matrix is thus removed. The transformed currents i_d and i_q are related to the original currents i_a and i_b by

$$\begin{bmatrix} i_a \\ i_b \end{bmatrix} = \begin{bmatrix} \cos\nu\theta & \sin\nu\theta \\ -\sin\nu\theta & \cos\nu\theta \end{bmatrix} \begin{bmatrix} i_d \\ i_q \end{bmatrix} \tag{5.42}$$

and similar relations hold for voltages also.

Two distinct advantages are gained by this transformation. First, it can be verified that the original torque equation

$$-T_e = \frac{1}{2}\tilde{i}\frac{\partial L}{\partial \theta}i \tag{5.43a}$$

can be reduced to

$$T_e = \nu[i_d i_q(L_d - L_q) + i_f i_q L^{sr}] \tag{5.43b}$$

which does not contain any angular dependence. This equation also puts the torque components due to saliency (the first term) and due to field excitation (the second term) in proper perspective. Secondly, the volt-ampere equations become linearized for constant-speed operation (see Problem 5-4).

Further advantages of the transformations become evident when the dynamics of synchronous machines, induction machines, and dc machines are studied in the following chapters.

The transformation given by the S_{dq} matrix is known as the *dq transformation.*

EXAMPLE 5–6

Given the 2-phase induction machine shown in Fig. 5E-6(a), it is desired to obtain a transformation which will linearize the equations of motion.

The method of the preceding example is not directly applicable here, because

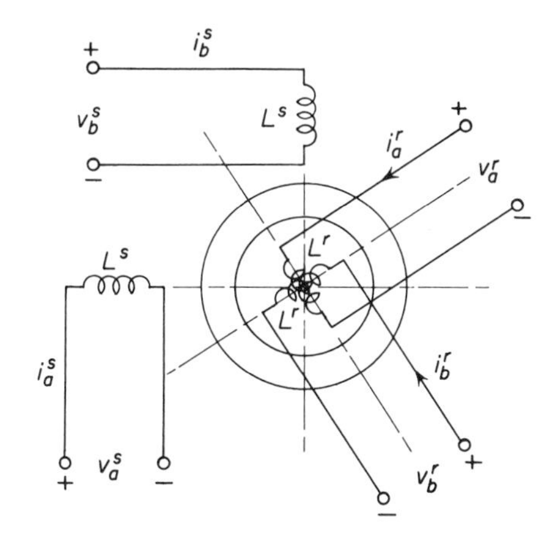

(a)

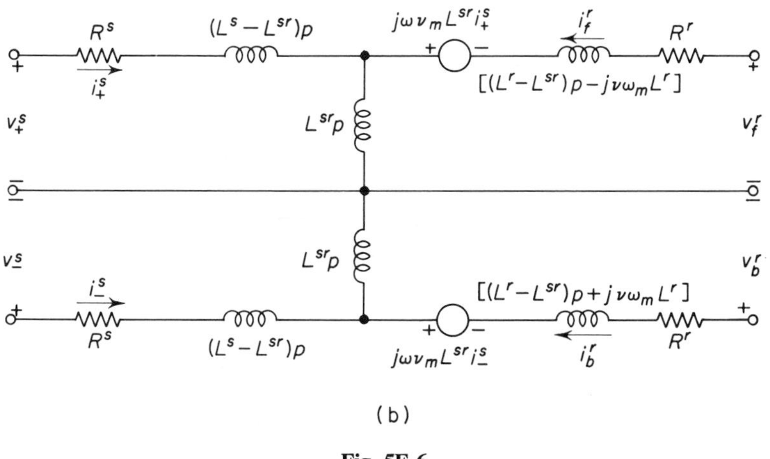

(b)

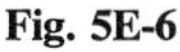
Fig. 5E-6

the stator-inductance submatrix is already in the diagonal form [see Eq. (5.22)]. Nevertheless, it is a cyclic-symmetric matrix, so that according to Theorem 3 a transformation matrix S can be found which keeps the L^{ss} submatrix in the diagonal form. Moreover, the elements of the L^{sr} and L^{rs} submatrices are of the form given in Theorem 4. When these submatrices are operated upon by the S matrix, they are also diagonalized. The entire matrix operation is summarized below.

The S matrix, which is denoted S_{+-} for this case, is

$$S_{+-} = \frac{1}{\sqrt{2}}\begin{bmatrix} 1 & 1 \\ -j & j \end{bmatrix}$$

and

$$S_{+-}^{-1} = \tilde{S}^*$$

The transformation denoted by S_{+-} is called the *symmetrical-component transformation.*

The original equations of motion are given by Eqs. (5.28a,b). The relationships between the original (unprimed) and transformed (primed) quantities are

$$\left.\begin{aligned} i &= S_{+-}i' \\ v &= S_{+-}v' \end{aligned}\right\} \tag{5.44}$$

and

The transformed equations of motion in expanded form are

$$\begin{bmatrix} v^s_+ \\ v^s_- \\ v^r_+ \\ v^r_- \end{bmatrix} = \begin{bmatrix} R^s+L^sp & 0 & L^{sr}e^{j\nu\theta}(p+j\nu\dot{\theta}) & 0 \\ 0 & R^s+L^sp & 0 & L^{sr}e^{-j\nu\theta}(p-j\nu\dot{\theta}) \\ L^{sr}e^{-j\nu\theta}(p-j\nu\dot{\theta}) & 0 & R^r+L^rp & 0 \\ 0 & L^{sr}e^{j\nu\theta}(p+j\nu\dot{\theta}) & 0 & R^r+L^rp \end{bmatrix} \begin{bmatrix} i^s_+ \\ i^s_- \\ i^r_+ \\ i^r_- \end{bmatrix} \tag{5.45a}$$

$$T_e = -j\nu L^{sr}[(i^{s*}_+ i^r_+)e^{j\nu\theta} - (i^s_+ i^{r*}_+)e^{-j\nu\theta}] \tag{5.45b}$$

It follows from the above equations that for a given speed ω_m, Eq. (5.45a) is a linear differential equation with time-varying coefficients. The torque equation, Eq. (5.45b), is still nonlinear. To overcome this difficulty, the rotor quantities are referred to the stator by a further transformation, called the forward–backward or *fb transformation.* The *fb* components are related to the $+-$ components by the equation

$$v^r_{+-} = S_{fb}v^r_{fb} \tag{5.46a}$$

where

$$S_{fb} = \begin{bmatrix} e^{-j\nu\theta} & 0 \\ 0 & e^{j\nu\theta} \end{bmatrix} \tag{5.46b}$$

When these transformations are carried out on the rotor variables, Eqs. (5.45a, b) become

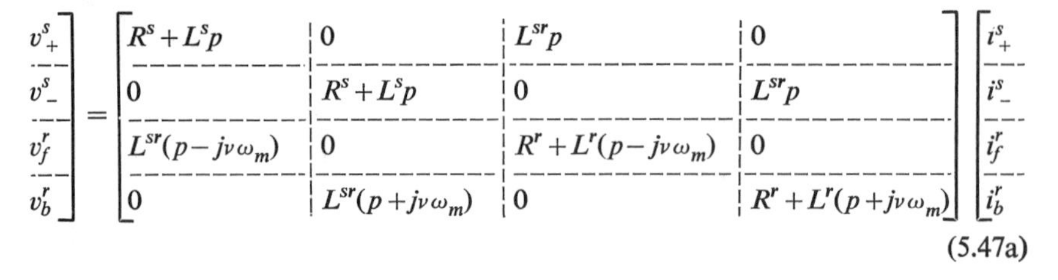

$$\begin{bmatrix} v^s_+ \\ v^s_- \\ v^r_f \\ v^r_b \end{bmatrix} = \begin{bmatrix} R^s+L^sp & 0 & L^{sr}p & 0 \\ 0 & R^s+L^sp & 0 & L^{sr}p \\ L^{sr}(p-j\nu\omega_m) & 0 & R^r+L^r(p-j\nu\omega_m) & 0 \\ 0 & L^{sr}(p+j\nu\omega_m) & 0 & R^r+L^r(p+j\nu\omega_m) \end{bmatrix} \begin{bmatrix} i^s_+ \\ i^s_- \\ i^r_f \\ i^r_b \end{bmatrix} \tag{5.47a}$$

$$-T_e = j\nu L^{sr}(i^{s*}_+ i^r_f - i^s_+ i^{r*}_f) \tag{5.47b}$$

where $p = d/dt$.

The above are simplified linear differential equations representing the quantitative behavior of an induction machine. The further usefulness of this transformation is also discussed in later chapters.

An equivalent circuit representing Eq. (5.47a) is shown in Fig. 5E-6(b).

5.7.5 *Geometrical Interpretations of Transformations*

Examples 5-5 and 5-6 illustrate two of the most commonly used transformations in the analysis of rotating electrical machines. We shall now consider these transformations and interpret them geometrically. First, we shall consider the dq transformation of Example 5-5 and then the $+-$ and fb transformations of Example 5-6.

The dq transformation is defined by Eq. (5.42), and is represented geometrically in Fig. 5-10. The two stator current sheets (in electrical space quadrature) i_a and i_b are in relative motion with respect to the rotor reference frame. If these currents are transformed according to Eq. (5.42), the resultant equivalent current sheet is such that there is no relative motion between the rotor and the current sheet. The dq transformation is thus a transformation between fixed and rotating axes. We shall consider this transformation in some detail in a later chapter on synchronous machines.

In Example 5-6 we used the $+-$ and fb transformations. These are complex-type transformations and do not have a true physical meaning. However, their geometrical relationships are illustrated in Figs. 5-11(a) and (b). In the study of static networks under steady-state conditions, the $+-$ transformation does have a physical meaning. For example, a 2-phase unbalanced network can be reduced to two equivalent balanced networks by the $+-$ transformation.

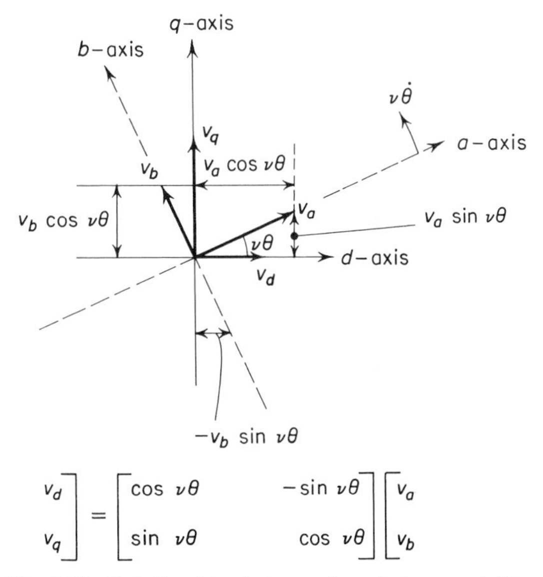

$$\begin{bmatrix} v_d \\ v_q \end{bmatrix} = \begin{bmatrix} \cos \nu\theta & -\sin \nu\theta \\ \sin \nu\theta & \cos \nu\theta \end{bmatrix} \begin{bmatrix} v_a \\ v_b \end{bmatrix}$$

Fig. 5-10 Relationships between dq and phase variables.

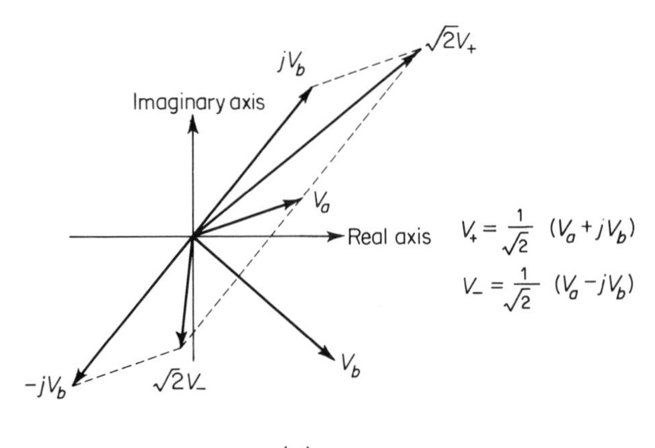

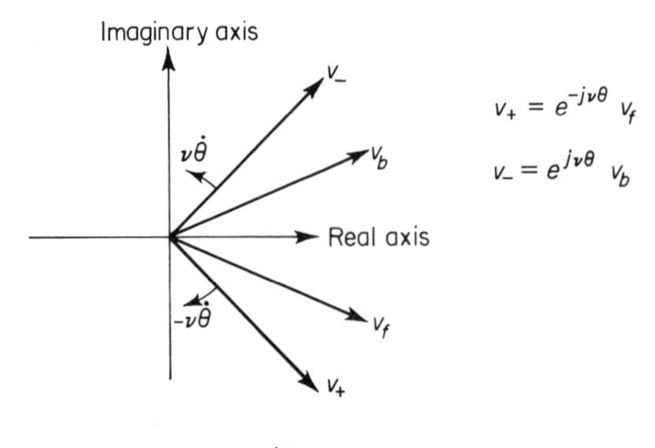

Fig. 5-11 (a) Relationship between two-phase symmetrical components and phase variables; (b) relationship between instantaneous $+-$ and fb components.

Turning now to the fb transformation, we notice from Fig. 5-11(b) that there is a relative motion between the $+-$ and fb components. If the fb components are assumed to be fixed the $+-$ components can be considered to rotate with an angular velocity $\nu\dot{\theta}$ in the directions shown in Fig. 5-11(b). We shall consider these transformations in more detail in connection with the study of induction machines in a later chapter.

5.7.6 Summary of Transformations

The transformations given in the last two examples are summarized in Fig. 5-12, where S and S^{-1} are also defined. The relationships between the

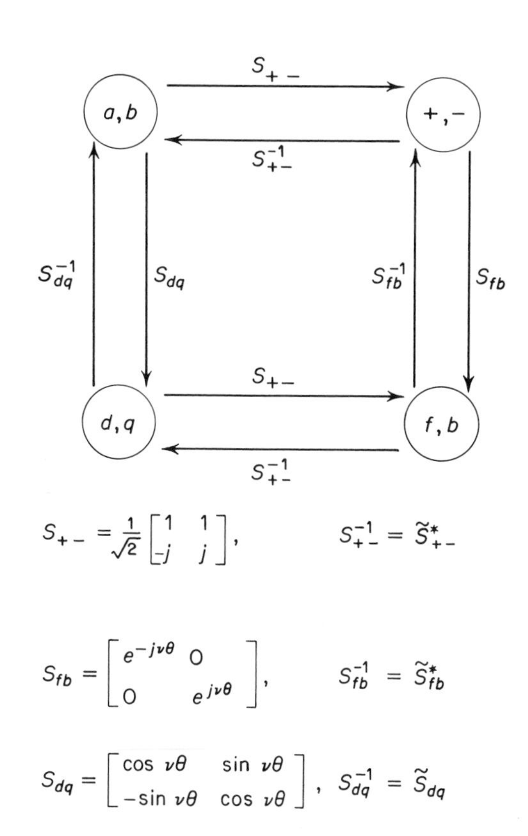

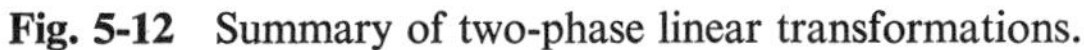

Fig. 5-12 Summary of two-phase linear transformations.

original (unprimed) and transformed (primed) variables are given by

$$x = Sx'$$

In the transformation matrices, the superscripts -1, $*$, and ~ respectively denote the inverse, complex conjugate, and transpose of the matrix.

Geometrical interpretations of the *dq*, $+-$, and *fb* transformations are shown in Figs. 5-10 and 5-11.

5.8 Summary

In this chapter, a resumé on rotating machines was presented. An idealized machine was described and its parameters determined. The equations of motion were derived, and it was shown that applications of appropriate constraints lead to a number of conventional machines. Linear transformations in electric-machine theory are derived from the concepts of linear algebra. These transformations are applied to synchronous and induction machines.

PROBLEMS

5-1. (a) Simplify Eqs. (5.21a–c) for the 2-phase machine if $g \ll \tau_r$. (b) Express the inductances of Eq. 5.22 in terms of those of Eqs. (5.21a–c) for the idealized 2-phase machine.

5-2. Derive the expressions for inductances using the permeance function for the machine for which the L matrix is given by Eq. (5.27).

5-3. Specify the constraints to be imposed on the idealized machine in order to develop a single-phase induction motor.

5-4. (a) Derive Eq. (5.43b) from Eqs. (5.39–5.43a). (b) Expand the volt-ampere equation for the salient-pole machine of Example 5-1.

5-5. A 60-cycle synchronous generator feeds an 8-pole induction motor running with a slip of 2%. What is the speed of the motor? At what speed must the generator run if it has two poles?

5-6. Derive the equation for the voltage induced in the elementary generator shown in Fig. 5-6(a).

5-7. Derive Eq. (5.32) from Eqs. (5.27) and (5.28b).

5-8. Reduce the equivalent circuit of Fig. 5E-6(b) for steady-state balanced operating conditions defined by $v_a^s = V\angle 0°$ and $v_b^s = V\angle -90°$.

5-9. Develop a Thevenin equivalent circuit for the simplified circuit of Problem 5-8.

5-10. For the 2-phase machine of Example 5-6 the following constraints hold:

$$\begin{aligned}
v_a^s &= V\cos\omega t \\
v_b^s &= V\sin\omega t \\
v_a^r &= v_b^r = 0; \qquad \omega^r = \omega - \omega_m \\
i_a^s &= I^s\cos(\omega t - \psi) \\
i_b^s &= I^s\sin(\omega t - \psi) \\
i_a^r &= I^r\cos(\omega^r t - \phi)
\end{aligned}$$

and

$$i_b^r = I^r\sin(\omega^r t - \phi)$$

Develop an equivalent circuit which represents the electrical equation of motion for the above machine (for steady-state operation) and derive the torque equation. Verify that, using the derived circuit, the energy-conservation principle also leads to identical torque equation.

5–11. Derive the frequency constraint given by Eq. (5.31) for average torque production in a uniform-airgap machine.

5–12. Derive Eqs. (5.33a–c) for a salient-pole machine.

REFERENCES

1. Saunders, R. M., "Electromechanical Energy Conversion in Double Cylindrical Structures," *IEEE Trans.*, Vol. PAS-82 ("Power Apparatus and Systems"), 1963, p. 631.

2. White, D. C., and H. H. Woodson, *Electromechanical Energy Conversion* (New York: John Wiley & Sons, Inc., 1959).

3. Nasar, S. A., "Electromagnetic Theory of Electrical Machines," *Proc. IEE* (London), Vol. 111, 1964, pp. 1123–1131.

4. Robinson, R. B., "Inductance Coefficients of Rotating Machines Expressed in Terms of Space Harmonics," *Proc. IEE* (London), Vol. 111, 1964, p. 769.

5. Bewley, L. V., *Two-Dimensional Fields in Electrical Engineering* (New York: Dover Publications, 1963).

6. Fukushima, K., S. A. Nasar, and R. M. Saunders, "Electromechanical Energy Conversion in Salient Pole Structures," *IEEE Trans.*, Vol. PAS-82 ("Power Apparatus and Systems"), 1963, p. 760.

7. Friedman, B., *Principles and Techniques of Applied Mathematics* (New York: John Wiley & Sons, Inc., 1956).

6

DC Machines

In the last chapter we made some introductory remarks relating to dc machines. To recapitulate, we discussed in particular the action of a 2-segment commutator and briefly commented upon the voltage and torque productions in dc machines. We also introduced the idealized machine and derived formulas for inductance coefficients. In this chapter we shall concentrate our attention on dc machines and shall obtain their dynamical characteristics from their equations of motion. Because dc machines find applications as elements of control systems, it is important that we understand their transient behavior. In this connection we shall use some of the concepts—such as the transfer functions, state equations, signal-flow graphs, etc.—discussed in Chapter 2. We shall also consider briefly the steady-state physical characteristics of dc machines with special reference to the effects of saturation, armature reaction, and commutation. But, first of all we shall consider the general equations of motion and see how much information about the machine can be obtained from them.

6.1 The 2-Axis dc Machine

Most of the commonly used dc machines have a set of two ports for electrical inputs, indicating one armature winding and one (or two, in case of compound machines) field winding. This type of machine is a *single-axis dc machine*. Here we consider a more general case and assume that the machine has a set of four independent windings, two on the stator and two on the rotor, as shown in Fig. 6-1. Assuming saliency on the stator, the inductances

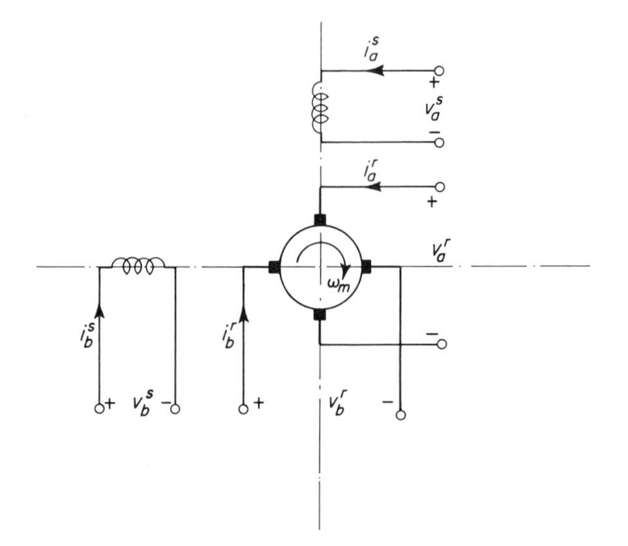

Fig. 6-1 A 2-axis dc machine.

of the four windings can be expressed in matrix notation as[1]

$$\left[\begin{array}{cc|cc} L^s+L^s_o & 0 & (L^{sr}+L^{sr}_o)\cos v\theta & -(L^{sr}+L^{sr}_o)\sin v\theta \\ 0 & L^s-L^s_o & (L^{sr}-L^{sr}_o)\sin v\theta & (L^{sr}-L^{sr}_o)\cos v\theta \\ \hline (L^{sr}+L^{sr}_o)\cos v\theta & (L^{sr}-L^{sr}_o)\sin v\theta & L^r+L^r_o\cos 2v\theta & -L^r_o\sin 2v\theta \\ -(L^{sr}+L^{sr}_o)\sin v\theta & (L^{sr}-L^{sr}_o)\cos v\theta & -L^r_o\sin 2v\theta & L^r-L^r_o\cos 2v\theta \end{array}\right]$$

$$= \left[\begin{array}{cc|cc} L^{ss}_{aa} & L^{ss}_{ab} & L^{sr}_{aa} & L^{sr}_{ab} \\ L^{ss}_{ba} & L^{ss}_{bb} & L^{sr}_{ba} & L^{sr}_{bb} \\ \hline L^{rs}_{aa} & L^{rs}_{ab} & L^{rr}_{aa} & L^{rr}_{ab} \\ L^{rs}_{ba} & L^{rs}_{bb} & L^{rr}_{ba} & L^{rr}_{bb} \end{array}\right] = \begin{bmatrix} L^{ss} & L^{sr} \\ L^{rs} & L^{rr} \end{bmatrix} = [L] \tag{6.1}$$

The volt-ampere and torque equations then take the form

$$v] = \frac{d}{dt}[L][i+[R][i \tag{6.2a}$$

and

$$T = -\frac{1}{2}\underline{i}\,\frac{\partial}{\partial\theta}[L][i \tag{6.2b}$$

where

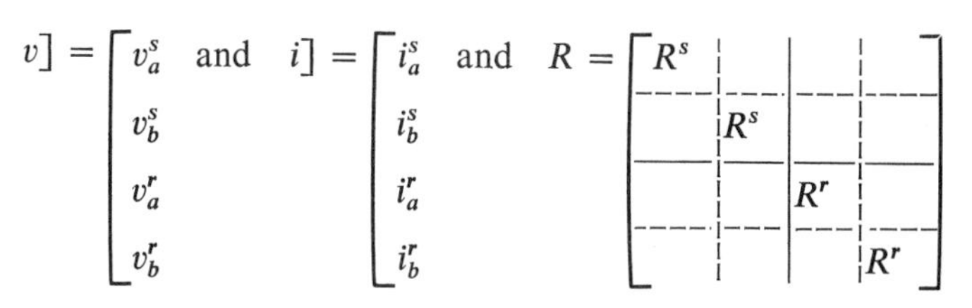

$[L]$ = inductance matrix given by Eq. (6.1), and

$\underline{i}$ = transpose of the matrix $i]$

6.1.1 The dq Transformation

In principle, Eqs. (6.2a,b) describe the behavior of the machine under all operating conditions. However, by expanding Eqs. (6.2a,b) we recognize that these constitute a set of five nonlinear θ-dependent equations. These equations can be considerably simplified if we introduce a change of variables. In dc machines we notice that the commutator physically brings about a change of variables on the rotor; specifically, the rotor (or *armature*) currents are different from the terminal currents. Thus, for dc machines it is quite natural that we mathematically take into account the change of variables introduced by the presence of the commutator. In this connection, we find that the *dq* transformation, first used in Example 5-5, is quite useful.

The *dq* transformation was introduced by Park[2,3] in 1929 in connection with the analysis of salient-pole synchronous machines. Since then, this transformation has been used in analyzing induction machines as well as dc machines. In the preceding chapter we gave a geometrical interpretation to the *dq* transformation. We found that the angular dependence was eliminated and that the equations of motion of a salient-pole machine were linearized for constant-speed operation. We considered the *dq* transformation as a transformation between fixed and rotating axes. This physical significance of the *dq* transformation becomes clearer when applied to the dc machine.

First, let us recall the S_{dq} transformation matrix developed in the last chapter and rewrite the relationships between the *ab* and *dq* currents on the rotor as follows:

$$\begin{bmatrix} i_d^r \\ i_q^r \end{bmatrix} = \begin{bmatrix} \cos \nu\theta & -\sin \nu\theta \\ \sin \nu\theta & \cos \nu\theta \end{bmatrix} \begin{bmatrix} i_a^r \\ i_b^r \end{bmatrix} \tag{6.3}$$

The above equation implies that a set of rotor currents i_a^r and i_b^r flowing in the actual rotor windings (moving with the rotor) can be transformed to the rotor currents i_d^r and i_q^r, which are stationary with respect to the stator reference frame. In other words, the airgap fields produced by the original currents i_a^r and i_b^r flowing in the actual rotor windings are the same as those produced by the fictitious currents i_d^r and i_q^r flowing in fictitious windings stationary with respect to the stator. The constraints on i_d^r and i_q^r are given by Eq. (6.3).

One way of accomplishing the above transformation physically is by injecting appropriate currents into the rotor windings by means of the brushes that are sliding on the rotor but fixed with respect to the stator. Specifically, the brushes slide on commutator segments to which are connected the leads of various rotor coils (forming the re-entrant type armature winding).

The transformation from moving-axis to fixed-axis operation can be visualized from Fig. 6-2(a). It is seen that the field produced by the rotor winding remains stationary with respect to the stator, because the direction

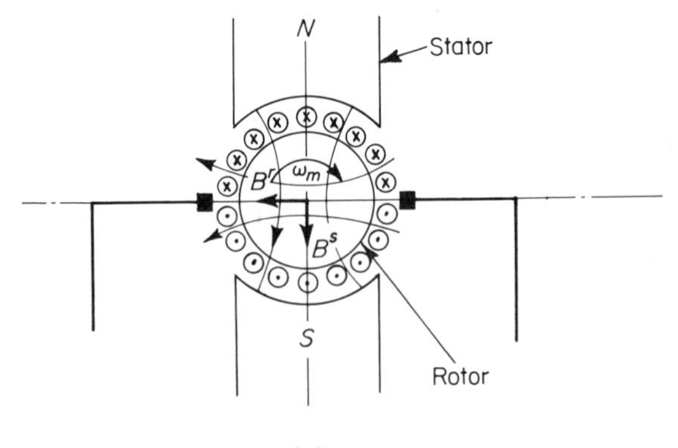

(a)

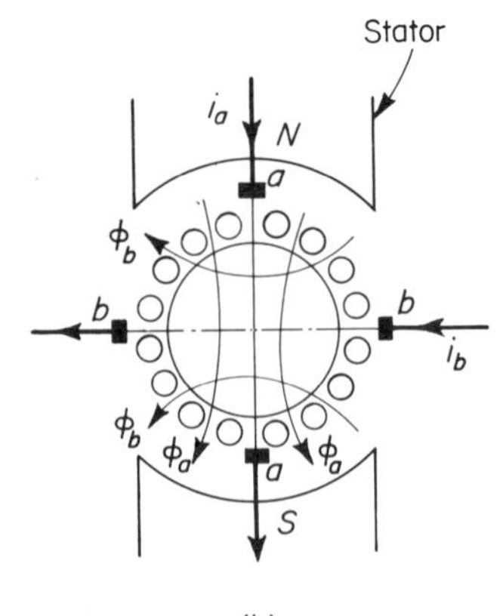

(b)

Fig. 6-2 Fluxes due to rotor currents in (a) a 2-brush machine; (b) a 4-brush machine.

of flow of currents in the rotor conductor depends on the position of the brushes and not on the position or velocity of the rotor.

If we mount a set of four brushes on the rotor, as shown in Fig. 6-2(b), and feed them with balanced 2-phase currents, the individual magnetic fields of the two currents will be stationary with respect to the stator. But, as a consequence of commutator action, the resultant is a field rotating relative to the stator but independent of the rotor position or velocity. Finally, if the brushes are mechanically rotated, the speed of rotation of the brushes will affect the speed of the rotating field. We shall now consider this aspect of commutator action.

6.1.2 *The Commutator as a Frequency Changer*

We may now recall the commutator action discussed in the last chapter. We see from Fig. 5-6 that the coil voltage goes through one complete cycle each time the coil makes one complete revolution with respect to the brushes. We may also recall that the magnitude of the voltage measured at the brushes is proportional to the angular velocity of the coil and the flux density (produced by the stationary field poles, Fig. 5-6). In other words, the frequency of variation of the voltage at the brushes depends on the velocity of the coil relative to the brushes, while the magnitude of this voltage is dependent on the velocity of the rotor coil relative to the airgap field. Thus, the magnitude of the brush voltage v_b can be expressed as

$$|v_b| = k_1(\omega_m - \omega^s) \tag{6.4a}$$

where k_1 = a constant, ω_m = mechanical velocity of the rotor coil, and ω^s = angular velocity of the airgap field (produced by the stator, or field poles). On the other hand, if we assume the brushes to be stationary, the frequency of the brush voltage is unchanged regardless of the relative motion between the rotor and the airgap field. The frequency of the brush voltage ω is given by

$$\omega = \omega_m - \omega_b \tag{6.4b}$$

where ω_m and ω_b are the angular velocities of the rotor and the brushes, respectively, relative to a stationary reference frame.

The above discussion suggests that the commutator is a kind of frequency changer. This interpretation becomes clear by considering an armature (rotor) winding having a set of four brushes and four sliprings, as shown in Fig. 6-3. The rotor is excited by two phase currents through the sliprings. If the frequency of these currents is ω, the resulting airgap field is a rotating field having a velocity ω relative to the stationary rotor. The frequency of the brush voltage is also ω. Now, if the rotor turns with a velocity ω_m, the resultant velocity of the rotating field as seen by the brushes is $(\omega - \omega_m)$.

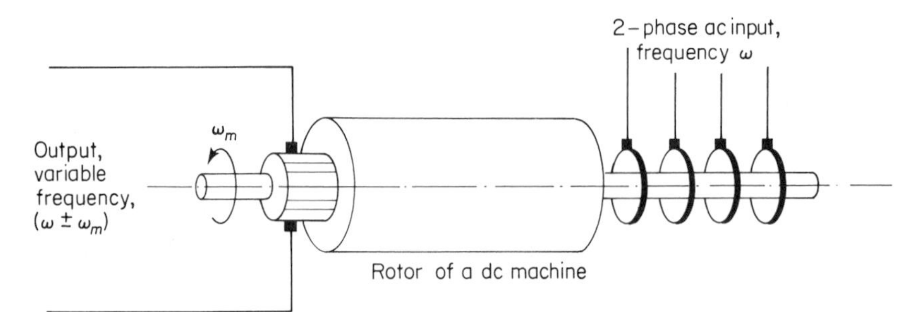

Fig. 6-3 Commutator as a frequency changer.

Consequently, the frequency of the brush voltage in this case is $(\omega - \omega_m)$, which illustrates the action of the commutator as a frequency changer. We shall illustrate this point further by considering frequency conversion by *dq* transformation in the next section.

6.1.3 Frequency Conversion by dq Transformation

Let us consider the 2-phase voltages

$$v_a^r = V \sin \omega t \tag{6.5a}$$

and

$$v_b^r = V \cos \omega t \tag{6.5b}$$

applied to the rotor coils of Fig. 6-1. From Eq. (6.3) the *dq* voltages are given by

$$v_d^r = V \cos \nu\theta \sin \omega t - V \sin \nu\theta \cos \omega t \tag{6.6a}$$

$$v_q^r = V \sin \nu\theta \sin \omega t + V \cos \nu\theta \cos \omega t \tag{6.6b}$$

Since $\theta = \omega_m t + \delta$, Eqs. (6.6a,b) can be expressed as

$$v_d^r = V \sin [(\omega - \nu\omega_m)t - \nu\delta] \tag{6.6c}$$

$$v_q^r = V \cos [(\omega - \nu\omega_m)t - \nu\delta] \tag{6.6d}$$

From Eqs. (6.6c,d) it is clear that the frequency for the *dq* voltages has changed from ω to $(\omega - \nu\omega_m)$, which is the same change accomplished by the commutator discussed in the preceding section.

We may then conclude that the commutator action can be described by the *dq* transformation, since both result in a frequency conversion.

6.1.4 The Transformed Equations of Motion

The dynamic equations of motion in terms of the original variables are given by Eqs. (6.2a,b). Transforming these equations in terms of dq variables, using Eq. (6.3), we have (with $p = d/dt$)

$$\begin{bmatrix} v_a^s \\ v_b^s \\ \hline v_d^r \\ v_q^r \end{bmatrix} = \left[\begin{array}{cc|cc} R^s + p(L^s + L_o^s) & 0 & p(L^{sr} + L_o^{sr}) & 0 \\ 0 & R^s + p(L^s - L_o^s) & 0 & p(L^{sr} - L_o^{sr}) \\ \hline p(L^{sr} + L_o^{sr}) & \nu\dot{\theta}(L^{sr} - L_o^{sr}) & R^r + p(L^r + L_o^r) & \nu\dot{\theta}(L^r - L_o^r) \\ -\nu\dot{\theta}(L^{sr} + L_o^{sr}) & p(L^{sr} - L_o^{sr}) & -\nu\dot{\theta}(L^r + L_o^r) & R^r + p(L^r - L_o^r) \end{array}\right] \begin{bmatrix} i_a^s \\ i_b^s \\ \hline i_d^r \\ i_q^r \end{bmatrix} \tag{6.7a}$$

$$T = -\nu L^{sr}(i_b^s i_d^r - i_a^s i_q^r) + \nu L_o^{sr}(i_b^s i_d^r + i_a^s i_q^r) + 2\nu L_o^r i_d^r i_q^r \tag{6.7b}$$

In Eq. (6.7a) the inductances occur as the sum or difference and function as coefficients of the speed $\dot{\theta}$. Defining $L_a^s = L^s + L_o^s$, $L_b^s = L^s - L_o^s$, $L_d^r = L^r + L_o^r$, $L_q^r = L^r - L_o^r$, $L_d^{sr} = L^{sr} + L_o^{sr}$, $L_q^{sr} = L^{sr} - L_o^{sr}$, $G_q^{rs} = \nu(L^{sr} + L_o^{sr})$, $G_d^{rs} = \nu(L^{sr} - L_o^{sr})$, $G_q^r = \nu(L^r + L_o^r)$, and $G_d^r = \nu(L^r - L_o^r)$, Eqs. (6.7a,b) can be abbreviated as

$$\begin{bmatrix} v_a^s \\ v_b^s \\ v_d^r \\ v_q^r \end{bmatrix} = \begin{bmatrix} R^s + pL_a^s & 0 & pL_d^{sr} & 0 \\ 0 & R^s + pL_b^s & 0 & pL_q^{sr} \\ pL_d^{sr} & \dot{\theta}G_d^{rs} & R^r + pL_d^r & \dot{\theta}G_d^r \\ -G\dot{\theta}_q^{rs} & pL_q^{sr} & -\dot{\theta}G_q^r & R^r + pL_q^r \end{bmatrix} \begin{bmatrix} i_a^s \\ i_b^s \\ i_d^r \\ i_q^r \end{bmatrix} \tag{6.8a}$$

$$T = -[G_d^{rs} i_b^s i_d^r - G_q^{rs} i_a^s i_q^r + (G_d^r - G_q^r) i_d^r i_q^r] \tag{6.8b}$$

The above equations are quite general and the equations of motion of various kinds of dc machines can be derived therefrom. We shall illustrate this by means of a number of examples. But at this stage it is worthwhile to discuss the physical and constructional features of some commonly encountered dc machines.

6.2 Physical Features of dc Machines

In order that we may subsequently consider some conventional dc machines, it is best that we briefly study the constructional features of their various parts and discuss the usefulness of these parts. A common large dc machine is shown in Fig. 6-4 which, along with the schematic shown in Fig. 6-5, shows most of the most important parts of machines. The *field poles*, mounted on the *stator*, carry windings called *field windings*. Some machines carry more than one separate field winding on the same core. The cores of the poles are built of sheet-steel laminations. Because the field windings carry direct current, it is not necessary to have the cores laminated. It is, however, necessary for the pole faces to be laminated because of their proximity to the armature windings. (Use of laminations for the cores as well as for the pole faces facilitates assembly.) The armature core, which carries the *armature windings*, is generally on the *rotor* and is made of sheet-steel laminations. The

Fig. 6-4 A dc machine. (Courtesy of General Electric Company.)

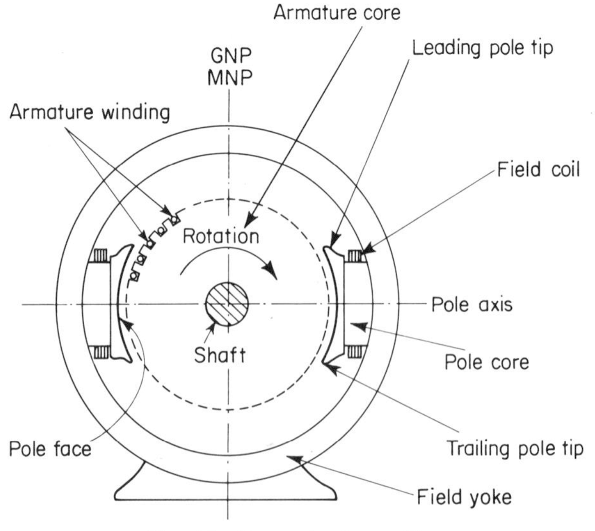

Fig. 6-5 Parts of a dc machine.

commutator is made of hard-drawn copper segments insulated from one another by mica. As shown in Fig. 6-6, the armature windings are connected to the commutator segments over which the carbon *brushes* slide and serve as leads for electrical connection.

The armature winding may be a *lap winding* [Fig. 6-6(a)] or a *wave winding* [Fig. 6-6(b)], and the various coils forming the armature winding may be connected in a series–parallel combination. It has been found that in a simplex lap winding the number of paths in parallel a is equal to the number of poles p; whereas in a simplex wave winding the number of parallel paths is always 2. (The interested reader should consult Reference 4 for details.)

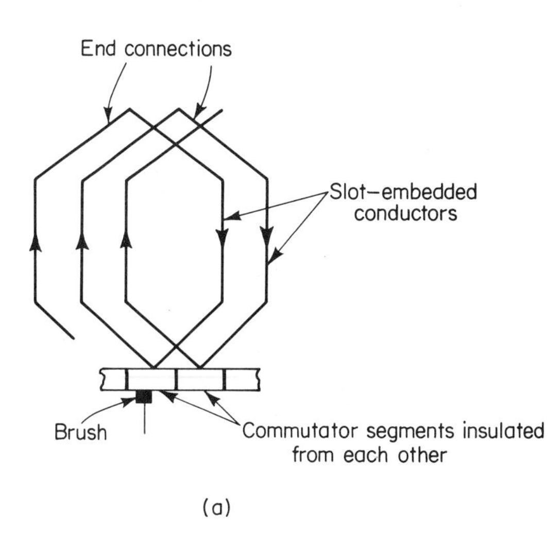

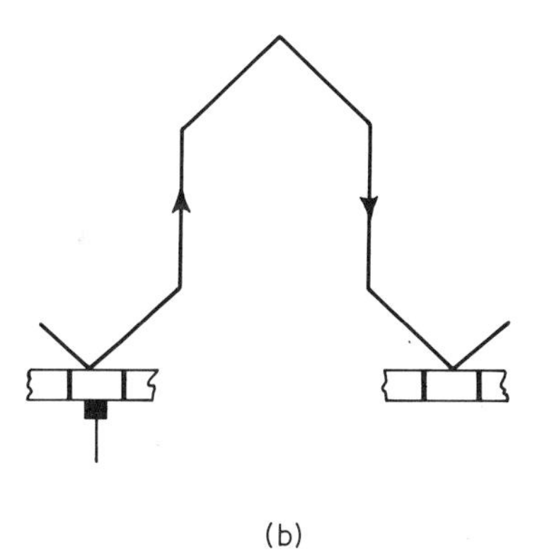

Fig. 6-6 Schematic representation of armature windings: (a) lap winding; (b) wave winding.

In addition to the armature and field windings, *commutating poles* and *compensating windings* are also found on large dc machines. These are used essentially to improve the performance of the machine, as we shall see in later sections (Secs. 6.2.3 and 6.2.4).

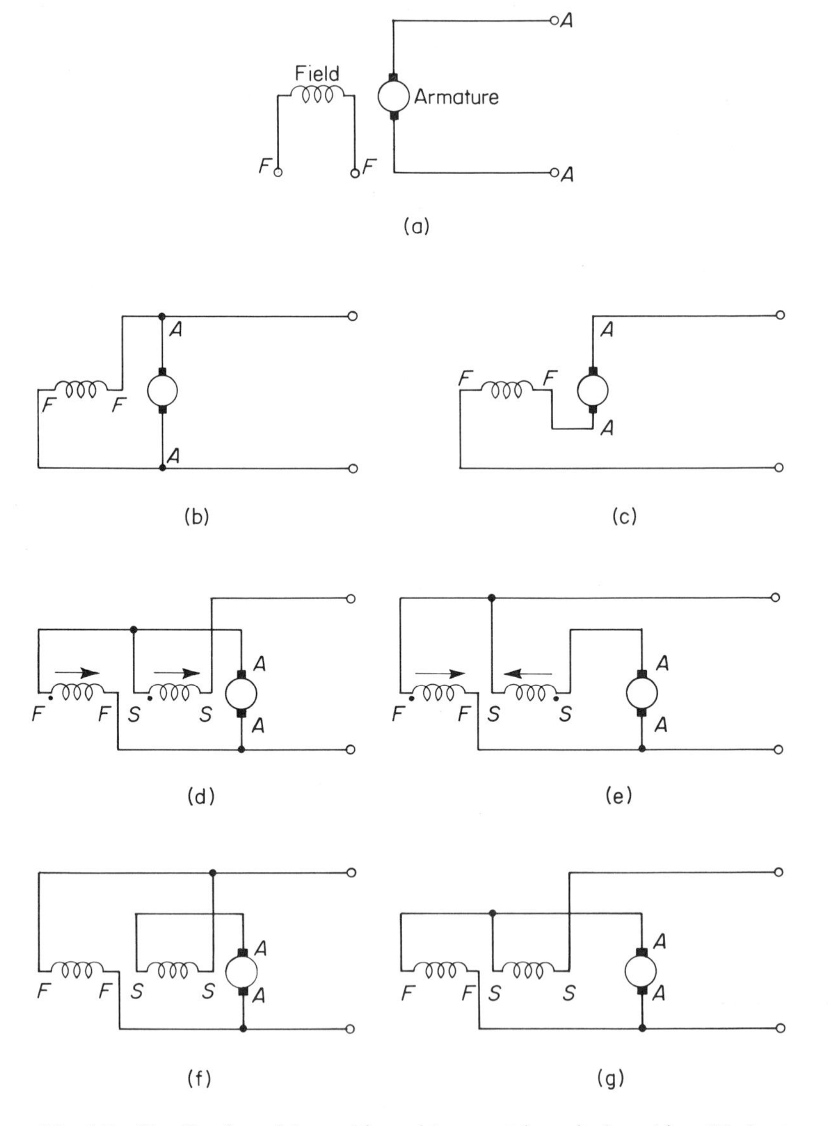

Fig. 6-7 Classification of dc machines: (a) separately excited machine; (b) shunt machine; (c) series machine; (d) cumulative compound machine; (e) differential compound machine; (f) long-shunt machine; (g) short-shunt machine.

6.2.1 Classification According to Forms of Excitation

We have repeatedly mentioned that electromechanical energy conversion results from the interaction of current-carrying conductors and electromagnetic fields. In a dc machine, the armature windings serve as "current-carrying conductors" and the magnetic field in the airgap of the machine is produced by the field windings or by a permanent magnet. In the latter case, the machine is called a permanent-magnet machine; we shall not consider this type machine here.

Conventional machines having a set of field windings and armature windings can be classified, on the basis of mutual electrical connections between the field and armature windings, as follows:

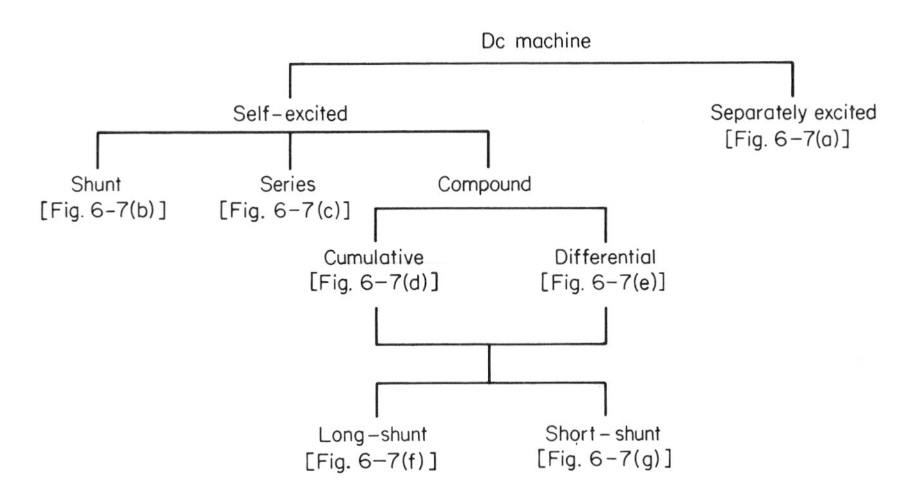

We can now apply the theory developed in Sec. 6.1 to the analysis of some conventional machines.

EXAMPLE 6–1

The dc shunt machine is shown in Fig. 6-7(b). It is a single-axis machine with one winding on the field (stator) and one on the armature (rotor). Its equations of motion and speed–torque characteristics are to be obtained.

For this case, windings "a" on the stator and "q" on the rotor are removed and Eqs. (6.8a,b) are simplified to

$$\begin{bmatrix} v_f \\ \hline v_a \end{bmatrix} = \begin{bmatrix} R_f + L_f p & 0 \\ \hline \dot{\theta} G_{af} & R_a + L_a p \end{bmatrix} \begin{bmatrix} i_f \\ \hline i_a \end{bmatrix} \tag{a}$$

and

$$T_e = G_{af} i_f i_a \tag{b}$$

where $R_f = R^s$, $L_f = L_b^s$, $R_a = R^r$, $L_a = L_d^r$, and $G_{af} = G_d^{rs}$ are substituted in Eqs. (6.8a,b).

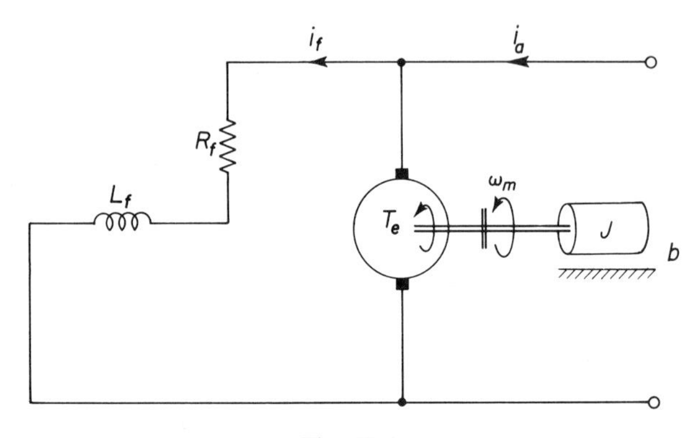

Fig. 6E-1

Under steady-state and constant-speed operation ($\dot{\theta} = \omega_m$), the above equations further reduce to

$$V_f = R_f I_f \tag{c}$$

and

$$V_a = \omega_m G_{af} I_f + R_a I_a \tag{d}$$

Because of shunt connection, $V_f = V_a = V$, so that

$$I_f = \frac{V}{R_f} \tag{e}$$

$$I_a = \frac{V}{R_a}\left(1 - \frac{\omega_m G_{af}}{R_f}\right) \tag{f}$$

and

$$T_e = \frac{G_{af} V^2}{R_a R_f}\left(1 - \frac{\omega_m G_{af}}{R_f}\right) \tag{g}$$

The above are the required equations. The torque–speed characteristics can be sketched, for the specified voltage V, from the last equation.

Reconsidering the torque equation

$$T_e = G_{af} I_f I_a \tag{6.9}$$

and the armature-voltage balance equation

$$V_a = G_{af} \omega_m I_f + R_a I_a \tag{6.10}$$

we notice that the speed coefficient G_{af} can be considered an electromechanical energy-conversion constant. This constant can be related to the number of poles, the armature-winding data, and other constants as discussed in the following section.

6.2.2 The EMF Equation

In Chapter 5 we obtained the expression for the voltage induced in an N-turn coil rotating at an angular velocity ω as

$$e = \omega N\phi \sin \omega t$$

The voltage given by this expression is sinusoidal, but in a dc machine a rectified voltage wave [as shown in Fig. 5-6(b)] is obtained at the brushes because of the commutator action. The average or dc voltage at the brushes is, therefore,

$$E = \frac{1}{\pi}\int_0^{\pi} \omega N\phi \sin \omega t \, d(\omega t) = \frac{2\omega}{\pi} N\phi \tag{6.11a}$$

But $\omega = 2\pi f$ and $f = pn/120$ [from Eq. (5.7)], so that Eq. (6.11a) takes the form

$$E = \frac{n}{60} 2p\phi N \tag{6.11b}$$

where n = speed of rotation in r/min. If Z is the total number of conductors connected in a parallel paths, Eq. (6.11b) takes the final form[5]

$$E = \phi n \frac{Zp}{60a} \tag{6.12a}$$

But $\omega_m = 2\pi n/60$, so Eq. (6.12a) can also be expressed as

$$E = \frac{\omega_m}{2\pi} \phi Z \frac{p}{a} \tag{6.12b}$$

If the core of the field poles is not saturated, the flux per pole ϕ is directly proportional to the field current I_f; that is,

$$\phi = k_f I_f \tag{6.12c}$$

where k_f is the proportionality constant. Combining Eqs. (6.12b) and (6.12c) we have

$$E = \frac{k_f}{2\pi} \frac{pZ}{a} \omega_m I_f \tag{6.12d}$$

The constant G_{af} in Eq. (6.10) can thus be identified as

$$G_{af} = \frac{k_f pZ}{2\pi a}$$

and the voltage $G_{af}\omega_m I_f$ is a voltage due to the rotation of the armature conductors in the magnetic field. This voltage is termed the *induced voltage*, *back EMF*, or *speed voltage*. Evidently, the equivalent circuit of the dc machine with one field winding follows from Eqs. (a) of Example 6-1 and takes the form shown in Fig. 6-8, where $v_a = \dot{\theta} G_{af} i_f$.

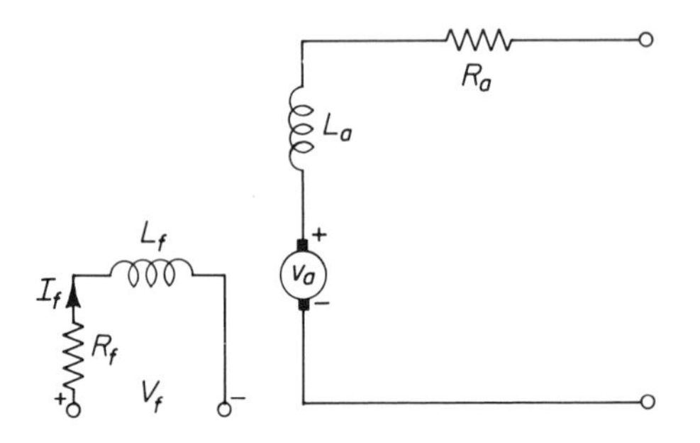

Fig. 6-8 Equivalent circuit of a dc machine.

6.2.3 Airgap Fields and Armature Reaction

In the discussions so far we have assumed no interaction between the fields produced by the field windings and by the current-carrying armature windings. In reality, however, the situation is quite different. Consider the 2-pole machine shown in Fig. 6-9(a). If the armature does not carry any current (that is, when the machine is on *no-load*), the airgap field takes the form shown in Fig. 6-9(b). The *geometric neutral plane* and *magnetic neutral plane* (GNP and MNP respectively) are coincident. (*Note:* Magnetic lines of force enter the MNP at right angles.) Noting the polarities of the induced voltages in the conductors, we see that the brushes are located at the MNP for maximum voltage at the brushes. We now assume that the machine is on "load" and that the armature carries current. The direction of flow of current in the armature conductors depends on the location of the brushes. For the situation in Fig. 6-9(b), the direction of the current flow is the same as the direction of the induced voltages. In any event, the current-carrying armature conductors produce their own magnetic fields, as shown in Fig. 6-9(c), and the airgap field is now the resultant of the fields due to the field and armature windings. The resultant airgap field is thus distorted and takes the form shown in Fig. 6-9(d). The interaction of the fields due to the armature and field windings is known as *armature reaction*. As a consequence of armature reaction the airgap field is distorted and the MNP is no longer coincident with the GNP. For maximum voltage at the terminals, the brushes have to be located at the MNP. Thus, one undesirable effect of armature reaction is that the brushes must be shifted constantly, since the shift of the MNP from the

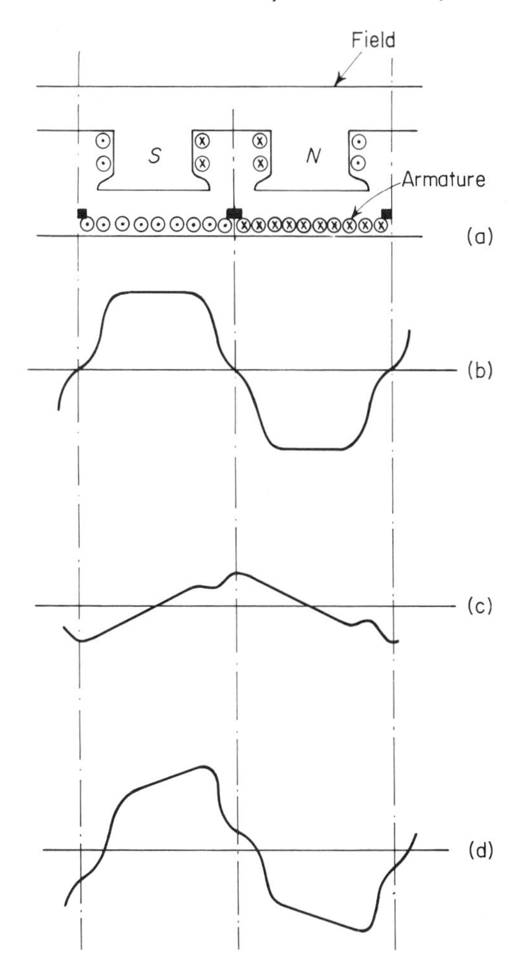

Fig. 6-9 Airgap fields in a dc machine: (a) A two-pole machine, showing armature and field MMF's; (b) flux-density distribution due to field MMF; (c) flux-density distribution due to armature MMF; (d) resultant flux-density distribution [Curve (b) + Curve (c)].

GNP depends on the load (which is presumably always changing). The effect of armature reaction can be analyzed in terms of *cross-magnetization* and *demagnetization*, as shown in Fig. 6-10(a). We just mentioned the effect of cross-magnetization resulting in the distortion of the airgap field and requiring the shifting of brushes according to the load on the machine. The effect of demagnetization is to weaken the airgap field. All in all, therefore, armature reaction is not a desirable phenomenon in a machine.

The effect of cross-magnetization can be neutralized by means of compensating windings, as shown in Fig. 6-10. These are conductors embedded

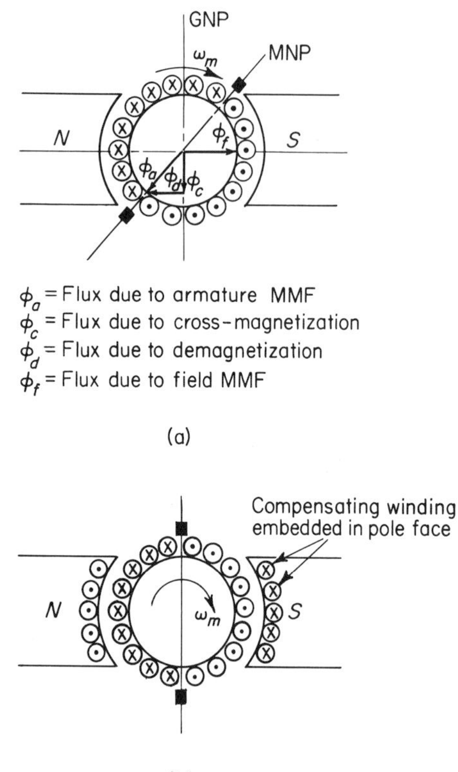

Fig. 6-10 (a) Armature reaction resolved into cross and demagnetizing components; (b) neutralization of cross-magnetizing component by compensating winding.

in pole faces, connected in series with the armature windings and carrying currents in an opposite direction to that flowing in the armature conductors under the pole face (Fig. 6-10). Once cross-magnetization has been neutralized, the MNP does not shift with load and remains coincident with the GNP at all loads. The effect of demagnetization can be compensated for by increasing the MMF on the main field poles. By neutralizing the net effect of armature reaction, we are justified in our preceding and following discussions where we assume no "coupling" between the armature and field windings.

6.2.4 *Reactance Voltage and Commutation*

In discussing the action of the commutator in the last chapter, we indicated that the direction of flow of current in a coil undergoing commutation reverses by the time the brush moves from one commutator segment to the other. This is schematically represented in Fig. 6-11. The flow of current in coil *a* for three different instants is shown. We have assumed that the current fed by a segment is proportional to the area of contact between the brush and the commutator segment. Thus, for satisfactory commutation, the direction of flow of current in coil *a* must completely reverse [Figs. 6-11(a) and (c)] by the time the brush moves from segment no. 2 to segment no. 3. The ideal situation is represented by the straight line in Fig. 6-12, and may be termed *straight-line commutation.* Because coil *a* has some in-

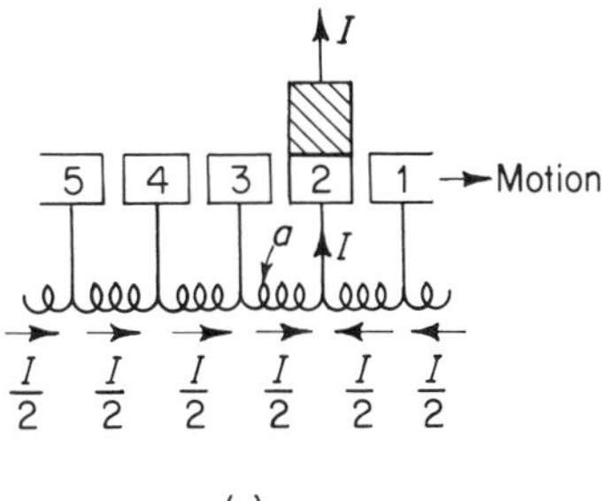

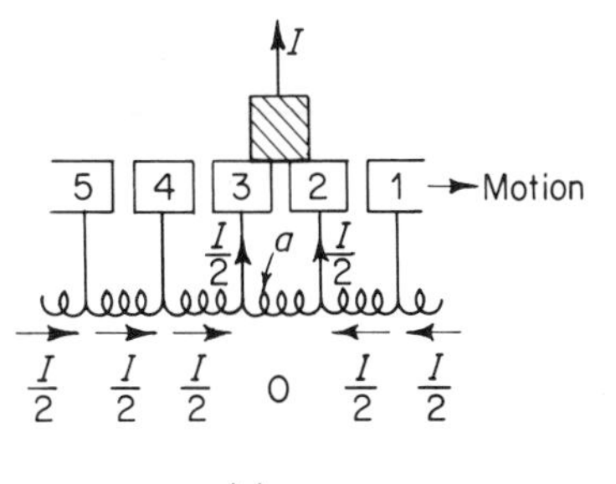

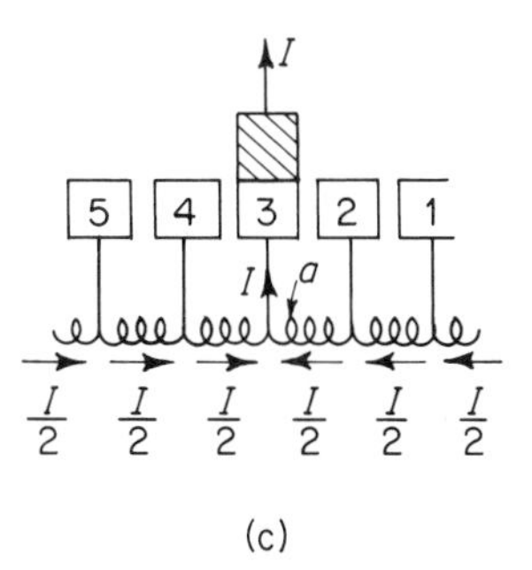

Fig. 6-11 Coil *a* undergoing commutation.

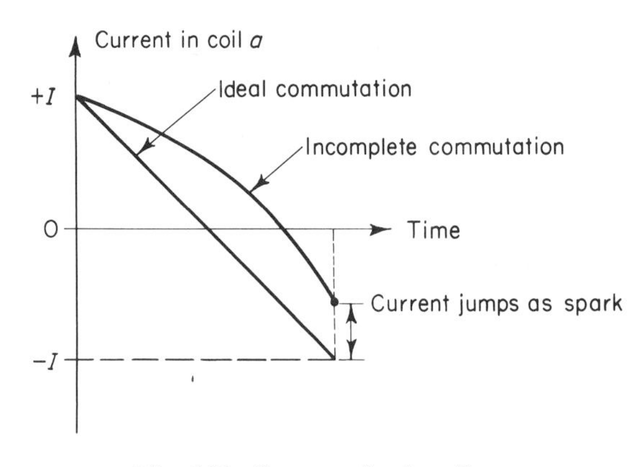

Fig. 6-12 Commutation in coil *a*.

ductance L, the change of current ΔI in a time Δt induces a voltage $L(\Delta I/\Delta t)$ in the coil. According to Lenz's law, the direction of this voltage, called *reactance voltage*, is opposite to the change (ΔI) which is causing it. As a result, the current in the coil does not completely reverse by the time the brush moves from one segment to the other. The balance of the "unreversed" current jumps over as a spark from the commutator to brush, and thereby the commutator wears out because of pitting. This departure from ideal commutation is also shown in Fig. 6-12.

The directions of the (speed-)induced voltage, current flow, and reactance voltage are shown in Fig. 6-13(a). Note that the direction of the induced voltage depends on the direction of rotation of the armature conductors and on the direction of the airgap flux. It is determined from the $\mathbf{u} \times \mathbf{B}$ (or the right-hand) rule. Next, the direction of the current flow depends on the location of the brushes (or tapping points). Finally, the direction of the reactance voltage depends on the change in the direction of current flow and is determined from Lenz's law. For the brush position shown in Fig. 6-13(a), observe that the reactance voltage opposes the induced voltage and retards the current reversal. If the brushes are advanced in the direction of rotation (for generator operation), we may notice, from Fig. 6-13(b), that the reactance voltage is in the same direction as the (speed-)induced voltage and therefore current reversal is not opposed. We may further observe that the coil undergoing commutation, being near the tip of the north pole, is under the influence of the field of a weak north pole. From this argument, we may conclude that commutation improves if we advance the brushes. But this is not a very practical solution. The same—perhaps better—results can be achieved if we keep the brushes at the GNP, or MNP, as in Fig. 6-13(a), but produce the "field of a weak north pole" by appropriately winding and connecting an auxiliary field winding, as shown in Fig. 6-13(c). The poles producing the desired field for better commutation are known as *commutating poles*.

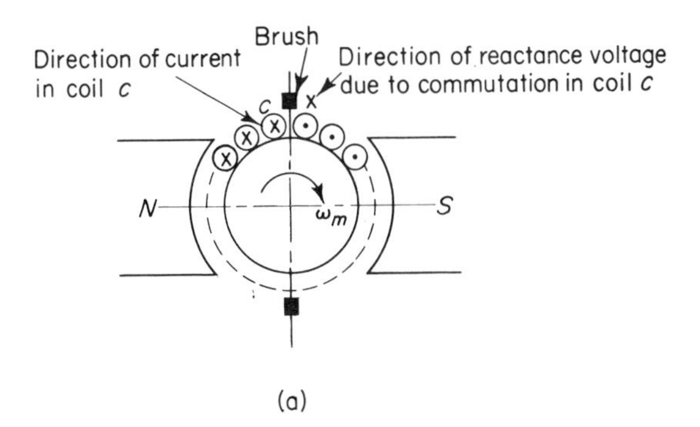

(a)

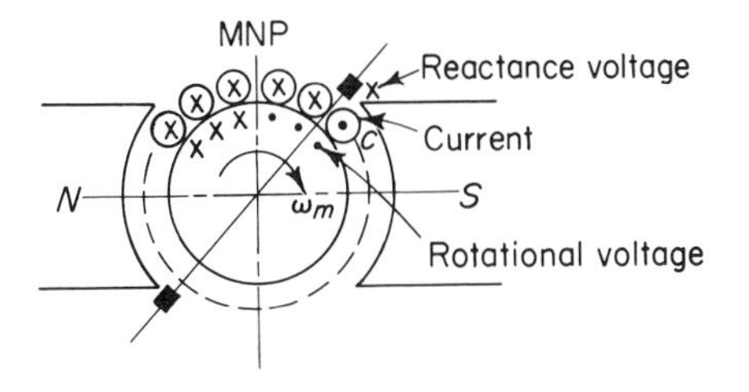

(b)

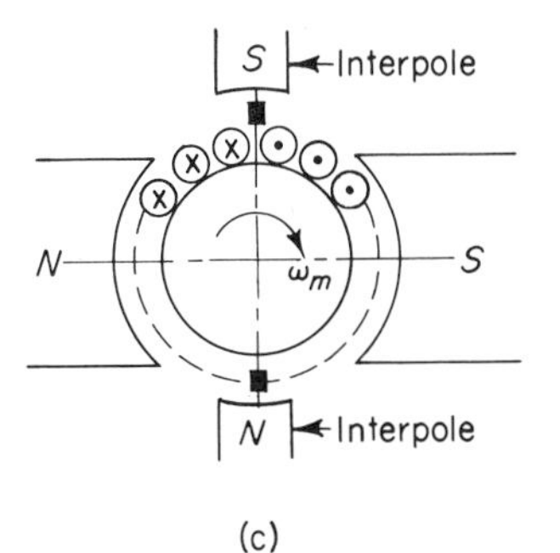

(c)

Fig. 6-13 Reactance voltage and its neutralization: (a) reactance voltage and current in coil c, rotational voltage $\simeq 0$; (b) reactance voltage, rotational voltage, and current in coil c; (c) interpoles.

6.2.5 *Effects of Saturation*

Saturation plays a very important role in governing the behavior of dc machines. It is extremely difficult to take into account the effects of satura-

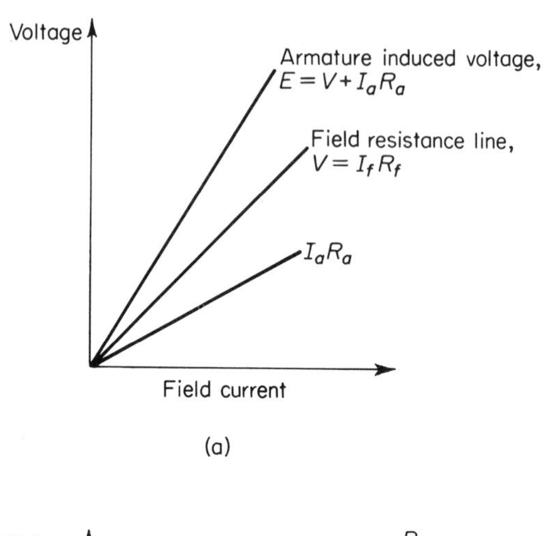

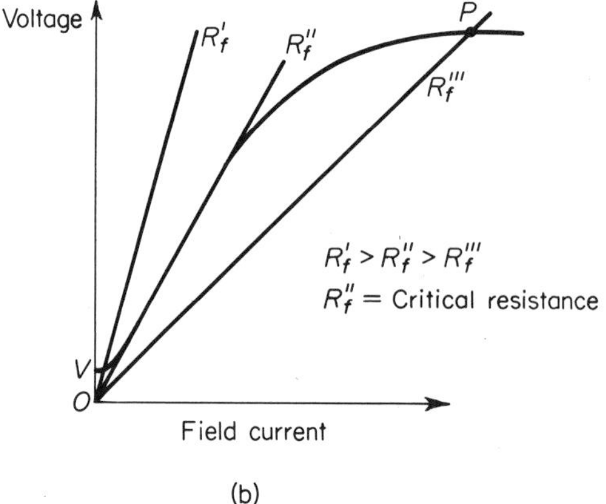

Fig. 6-14 No-load characteristic of a shunt generator: (a) no stable operating point for the shunt generator; (b) stable no-load voltage of a shunt generator.

tion in the dynamical equations of motion. Recently, an analysis based on differential geometry has been presented which considers saturation in the equations of motion.[6] The details are quite involved and will not be considered here. Generally a graphical or semiempirical method is used to investigate saturation. For certain purposes, saturation is taken into account by using the concept of the inverse saturation factor.[1] For the time being, let us consider qualitatively the consequences of saturation on the operation of a self-excited shunt generator. Then we shall take up an example of a motor–generator system.

A self-excited shunt machine is shown in Fig. 6-7(b). Its equations of motion were derived in Example 6-1. For convenience we rewrite the steady-state equations for the operation of the machine as a generator. From the circuit shown in Fig. 6-8, rewriting Eqs. (c) and (d) of Example 6-1 we have

$$\left.\begin{aligned} V &= R_f I_f \\ \text{and} \qquad E &= V + I_a R_a = I_f R_f + I_a R_a \end{aligned}\right\} \tag{6.13}$$

These equations are represented by the straight lines shown in Fig. 6-14(a). Notice that the voltages V and E will keep building up and no equilibrium point can be reached. On the other hand, if we include the effect of saturation, as in Fig. 6-14(b), the point P defines the equilibrium, because at this point the field-resistance line intersects the saturation curve. A deviation from P to P' or P'' would immediately show that at P' the voltage drop across the field is greater than the induced voltage, which is not possible; and at P'' the induced voltage is greater than the field-circuit voltage drop.

In Fig. 6-14(b) is shown some residual magnetism as measured by the small voltage OV. Evidently, without it the shunt generator will not build up any voltage. Also shown in Fig. 6-14(b) is the *critical resistance.* A field resistance greater than the critical resistance (for a given speed) would not let the shunt generator build up any appreciable voltage. Finally, we should ascertain that the polarity of the field winding is such that a current through it produces a flux which aids the residual flux. If it does not, the two fluxes tend to neutralize and the machine voltage will not build up. To summarize, the conditions for the building-up of a voltage in a shunt generator are: the presence of residual flux, field-circuit resistance less than the critical resistance, and appropriate polarity of the field winding.

EXAMPLE 6–2

Let us consider the separately excited motor shown in Fig. 6E-2 to be fed by a

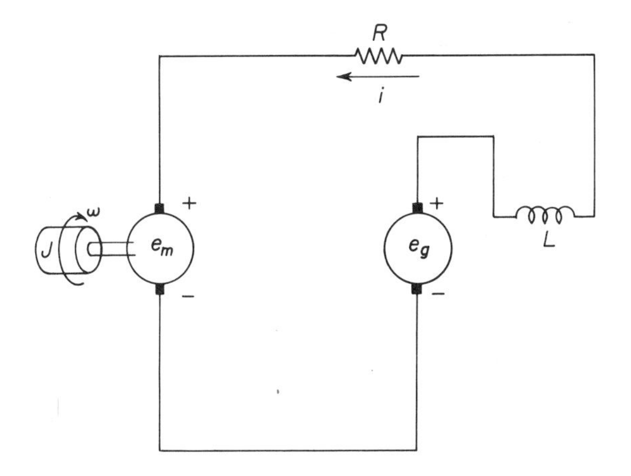

Fig. 6E-2

series generator. The motion of the motor–generator system is to be discussed.

This is an example wherein the effects of saturation result in sustained nonlinear oscillatory behavior of the motor, if certain conditions are met. Qualitatively, as soon as the switch is closed the generator voltage e_g starts building up. The current i flows into the motor armature, which starts rotating in one direction. However, as the current increases the core of the generator begins to saturate, and at the same time the motor back EMF e_m begins to increase as the motor speeds up. Because e_g and e_m act in opposition and begin to increase simultaneously, it is possible that the current i (which depends on $e_g - e_m$) will stop flowing at some instant. At that instant the voltage e_g collapses and, because of the inertia of the rotor, the motor temporarily acts as a generator, although the voltage e_m tends to go down. This generator action of the motor forces the current to flow in the opposite direction. When the current in the series field of the generator is thus reversed, the generator voltage builds up in the opposite direction and forces the current to flow in the opposite direction. Now that the current is reversed in the armature of the motor, it develops torque in a direction to stop the original motion. After the rotor stops, it picks up speed in the opposite direction, and the cycle repeats. Consequently, the motor reverses the direction of rotation periodically.

The equations describing the motion of the system are

$$e_g - e_m = L\frac{di}{dt} + Ri \tag{a}$$

$$k_m i = J\dot{\omega} \tag{b}$$

$$e_g = f(i) \tag{c}$$

where k_m = motor electromechanical-conversion constant. Other constants are as shown in Fig. 6E-2. In Eq. (c), the functional relationship $f(i)$ depends on the saturation characteristics of the generator. It is sometimes expressed as a polynomial, or by the *Froelich equation*:

$$e_g = \frac{k_1 i}{k_2 + i} \tag{d}$$

where k_1 and k_2 are constants. Another commonly used expression is

$$e_g = a_1 \tanh i \tag{e}$$

where a_1 is a constant.

In any event, we see that if either (c), (d), or (e) is substituted in Eqs. (a) and (b), the resulting equations are nonlinear and yield nonlinear oscillations as possible solutions.

The last two examples—the stable self-excited operation of a shunt generator and nonlinear sustained oscillations of the motor–generator system—show qualitatively the effects of saturation on the behavior of dc machines. To summarize, the effects of saturation are to decrease the field inductance as well as the induced voltage. As mentioned earlier, the effects of saturation cannot explicitly be taken into account in the equations of motion. However, approximate results can be obtained by graphical methods or by using Eqs. (d) and (e) of Example 6-2. An alternative way to include saturation in analog-computer representation is by using a nonlinear-function generator to generate what has been defined as the inverse saturation factor.[1]

6.3 The Dynamics of dc Machines

In the preceding section we considered the diverse physical aspects of dc machines. We now return to the equations of motion and attempt to obtain the dynamical (linearized) behavior of different kinds of dc machines. In this connection, let us recall Eqs. (a) and (b) of Example 6-1. These are rewritten as (see Fig. 6E-1)

$$\begin{bmatrix} v_f \\ \hline v_a \end{bmatrix} = \left[\begin{array}{c|c} R_f + L_f p & 0 \\ \hline \dot{\theta} G_{af} & R_a + L_a p \end{array}\right] \begin{bmatrix} i_f \\ \hline i_a \end{bmatrix} \tag{6.14}$$

$$T_e = G_{af} i_f i_a = J\ddot{\theta} + b\dot{\theta} \tag{6.15}$$

Notice that these equations are nonlinear. Also, we have assumed the mechanical load on the machine as inertia J and friction b. However, they can be linearized by the methods of Sec. 2.5.7. Consequently, we assume

$$\begin{aligned} v_f &= V_{f0} + v_{f1} \\ i_f &= I_{f0} + i_{f1} \\ \omega &= \Omega_0 + \omega_1 = \dot{\theta} \\ i_a &= I_{a0} + i_{a1} \\ v_a &= V_{a0} + v_{a1} \\ T_e &= T_{e0} + T_{e1} \end{aligned} \tag{6.16}$$

where the lower-case symbols with subscripts 1 denote small perturbations about the steady-state operating point. When Eqs. (6.16) are substituted in Eqs. (6.14) and (6.15), the steady-state behavior is given by

$$\begin{aligned} V_{f0} &= I_{f0} R_f \\ V_{a0} &= I_{a0} R_a + G_{af} \Omega_0 I_{f0} \\ T_{e0} &= G_{af} I_{f0} I_{a0} = b\Omega_0 \end{aligned} \tag{6.17}$$

The complete linearized block diagram for the machine is shown in Fig. 6-15, which yields the dynamics of the machine for incremental motions about a quiescent operating point defined by Eqs. (6.17).

In the above discussion, we considered linearization by using Eqs. (6.16). There exist cases where the machine operates under constraints which reduce

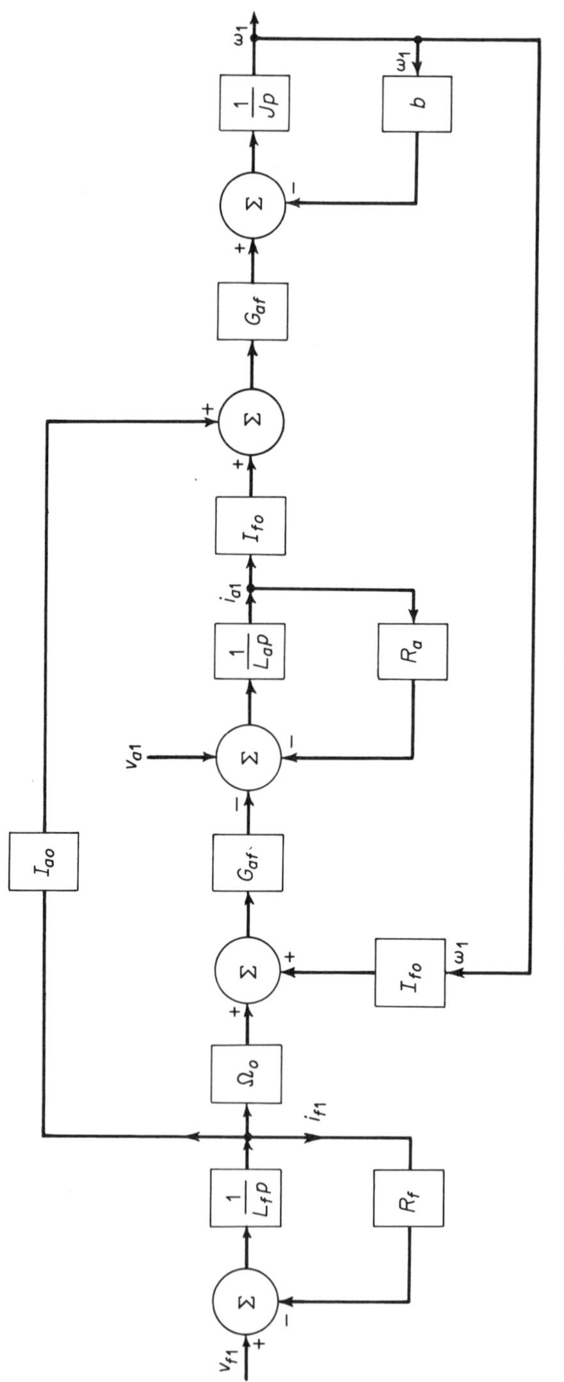

Fig. 6-15 Linearized block diagram of a dc machine.

the equations of motion to linear equations. We shall consider these in the following sections.

6.3.1 The Shunt Motor

A shunt motor is shown in Fig. 6-16. The general equations of motion are, from Eqs. (6.14) and (6.15),

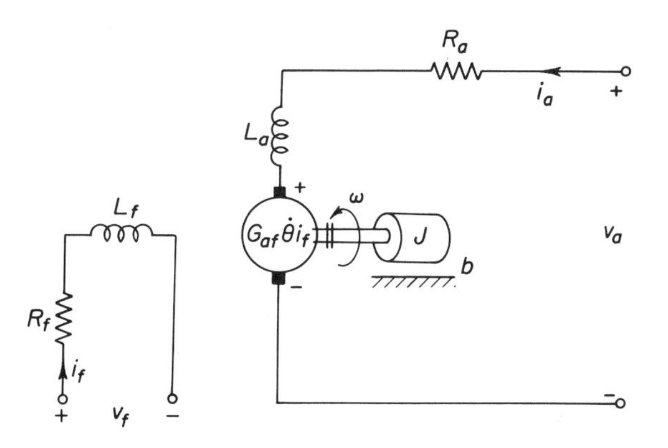

Fig. 6-16 An idealized shunt motor.

$$v_f = R_f i_f + L_f p i_f \tag{6.18a}$$

$$v_a = R_a i_a + L_a p i_a + G_{af}\dot{\theta} i_f \tag{6.18b}$$

and

$$G_{af} i_f i_a = J\ddot{\theta} + b\dot{\theta} \tag{6.18c}$$

In practice, the motor is either (1) field-controlled, in which $i_a = I_a =$ constant, or (2) armature-controlled, in which $i_f = I_f =$ constant. We consider the two cases separately.

Case (1): If $i_a = I_a$, we need to consider only Eqs. (6.18a,c). These become

$$v_f = R_f i_f + L_f p i_f \tag{6.19a}$$

and

$$G_{af} I_a i_f = J\ddot{\theta} + b\dot{\theta} = J\dot{\omega} + b\omega \tag{6.19b}$$

Both of these equations are linear. Defining the speed ω as an output and the field voltage as an input, the transfer function for the motor becomes

$$\frac{\Omega(s)}{V_f(s)} = \frac{G_{af} I_a}{(Js+b)(R_f+L_f s)} \tag{6.20}$$

The block diagram and the signal-flow graph are shown in Figs. 6-17(a) and (b) respectively.

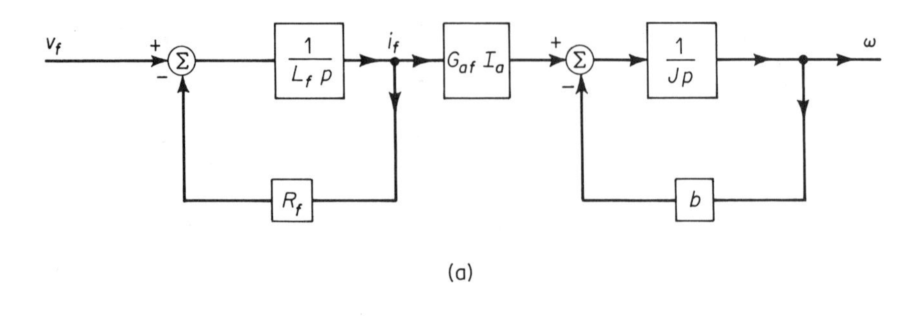

(a)

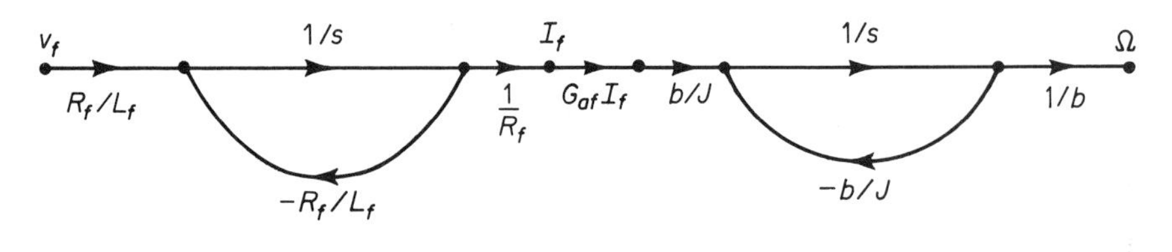

(b)

Fig. 6-17 (a) Block diagram; (b) signal-flow graph of a field-controlled dc motor.

Case (2): If $i_f = I_f$, we have, from Eqs. (6.18b,c),

$$v_a = R_a i_a + L_a p i_a + G_{af} I_f \omega \tag{6.21a}$$

and

$$G_{af} I_f i_a = J\dot{\omega} + b\omega \tag{6.21b}$$

As in case (1), Eqs. (6.21a,b) are linear and the techniques of solving the linear equations discussed in Chapters 2 and 3 can be used to solve these equations. In particular, the transfer function and the state equations are given by Eqs. (6.22) and (6.23) respectively:

$$\frac{\Omega(s)}{V_a(s)} = \frac{G_{af} I_f}{(R_a + L_a s)(b + Js) + (G_{af} I_f)^2} \tag{6.22}$$

$$\begin{bmatrix} \dfrac{di_a}{dt} \\[2ex] \dfrac{d\omega}{dt} \end{bmatrix} = \begin{bmatrix} \dfrac{-R_a}{L_a} & \dfrac{-G_{af} I_f}{L_a} \\[2ex] \dfrac{G_{af} I_f}{J} & \dfrac{-b}{J} \end{bmatrix} \begin{bmatrix} i_a \\[2ex] \omega \end{bmatrix} + \begin{bmatrix} \dfrac{1}{L_a} \\[2ex] 0 \end{bmatrix} v_a \tag{6.23}$$

The block diagram and signal-flow graph are shown in Figs. 6.18(a) and (b) respectively.

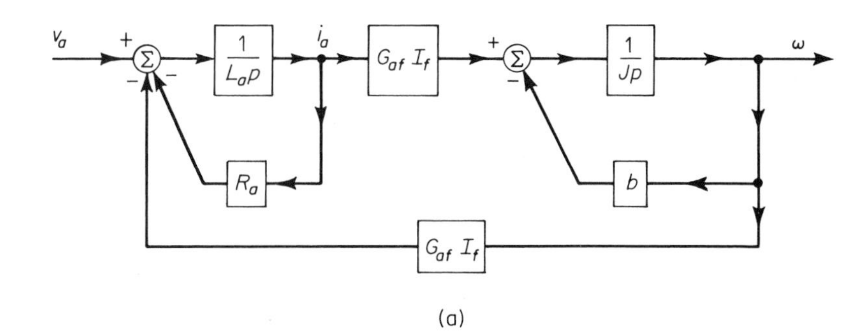

(a)

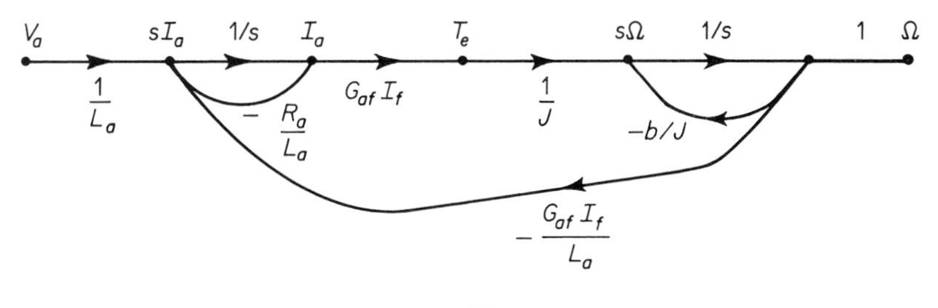

(b)

Fig. 6-18 (a) Block diagram; (b) signal-flow graph of an armature-controlled dc motor.

From Eqs. (6.21a,b) we can obtain an equivalent circuit of the motor (see Problem 6-4). We consider a somewhat simplified, but realistic, case in which we let $L_a = b = 0$ and $G_{af}I_f = k_m$ in Eqs. (6.21a,b), so that these reduce to

$$v_a = R_a i_a + k_m \omega \tag{6.24a}$$

and

$$J\dot{\omega} = k_m i_a \tag{6.24b}$$

Combining Eqs. (6.24a,b) we have

$$v_a = R_a i_a + \frac{k_m^2}{J}\int_0^t i_a \, dt \tag{6.24c}$$

which is analogous to

$$v = Ri + \frac{1}{C}\int_0^t i \, dt \tag{6.24d}$$

and can be represented by the circuit shown in Fig. 6-19, where $C = J/k_m^2$. Applying a constant voltage V_a at the circuit terminals, we wish to study the energy balance in the network. From Eq. (6.24c) we have

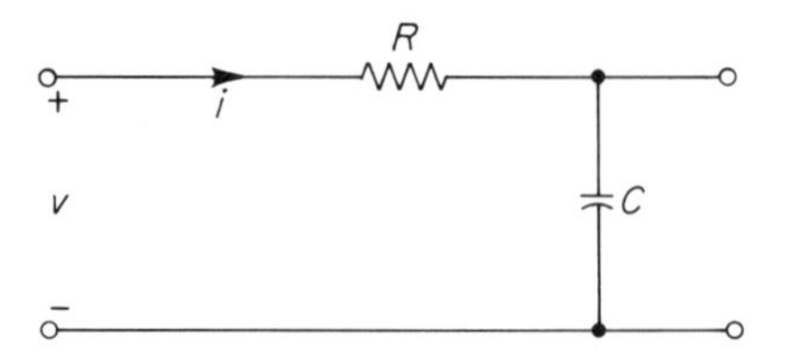

Fig. 6-19 Equivalent circuit of a dc motor.

$$\frac{di_a}{dt} = -\frac{k_m^2}{JR_a} i_a$$

Or,

$$i_a = I_{a0} \exp\left(-\frac{k_m^2}{JR_a} t\right) \tag{6.25}$$

where $I_{a0} = V_a/R_a$, the current at $t = 0$. The energy lost, W_e, in the resistance R_a is, therefore,

$$W_e = \int_0^\infty i_a^2 R_a \, dt = R_a I_{a0}^2 \int_0^\infty \exp\left(-\frac{k_m^2}{JR_a} 2t\right) dt$$

Or,

$$W_e = \frac{J}{2}\left(\frac{I_{a0}R_a}{k_m}\right)^2 \tag{6.26}$$

Now, from Eq. (6.24b) we have

$$\omega = \frac{k_m}{J} \int_0^t i_a \, dt$$

from which the final speed, ω_f, is given by

$$\omega_f = \frac{k_m}{J} \int_0^\infty i_a \, dt \tag{6.27}$$

Substituting Eq. (6.25) in Eq. (6.27),

$$\omega_f = \frac{k_m}{J} I_{a0} \int_0^\infty \exp\left(-\frac{k_m^2}{JR_a} t\right) dt = I_{a0} \frac{R_a}{k_m^2}$$

Therefore, the energy stored in the rotating parts, W_s, is

$$W_s = \frac{1}{2} J\omega_f^2 = \frac{J}{2}\left(\frac{I_{a0}R_a}{k_m}\right)^2 \tag{6.28}$$

Notice, from Eqs. (6.26) and (6.28), that the energy lost in the resistance is equal to the energy stored in rotating parts. Or, for the RC circuit, the total energy supplied at the terminals divides equally between the resistor and the capacitor.

6.3.2 The Series Motor

A series motor is shown in Fig. 6-20. Because the armature and field are

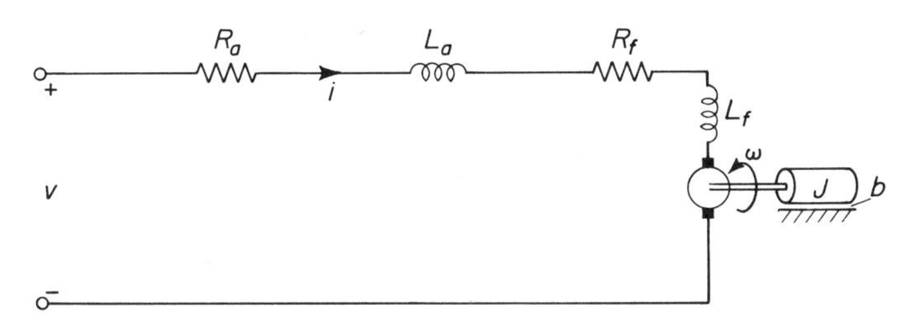

Fig. 6-20 A dc series motor.

connected in series, the total circuit inductance is $L = L_f + L_a$ and the resistance is $R = R_f + R_a$. The equations of motion are, from Eqs. (6.14) and (6.15)—with the constraint $i_f = i_a = i$:

$$L\frac{di}{dt} + Ri + G_{af}\dot{\theta}i = v \tag{6.29a}$$

and

$$T_e = G_{af}i^2 = J\dot{\omega} + b\omega \tag{6.29b}$$

Notice that both equations are nonlinear. For steady-state operation, however, it can be shown (see Problem 6-2) that

$$I = \frac{V}{(\omega G_{af} + R)} \tag{6.30a}$$

and

$$T_e = \frac{G_{af}V^2}{(\omega G_{af} + R)^2} \tag{6.30b}$$

Using Eqs. (6.16) we can linearize Eqs. (6.29a,b), and the linearized dynamical equations become (neglecting friction, $b = 0$)

$$L\frac{di_1}{dt} + (R + G_{af}\Omega_0)i_1 + G_{af}I_0\omega_1 = v_1 \tag{6.31a}$$

and

$$2G_{af}I_0i_1 = J\dot{\omega}_1 \tag{6.31b}$$

with the steady-state operating point defined by

$$(R+G_{af}\Omega_0)I_0 = V_0 \tag{6.32a}$$

and

$$G_{af}I_0^2 = b\Omega_0 = 0 \tag{6.32b}$$

Equations (6.31a,b) can be represented by the circuit shown in Fig. 6-21,

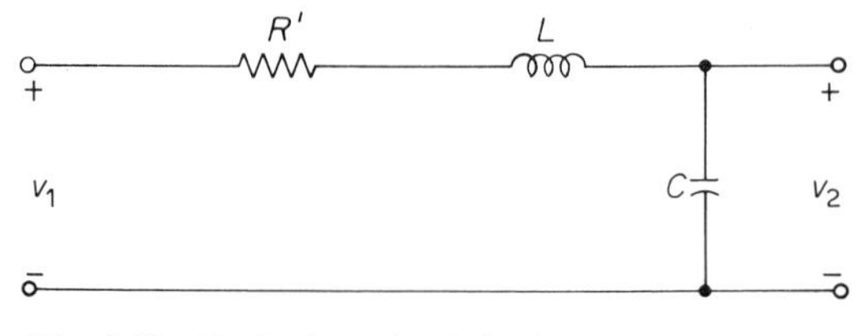

Fig. 6-21 Equivalent circuit for incremental motion.

in which $v_2 = G_{af}I_0\omega_1$, $R' = (R+G_{af}\Omega_0)$ and $C = J/2(G_{af}I_0)^2$. Notice again that this circuit is similar to that shown in Fig. 6-19.

6.3.3 The Separately Excited Generator

In the last two sections we considered the operation of dc machines as motors. The general equations for the single-axis machine, Eqs. (6.14) and (6.15), are equally applicable to the generator mode of operation also. For the sake of illustration, consider the separately excited generator shown in Fig. 6-7(a). The generator is driven by a prime mover at a constant speed Ω_0. The equations of motion are then

$$v_f = R_f i_f + L_f p i_f \tag{6.33a}$$

and

$$G_{af}\Omega_0 i_f = R_a i_a + L_a p i_a + Z_L(p) i_a \tag{6.33b}$$

where $Z_L(p)$ denotes the load impedance connected across the armature terminals. These equations can be represented by the block diagram shown in Fig. 6-22. The transfer function can be derived either from the block diagram or from Eqs. (6.33a,b).

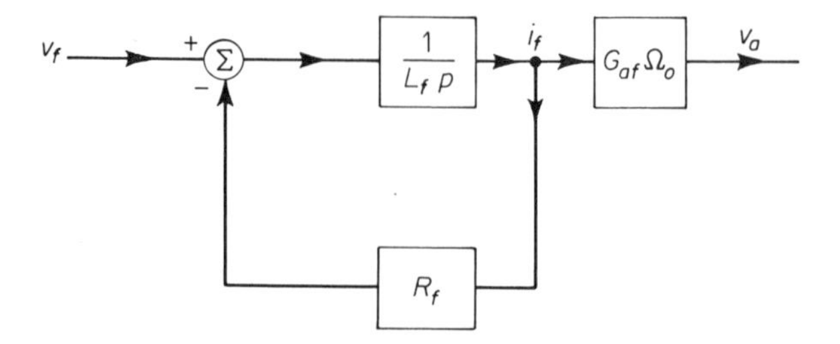

Fig. 6-22 Block diagram of a separately excited dc generator on no-load.

6.3.4 A 2-Axis Machine—The Amplidyne

So far we have considered the various types of single-axis dc machines, although in Sec. 6.1 we began with a general 2-axis machine. We shall now reconsider the 2-axis machine and discuss a practical form of such a machine.

In Sec. 6.2.3 we discussed the effect of armature reaction on the airgap fields in a dc machine. Recall that the brushes are located at the MNP and that, if these brushes are short-circuited [Fig. 6-23(a)], a cross-field results

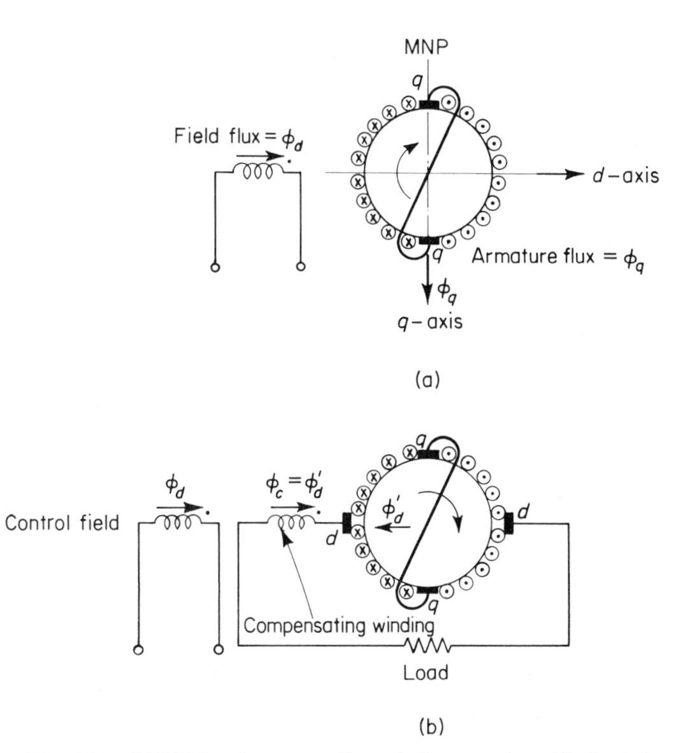

Fig. 6-23 (a) Field and armature fluxes (a dc generator with shorted brushes); (b) an amplidyne.

from the armature MMF. This machine can be considered a 2-axis machine, the main field being along the d axis and the cross-field (due to the armature) along the q axis, as shown in Fig. 6-23(a). If we locate another set of brushes along the d axis, connect it to a load, and run the machine as a generator, we observe the following facts: the main field produces a flux ϕ_d along the d axis; ϕ_d induces a voltage at the brushes qq and, these brushes being shorted, an armature current flows; the armature MMF produces a cross-flux ϕ_q

along the q axis; ϕ_q induces a voltage at the brushes dd, and a current flows through the load. However, as soon as the load current flows through the armature conductors, it also produces a flux ϕ_d' along the d axis. Notice that ϕ_d' opposes ϕ_d and weakens the resultant field considerably. To compensate for the effect of ϕ_d', another field winding is connected in series with the load [Fig. 6-23(b)]. This field winding is known as the compensating field and produces a flux ϕ_c which neutralizes ϕ_d' and thereby restores the original flux. A cross-field machine with the compensating winding is known as an *amplidyne*; without the compensating winding it is sometimes called the *metadyne*. Although we have shown only one winding on the main field, in practice the amplidyne is a multifield machine and carries a number of windings on the same field structure.

An amplidyne—and for that matter a conventional dc generator—is considered a rotating power amplifier. The amplidyne can be considered two generators in cascade, as shown in Fig. 6-24, where the brushes qq and dd

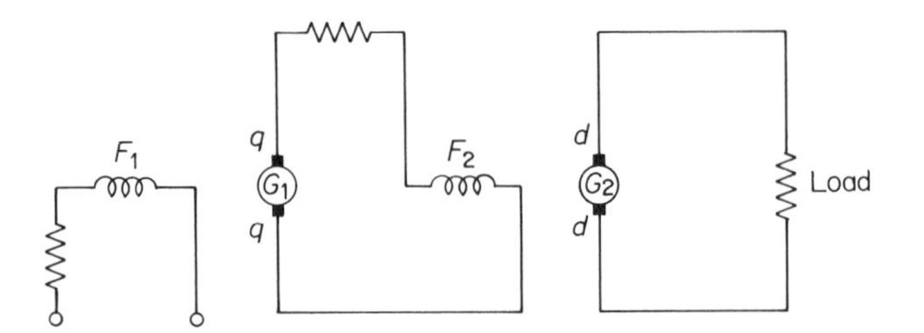

Fig. 6-24 An amplidyne as two generators in cascade.

identify the generators. If we define the power-amplification factor as the ratio of power output at the load to power input at the field (commonly known as the *control field*), we can immediately verify that the overall amplification factor of the amplidyne is the product of the individual amplification factors of the generators Q and D.

An equivalent circuit of the amplidyne is shown in Fig. 6-25, from which its voltage-balance equations can be written as

$$\begin{bmatrix} v_f \\ v_q \\ v_d \end{bmatrix} = \begin{bmatrix} R_f+pL_f & 0 & 0 \\ -\omega G_{qf} & R_q+pL_q & 0 \\ 0 & \omega G_{dq} & -(R_d+pL_d) \end{bmatrix} \begin{bmatrix} i_f \\ i_q \\ i_d \end{bmatrix} \tag{6.34}$$

It should be emphasized that Eq. (6.34) is for a compensated amplidyne. This is reflected by the fact that there is no coupling between the d-axis voltage and the main field.

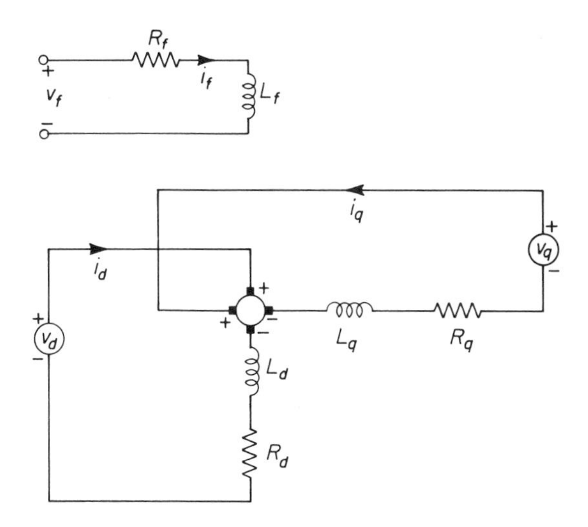

Fig. 6-25 Equivalent circuit of an amplidyne.

We now consider two examples with typical numerical values to determine the power-amplification factor for an uncompensated amplidyne (metadyne) and a (fully) compensated amplidyne.

EXAMPLE 6–3

An equivalent circuit of a metadyne is shown in Fig. 6E-3. It supplies a load of 1250 W at 250 V under steady-state conditions. The resistances of various circuits are as labeled in Fig. 6E-3. The constants $\omega G_{fq} = 300$ and $\omega G_{fd} = 200$. The power gain is to be calculated.

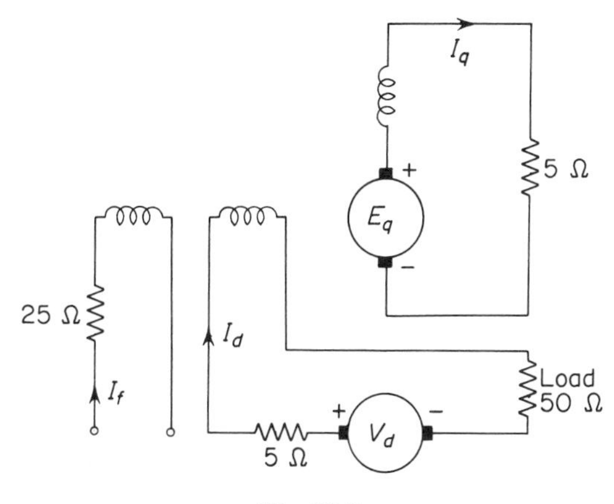

Fig. 6E-3

$$\text{power gain factor} = \frac{P_o}{P_i} = \frac{\text{output power to the load}}{\text{input power to the field}}$$

The output power $P_o = 1250$ W (given). To find the input power we have to calculate the field current I_f, since $P_i = I_f^2 R_f = I_f^2 \times 25$. From the circuit we have

$$V_d = V_L + R_d I_d = 250 + 5 \times 5 = 275 \quad \text{V}$$

$$I_q = \frac{V_d}{\omega G_{dq}} = \frac{275}{200} = 1.375 \quad \text{A}$$

$$E_q = R_q I_q = 1.375 \times 5 = 6.875 \quad \text{V}$$

But

$$E_q = \omega G_{fq} I_f - \omega G_{fd} I_d = 6.875 \quad \text{V}$$

Or,

$$300 I_f - 1000 = 6.875 \quad \text{V}$$

Or,

$$I_f = 3.36 \quad \text{A}$$

So that

$$P_i = (3.36)^2 \times 25 = 280 \quad \text{W}$$

and,

$$\text{power gain} = \frac{P_o}{P_i} = \frac{1250}{280} = 4.45$$

Notice that the gain is quite low as compared with that for a compensated amplidyne (see next example).

EXAMPLE 6–4

In the preceding example we observe that the uncompensated state is shown by the presence of the differential field (Fig. 6E-3), which is numerically expressed as $\omega G_{df} = 200$. If the metadyne is fully compensated and operated as an amplidyne, then $\omega G_{df} = 0$. In this case we have

$$300 I_f = 6.875$$

Or,

$$I_f = 0.023 \quad \text{A}$$

$$P_i = (0.023)^2 \times 25 = 0.0132 \quad \text{W}$$

and

$$\text{power gain} = \frac{1250}{0.0132} = 94{,}500$$

Although the difference in gains calculated in the last two examples might appear to be exaggerated, the effect of compensation is clearly illustrated.

Turning now to the linearized analysis of an amplidyne, we can show, from Eq. (6.34), that the transfer function relating to the no-load output voltage v_o to the field voltage v_f is given by

$$\frac{V_o(s)}{V_f(s)} = \frac{\omega^2 G_{dq} G_{qf}}{(R_f + sL_f)(R_q + sL_q)} \tag{6.35}$$

We shall have occasion to refer to this expression in a later example which illustrates the application of the amplidyne as an element in a closed-loop system.

6.4 Starting and Controlling the Speed of dc Motors

As compared to other motors, the speed of dc motors can be very flexibly controlled. In order to study the steady-state adjustable characteristics of dc motors we shall refer to the equations of motion derived in Secs. 6.3.1 and 6.3.2. For instance, from Eq. (6.18b) the steady-state speed ω_m of the dc shunt motor is given by

$$\omega_m = \frac{V_a - I_a R_a}{G_{af} I_f} \tag{6.36}$$

This equation is known as the *speed equation* of dc motors, and suggests three possible methods of speed control: (1) by varying the armature-terminal voltage V_a; (2) by changing the armature-circuit resistance R_a; and (3) by changing the field current I_f. The so-called *Ward Leonard system*, illustrated by the following example, provides a very wide range of speed control for a dc motor.

EXAMPLE 6–5

A scheme for the speed control of a dc motor M is shown in Fig. 6E-5(a). We make it an automatic speed-control system by feeding back the speed deviations to the second control field of the amplidyne. The dynamic characteristics of this system are to be studied.

In order to obtain the performance of the overall system, we first determine the small-signal transfer characteristic, about an operating point, for each component. This is done in the following order.

The Amplidyne [Fig. 6E-5(b)]

The transfer function of the amplidyne was derived in the preceding section. Because in the example under consideration there are two active fields on the amplidyne, the transfer function, as given by Eq. (6.35), is modified to

$$V_o(s) = \frac{k_A}{(R_f + sL_f)(R_q + sL_q)} [V_{f1}(s) - V_{f2}(s)] \tag{a}$$

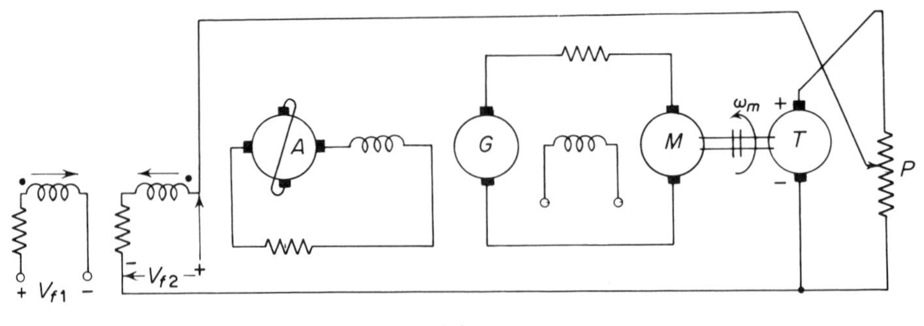

(a)

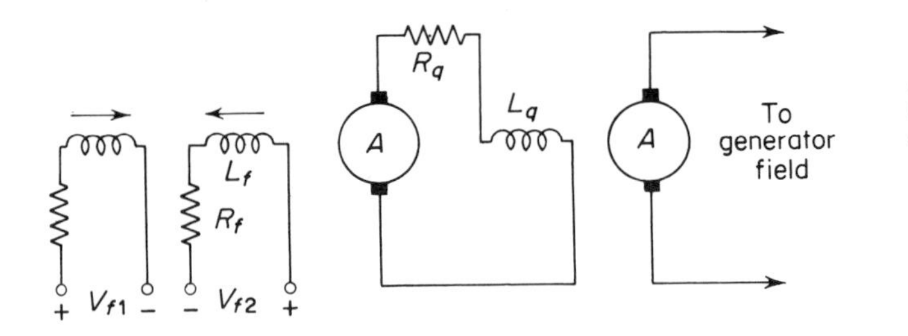

A = Amplidyne
G = Generator
M = Motor
T = Tachometer
P = Potentiometer

(b)

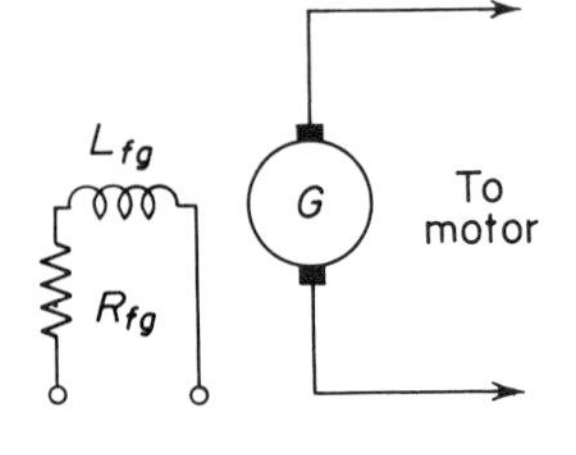

(c)

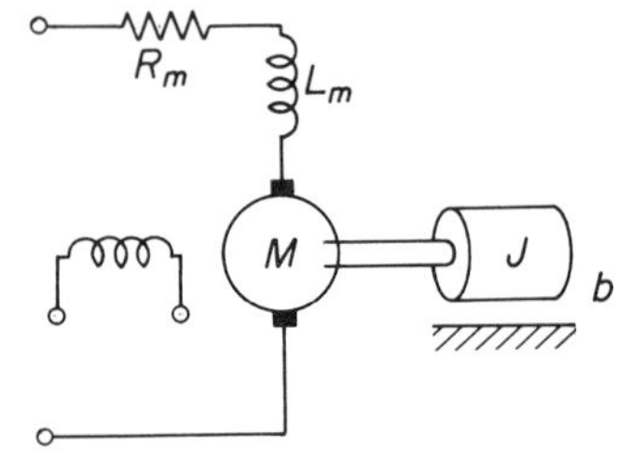

(d)

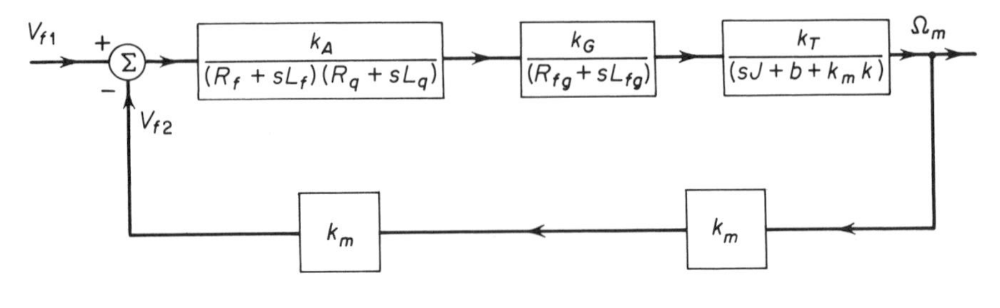

(e)

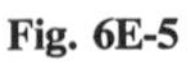
Fig. 6E-5

where k_A = a constant and we assume the same values of R_f and L_f for the two fields of the amplidyne.

The dc Generator [Fig. 6E-5(c)]

From Fig. 6E-5(c), we obtain the transfer function of the generator, and thus

$$V_g(s) = \frac{k_G}{R_{fg}+sL_{fg}} V_{fg}(s) \tag{b}$$

where R_{fg} and L_{fg} are the generator field resistance and inductance respectively, and k_G = a constant.

The dc Motor [Fig. 6E-5(d)]

The motor carries a constant field current I_{FM}, and is separately excited. The motor-armature inductance L_m and resistance R_m also include the generator-armature inductance and resistance, respectively. From Fig. 6E-5(d), therefore,

$$\Omega_m(s) = \frac{k'_m}{(R_m+sL_m)(b+sJ)} [V_g(s) - V_m(s)] \tag{c}$$

But $V_m(s) = k\Omega_m(s)$ and the armature-circuit time constant L_m/R_m is assumed negligible, so that Eq. (c) simplifies to

$$\Omega_m(s) = \frac{k_m}{sJ+b+k_m k} V_g(s) \tag{d}$$

where $k_m = k'_m/R_m$.

The Tacho-generator and the Potentiometer [Fig. 6E-5(e)]

The tacho-generator and the potentiometer transform as

$$V_T(s) = k_T\Omega_m(s) \tag{e}$$

and

$$V_{f2}(s) = kPV_T(s) \tag{f}$$

The overall open-loop transfer function of the system is, from Eqs. (a–f),

$$G(s)H(s) = \frac{k_A k_G k_T k P k_m}{(R_f+sL_f)(R_q+sL_q)(R_{fg}+sL_{fg})(sJ+b+k_m k)} \tag{g}$$

The block diagram for the system is shown in Fig. 6E-5(e). Knowing the transfer function, we can obtain the response of the system for specified inputs. We might recall another application of the amplidyne was mentioned in Chapter 1 (Fig. 1-3).

Of course, other than by utilizing the closed-loop system described above, the speed of the dc motor is commonly varied by varying the field-circuit resistance. By this method the speed of motor cannot be decreased below a certain value, determined by the inherent field-winding resistance. The speed can only be decreased by decreasing the armature input voltage or by inserting external resistance in the armature circuit. The latter method is inefficient.

6.4.1 Speed Control by Thyristors

We mentioned that the speed of a motor can be controlled by varying the armature input voltage. One of the methods of achieving this objective is by means of thyristors (or silicon-controlled rectifiers, SCR's), as shown in Fig. 6-26. The gate firing signal makes the thyristor conduct until the

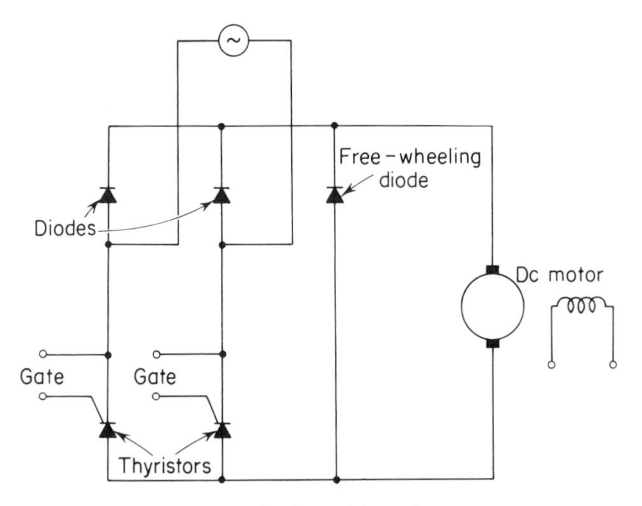

Fig. 6-26 Thyristor-driven dc motor.

forward voltage is removed. The thyristor then shuts off. By adjusting the timing of the firing signal, the armature input voltage can be varied and thereby the speed of the motor controlled. The waveforms illustrated in Fig. 6-27 indicate the possible variations in armature voltage. This method is not very suitable for large motors.

6.4.2 Starting

In addition to certain operational conveniences, the basic requirements for satisfactory starting of a dc motor are (1) sufficient starting torque, and (2) armature current, within safe limits, for successful commutation and

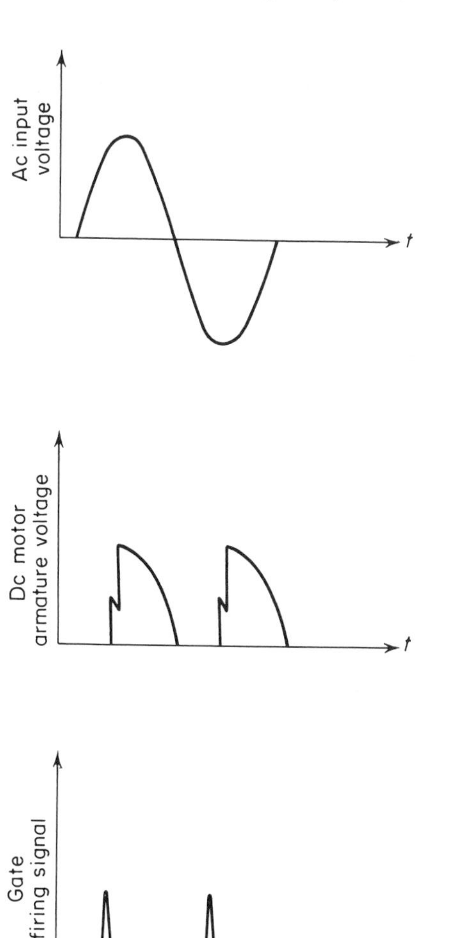

Fig. 6-27 Input-output waveforms for circuit shown in Fig. 6-26.

for preventing the armature from overheating. The second requirement becomes clear from Eq. (6.36), which shows that the armature current is given by $I_a = V_a/R_a$, when $\omega_m = 0$ (that is, when the motor is at rest) and the armature inductance is negligible. A typical 50hp, 230V motor might have an armature resistance of 0.05 Ω and if connected across 230V shall draw 4600A current. This current is evidently too large for this motor, which might be rated to take 180A full-load current. Commonly, double the full-load current is allowed to flow through the armature at the time of starting. For the motor under consideration, therefore, an external resistance $R_{\text{ext}} = 230/(2 \times 180) - 0.05 = 0.59\ \Omega$ must be inserted in the series with the armature to limit the starting current to double the rated value.

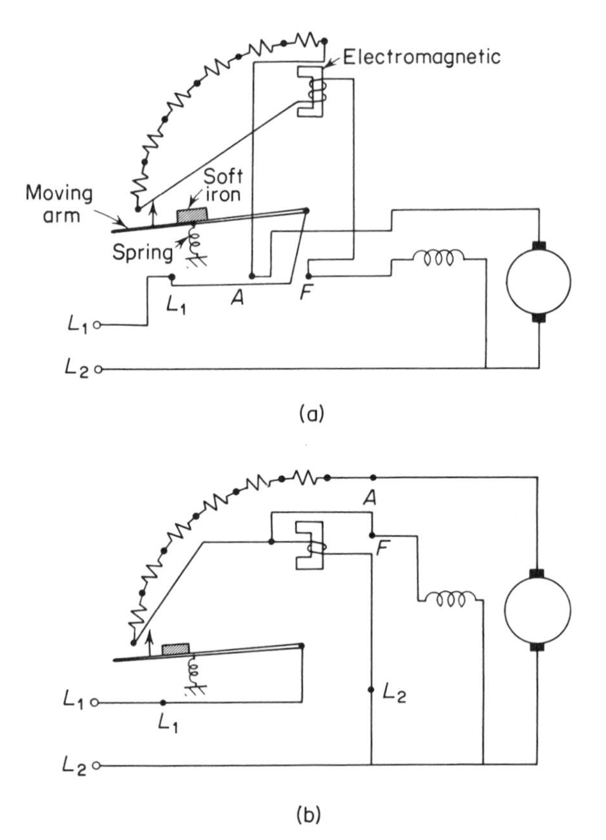

Fig. 6-28 (a) A three-point starter; (b) a four-point starter.

In practice, the necessary starting resistance is provided by means of a starter. Typical 3-point and 4-point starters are shown in Figs. 6-28(a) and (b). Notice that at the time of starting, the resistance R_{ext} comes in series with the armature. As the motor speeds up this resistance is cut out in steps. When the entire resistance is cut out, the starter arm is held by the electromagnet M. When the supply is turned off the starter arm is pulled back to the "off" position by the spring. In addition to the manual starter just described, there also exist pushbutton (or automatic) starters. (The interested reader may consult Reference 4 for an introduction to this topic.)

6.5 Losses and Efficiency

Most of the time, machines operate under steady-state conditions. Therefore, it is important that we understand the steady-state behavior as well as the dynamic behavior. Fortunately, the steady-state behavior can be easily obtained from the dynamical equations by setting the d/dt terms equal to zero. In obtaining the equations of motion, however, certain simplifying assumptions were made. One such assumption was that we generally neglect the effects of saturation. In determining the steady-state characteristics, the effects of saturation, armature reaction, etc. can be conveniently taken into account by graphical and semiempirical methods. Example 6-2 and Sec. 6.2.5 illustrate this point.

Besides the volt-ampere or speed–torque characteristics, the performance of a dc machine is also measured by its efficiency. The usual definition of *efficiency* is

$$\text{efficiency} = \frac{\text{power output}}{\text{power input}} = \frac{\text{power output}}{\text{power output} + \text{losses}}$$

Efficiency can, therefore, be determined either from load tests or by determination of losses. The various losses are classified as follows:

1. *Electrical*: (a) Copper losses in various windings such as the armature winding and different field windings. (b) Loss due to the contact resistance of the brush (with the commutator).
2. *Magnetic*: These are the iron losses and include the hysteresis and eddy-current losses in the various magnetic circuits, primarily the armature core and pole faces.
3. *Mechanical*: These include the bearing friction, windage and brush-friction losses.
4. *Stray-load*: These are other load losses not covered above. They are taken as 1 per cent of the output (as a rule of thumb).

Knowing the various losses, the efficiency of the machine can easily be calculated.

EXAMPLE 6–6

A 10 hp, 230 V shunt motor takes a full-load line current of 40 A. The armature and field resistances are 0.25 Ω and 230 Ω respectively. The total brush-contact drop is 2 V and the core and friction losses are 380 W. Calculate the efficiency of the motor if it delivers rated power.

$$\text{power output} = 10 \times 746 = 7460 \quad \text{W}$$

$$I_f^2 R_f \text{ loss} = \left(\frac{230}{230}\right)^2 \times 230 = 230 \quad \text{W}$$

$$I_a^2 R_a \text{ loss} = (40-1)^2 \times 0.25 = 380 \quad \text{W}$$

$$\text{core loss and friction loss} = 380 \quad \text{W (given)}$$

$$\text{brush-contact loss} = 2 \times 39 = 78 \quad \text{W}$$

$$\text{stray-load loss} = 7460 \times \tfrac{1}{100} = 75 \quad \text{W (assumed)}$$

$$\text{total losses} = 1143 \quad \text{W}$$
$$\text{power input} = 8603 \quad \text{W}$$
$$\text{efficiency} = \frac{7460}{8603} \times 100 = 86\%$$

6.6 Summary

In this chapter we discussed the steady-state and dynamic behavior of dc machines. The general equations of motion were obtained for a 2-axis machine using the *dq* transformation. The commutator action was discussed in terms of frequency conversion. The physical aspects of dc machines were then considered, and the effects of armature reaction, reactance voltage, and saturation described. A number of conventional dc generators and motors and the amplidyne were studied. Finally, the starting and speed controls of dc motors were considered and the various losses and efficiency calculations outlined.

PROBLEMS

6-1. Plot the speed–torque curve for the shunt machine of Example 6-1 and identify its range of mode of operation as a motor and as a generator.

6-2. Using the constraint $I_f = I_a = I$ for a series machine [Fig. 6-7(c)] and using the equations of Example 6-1 show that

$$I = \frac{V}{\omega_m G_{af} + R_a + R_f} \quad \text{and} \quad T_e = \frac{G_{af} V^2}{(\omega_m G_{af} + R_a + R_f)^2}$$

6-3. A separately excited dc motor is initially at rest, and its field winding carries a steady current I_f. It is started by inserting an external resistance in series with the armature and then connecting it across a voltage V_a. As the motor picks up speed the external resistance is gradually cut off. If the armature-circuit resistance is R and its inductance is negligible, show that the energy dissipated in this resistor is a constant and is independent of the variations of the armature-circuit resistance during the period of starting. Assume that the load on the motor is pure inertia J and that it takes a time τ to come to the final speed ω_f.

6-4. Obtain an equivalent circuit of the motor described by Eqs. (6.21a,b).

6-5. A separately excited dc generator, running at constant speed, supplies a load having a resistance of 0.9 Ω in series with 1 H inductance. The armature resistance is 0.1 Ω and its inductance is negligible. The field, having a resistance of 50 Ω and an inductance of 5 H, is suddenly connected to a 100 V source.

Calculate the build-up of the armature current. The armature-induced voltage/field current is 40 V/A for the speed at which the generator is running.

6–6. A separately excited motor carries a load given by $T_L = 2\dot{\omega}_m + \omega_m$. Its field current is constant. For a 100 V armature input, determine the transfer function, taking the speed ω_m as the output. The armature resistance is 1 Ω and its inductance is negligible, and the electromechanical-conversion constant is 10 N·m/A armature current.

6–7. For the motor of Problem 6-6, (a) develop its block diagram; (b) draw the signal-flow graph; and (c) plot its frequency response.

6–8. For the voltages specified in Sec. 6.1.3, using Eqs. (6.8a,b), determine the steady-state torque and the rotor currents I_d^r and I_q, if $i_a^s = i_b^s = 0$ and (a) $\omega = \omega_m$; (b) $\omega < \omega_m$; and (c) $\omega > \omega_m$.

6–9. Using appropriate symbols and constants, find the open-loop transfer function of the system shown in Fig. 1-3 by the method of Example 6-5.

6–10. The motor of Problem 6-6 is running at a constant speed of n r/min. Find, approximately, the effect of the following changes on the speed of the motor:

a. The armature resistance is doubled, armature-terminal voltage and field current remain constant.
b. The armature-terminal voltage is halved, armature resistance (of 1 Ω) and field current remain unchanged.
c. The field current is halved, armature-terminal voltage (of 100 V) and the armature resistance (of 1Ω) remain constant.

6–11. How does the electromagnetic torque of the motor change in parts (a), (b), and (c) in Prob. 6-10?

6–12. Using Eqs. (5.4) and (6.12b), derive the torque equation of a dc motor.

6–13. If the motor of Example 6-6 runs at 1100 r/min at full-load, calculate its efficiency and speed at half-load. The brush-contact drop and the core and friction losses remain unchanged.

6–14. It is claimed that "approximately, the metadyne can be used as a constant-current generator over a certain operating range." Prove or disprove this statement.

6–15. A large, separately excited dc motor having a total inertia of J is brought to rest by dynamic braking (by disconnecting the armature from the supply and then suddenly connecting it across a resistance R). If the steady speed of the motor was Ω_o, derive an expression for the speed $\omega_m(t)$. Assume appropriate symbols, such as those used in Sec. 6.3.1, and constants not specified above.

6–16. A scheme for the speed control of small dc motors (of say 5 hp) rating is proposed in Sec. 6.4.1. Derive the dynamical equations of motion for the system.

6–17. Adding an excitation winding on the field poles of a self-excited shunt generator converts it into a control-type tuned generator known as the *rototrol*. The generator is shown in Fig. 6P-17(a); the saturation characteristic

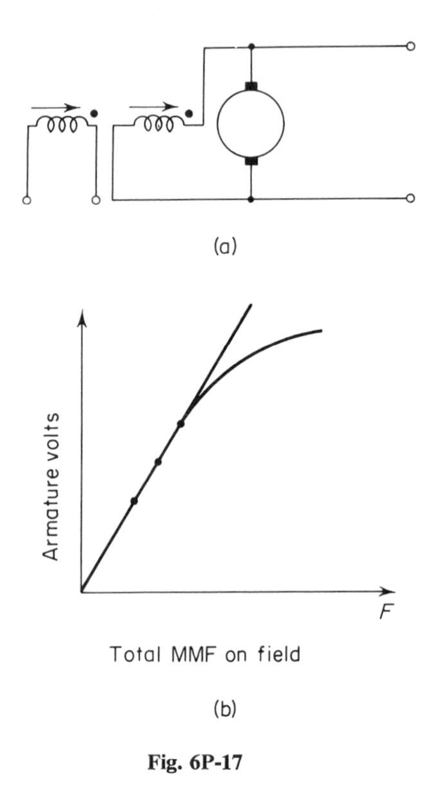

Fig. 6P-17

with "tuned" field resistance, shown in Fig. 6P-17(b), is a tangent to the saturation curve. A small power change in the control winding brings about a large change in the power output. Determine the voltage- and power-amplification factors in terms of the constants shown in Figs. 6P-17(a) and (b) and derive the dynamical equations of motion for small disturbances about an operating point.

REFERENCES

1. White, D. C., and H. H. Woodson, *Electromechanical Energy Conversion*. New York: John Wiley & Sons, Inc., 1959.

2. Park, R. H., "Two-Reaction Theory of Synchronous Machines, Part I," *AIEE Trans.*, Vol. 48, 1929, pp. 716–727.

3. Park, R. H., "Two-Reaction Theory of Synchronous Machines, Part II," *AIEE Trans.*, Vol. 52, 1933, pp. 352–355.

4. Kloeffler, R. C., R. M. Kerchner, and J. L. Brenneman, *Direct-Current Machinery*. New York: The Macmillan Company, 1950.

5. Fitzgerald, A. E., and C. Kingsley, *Electric Machinery*, 2nd ed. New York: McGraw-Hill Book Company, 1961.

6. von der Embse, Urban A., "A New Theory of Nonlinear Commutating Machines," *IEEE Trans*, Vol. PAS-87 ("Power Apparatus and Systems"), pp. 1804–1809.

7

Induction Machines

In Chapters 4 and 5 we introduced the principle of action of induction machines. In Chapter 4, we discussed in some detail application of the field equations and of the Lorentz force equation in determining the electromagnetic forces developed (by induction action) in continuous media (of finite conductivity). In this connection, we also discussed the production of the traveling field, the traveling-field theory, and the concept of slip (Secs. 4.3 and 4.4). In Chapter 5 we outlined some constructional features of the conventional induction machine (Sec. 5.1.2). We discussed the frequency constraints for energy conversion in Sec. 5.6.1. In Example 5-6 we obtained the equations of motion of a 2-phase cylindrical-rotor machine with mutual coupling between the stator and rotor, both having voltages applied at the terminals. The constraints for the operation of the machine as a 2-phase induction motor are given in Table 5-1. Thus, in principle, the dynamic characteristics of the induction machine follow directly from the equations of Example 5-6 together with the appropriate constraints given in Table 5.1. On the other hand, the *dq* transformation of Example 5-5 can also be used to analyze the induction machine. In the following discussions we shall use the two transformations (of Examples 5-5 and 5-6) to obtain the dynamical characteristics of the induction machine. We shall analyze the 2-phase induction machine and subsequently demonstrate the equivalence of the 2- and 3-phase machines. We shall thereby show that the theory developed for the 2-phase machine is also applicable to the 3-phase machine (which is more commonly encountered than the 2-phase machine).

7.1 The Equations of Motion of a 2-Phase Machine[1, 2]

Let us consider the uniform-airgap machine of Chapter 5 (Example 5-6). The machine has two windings on the stator and two on the rotor. The equations of motion in terms of the coil currents, voltages, and other constants are, from Eqs. (5.28a) and (5.28b),

$$\begin{bmatrix} v^s \\ \hline v^r \end{bmatrix} = \begin{bmatrix} R^s + pL^{ss} & pL^{sr} \\ \hline pL^{rs} & R^r + pL^{rr} \end{bmatrix} \begin{bmatrix} i^s \\ \hline i^r \end{bmatrix} \tag{7.1a}$$

$$T_e = -\frac{1}{2}[i^s \quad i^r]\frac{\partial}{\partial\theta}\begin{bmatrix} L^{ss} & L^{sr} \\ \hline L^{rs} & L^{rr} \end{bmatrix} \begin{bmatrix} i^s \\ \hline i^r \end{bmatrix} \tag{7.1b}$$

where v^s, i^s, v^r, and i^r are 2×1 column submatrices denoting stator and rotor voltages and currents; and R^s, L^{ss}, R^r, L^{rr}, L^{sr}, and L^{rs} are 2×2 submatrices of machine constants. From Table 5-1, the constraints for the operation of the machine as a 2-phase induction machine are as follows: $v_a^r = v_b^r = 0$; $\dot{\theta} = \omega_m$ (for constant-speed operation); $\omega^r = \omega^s - \nu\omega_m$ (for possible energy conversion, as mentioned in Sec. 5.6.1); and, if we assume symmetrical windings such that $R_a^s = R_b^s = R^s$, $R_a^r = R_b^r = R^r$, the resistance and inductance submatrices are:

$$R^s = \begin{bmatrix} R^s & 0 \\ 0 & R^s \end{bmatrix} \qquad R^r = \begin{bmatrix} R^r & 0 \\ 0 & R^r \end{bmatrix} \tag{7.2a}$$

$$L^{ss} = \begin{bmatrix} L^s & 0 \\ 0 & L^s \end{bmatrix} \qquad L^{rr} = \begin{bmatrix} L^r & 0 \\ 0 & L^r \end{bmatrix} \tag{7.2b}$$

and

$$L^{sr} = \begin{bmatrix} L^{sr}\cos\nu\theta & -L^{sr}\sin\nu\theta \\ L^{sr}\sin\nu\theta & L^{sr}\cos\nu\theta \end{bmatrix} \tag{7.2c}$$

with $L^{rs} = \tilde{L}^{sr}$, and $\theta = \omega_m t + \delta$ (since $\dot{\theta} = \omega_m$).

Next, we assume balanced 2-phase excitation on the stator such that

$$\left.\begin{aligned} v_a^s &= V^s \cos\omega^s t \\ v_b^s &= V^s \sin\omega^s t \end{aligned}\right\} \tag{7.2d}$$

and

$$\left.\begin{aligned} i_a^s &= I^s \cos(\omega^s t - \phi^s) \\ i_b^s &= I^s \sin(\omega^s t - \phi^s) \\ i_a^r &= I^r \cos(\omega^r t - \phi^r) \\ i_b^r &= I^r \sin(\omega^r t - \phi^r) \end{aligned}\right\} \tag{7.2e}$$

When Eqs. (7.2a–e) and the constraints mentioned above are substituted in Eq. (7.1a), we obtain

$$\left.\begin{aligned} V^s \cos \omega^s t + \omega^s L^{sr} I^r \sin(\omega^s t + \nu\delta - \phi^r) &= (R^s + L^s p) i_a^s \\ V^s \sin \omega^s t - \omega^s L^{sr} I^r \cos(\omega^s t + \nu\delta - \phi^r) &= (R^s + L^s p) i_b^s \\ \omega^r L^{sr} I^s \sin(\omega^r t - \nu\delta - \phi^s) &= (R^r + L^r p) i_a^r \\ -\omega^r L^{sr} I^s \cos(\omega^r t - \nu\delta - \phi^s) &= (R^r + L^r p) i_b^r \end{aligned}\right\} \tag{7.3}$$

where $p = d/dt$. Because the currents and voltages are sinusoidally varying quantities, we can express Eq. (7.3) in terms of complex exponentials by assuming

$$\mathbf{V}^s = V^s e^{j0}$$

$$\mathbf{I}^s = I^s e^{-j\phi^s}$$

and

$$\mathbf{I}^r = I^r e^{-j(\phi^r - \nu\delta)}$$

in which case Eqs. (7.3) reduce to

$$\mathbf{V}^s - j\omega^s L^{sr} \mathbf{I}^r = (R^s + j\omega^s L^s)\mathbf{I}^s \tag{7.4}$$

and

$$-j\omega^r L^{sr} \mathbf{I}^s = (R^r + j\omega^r L^r)\mathbf{I}^r \tag{7.5a}$$

Notice that Eq. (7.5a) is in terms of the rotor frequency ω^r. Recalling the definition of slip, $s = (\omega^s - \omega^r)/\omega^s$, we can rewrite Eq. (7.5a) as

$$-j\omega^s L^{sr} \mathbf{I}^s = \left(\frac{R^r}{s} + j\omega^s L^r\right)\mathbf{I}^r \tag{7.5b}$$

Equations (7.4) and (7.5b) can now be represented by the equivalent circuit shown in Fig. 7-1. Here the inductance $(L^s - L^{sr})$ is the *leakage inductance*

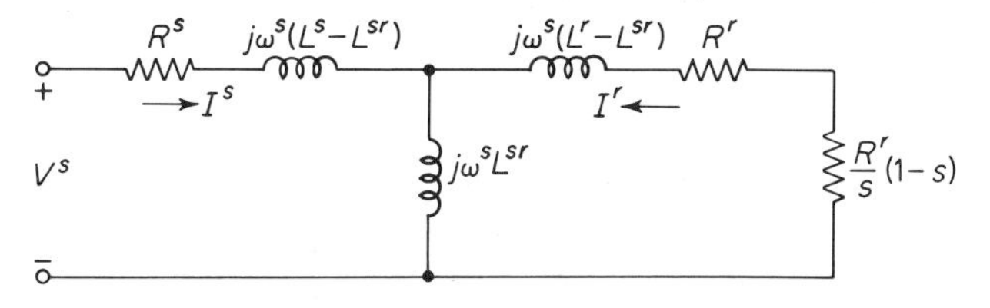

Fig. 7-1 Per-phase equivalent circuit of a two-phase balanced induction machine.

and L^{sr} is the *magnetizing inductance*. We notice that a variable resistance $R^r(1-s)/s$ is connected across the rotor terminals. This resistance is a measure of the power converted from electrical to mechanical forms.

7.1.1 Torque Characteristics

Turning our attention now to Eq. (7.1b) and substituting Eqs. (7.2b), (7.2c), and (7.2e), we have

$$T_e = \nu L^{sr} I^s I^r \sin(\phi^r - \phi^s - \nu\delta) \tag{7.6}$$

But $(1-s) = \nu\omega_m/\omega^s$, so that Eq. (7.6) can also be expressed as

$$T_e = \frac{1-s}{\omega_m} I^r[\omega^s L^{sr} I^s \sin(\phi^r - \phi^s - \nu\delta)] \tag{7.7}$$

This expression for torque can be considerably simplified by making use of the phasor diagram shown in Fig. 7-2, from which it follows that

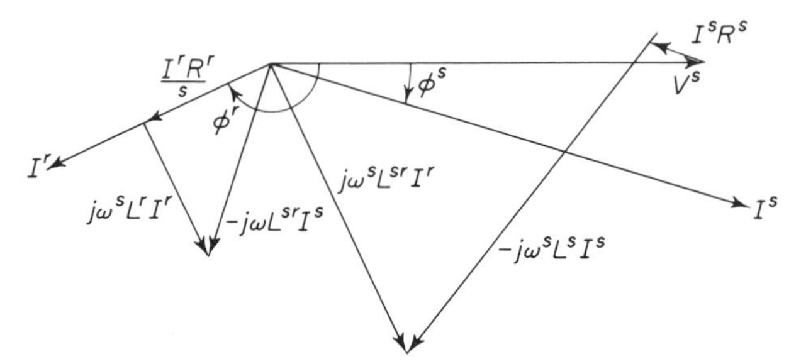

Fig. 7-2 Phasor diagram for the circuit shown in Fig. 7-1.

$$I^r \frac{R^r}{s} = \omega^s L^{sr} I^s \cos(\phi^r - \phi^s - \nu\delta - 90^\circ) = \omega^s L^{sr} I^s \sin(\phi^r - \phi^s - \nu\delta) \tag{7.8}$$

From Eqs. (7.7) and (7.8), therefore,

$$T_e = \frac{(I^r)^2}{\omega_m} \frac{(1-s)}{s} R^r \tag{7.9}$$

From Eq. (7.9) we see that the power converted is

$$T_e\omega_m = (I^r)^2 R^r \frac{(1-s)}{s} \tag{7.10}$$

The quantity on the right-hand side of Eq. (7.10) is the power absorbed in a fictitious resistance $R^r(1-s)/s$ through which the rotor current I^r flows. This further validates the statement made earlier that the resistance $R^r(1-s)/s$ is a measure of the power conversion.

A plot of Eq. (7.9) in Fig. 7-3 shows the three possible modes of operation of the machine. Depending upon the mechanical speed of the rotor, the machine can operate as a motor (for $0 < \nu\omega_m < \nu\omega^s$), as a generator (for $\nu\omega_m > \omega^s$), or as a brake (for $\nu\omega_m < 0$). Usually, the induction machine is operated as a motor; the speed range $0 < \nu\omega_m < \omega^s$ is therefore of greatest interest. We see, from Fig. 7-3, that the effect of increasing the rotor resistance is an increase in the starting torque. However, the maximum torque remains unchanged, although the speed (or slip) at which the maximum torque

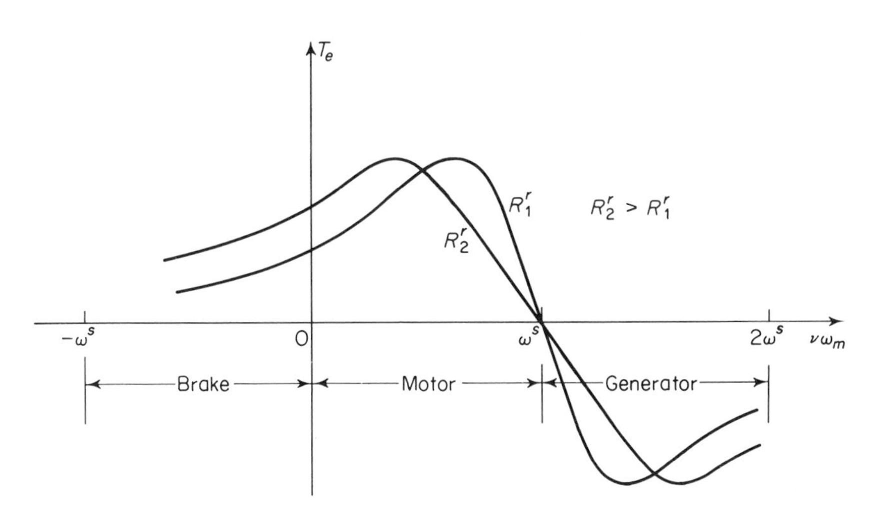

Fig. 7-3 Speed–torque curves for an induction machine.

occurs does depend on the rotor resistance. The maximum torque $(T_e)_m$ and the slip $s = s_m$ at which $(T_e)_m$ occurs are found from

$$\left.\frac{\partial T_e}{\partial s}\right|_{s=s_m} = 0 \tag{7.11}$$

where T_e is given by Eq. (7.9). But Eq. (7.9) cannot be used directly because I^r is also a function of s and is not an independent variable. The current I^r can be eliminated from Eq. (7.9) by using Eqs. (7.4) and (7.5b), from which

$$\mathbf{I}^r = \frac{-j\omega^s L^{sr} V^s}{\left[\frac{R^s R^r}{s} - (\omega^s)^2(L^s L^r - L^{sr} L^{sr})\right] + j\omega^s\left(\frac{L^s R^r}{s} + L^r R^s\right)} \tag{7.12a}$$

Or

$$|I^r| = \frac{\omega^s L^{sr} V^s}{\left\{\left[\frac{R^s R^r}{s} - (\omega^s)^2(L^s L^r - (L^{sr})^2)\right]^2 + (\omega^s)^2\left(\frac{L^s R^r}{s} + L^r R^s\right)^2\right\}^{1/2}} \tag{7.12b}$$

Equations (7.9) and (7.12b) yield

$$T_e = \frac{\frac{1}{2}\nu\omega^s(L^{sr} V^s)^2(R^r/s)}{\left\{\frac{R^s R^r}{s} - (\omega^s)^2[L^s L^r - (L^{sr})^2]\right\}^2 + (\omega^s)^2\left(\frac{L^s R^r}{s} + L^r R^s\right)^2} \tag{7.13}$$

From Eqs. (7.11) and (7.13) we have

$$s_m = \pm R^r\left[\frac{(R^s)^2 + (\omega^s L^s)^2}{(\omega^s L^r R^s)^2 + (\omega^s)^4[L^r L^s - (L^{sr})^2]^2}\right]^{1/2} \tag{7.14a}$$

and

$$(T_e)_m = \pm \frac{(\nu/2)(L^{sr}V^s)^2}{\{[(L^rR^s)^2+(\omega^s)^2(L^rL^s-(L^{sr})^2)][(R^s)^2+(\omega^sL^s)^2]+R^s\omega^s(L^{sr})^2\}^{1/2}} \tag{7.14b}$$

We observe from Eq. (7.14a) that the slip at which maximum torque occurs is directly proportional to the rotor resistance; but the value of the maximum torque, as obtained from Eqs. (7.9) and (7.14a), is independent of the rotor resistance. Evidently, the starting torque ranges from zero to $(T_e)_m$, depending upon the rotor resistance R^r. We thus conclude that the steady-state characteristics of the induction machine can be derived from the equations of motion and the various constraints. We now consider an example to illustrate the application of the preceding analysis in obtaining the performance characteristics of an induction motor.

EXAMPLE 7–1

A 2-phase, 60-cycle, 4-pole induction motor operates at 0.8 power factor at $\sqrt{2}\times 110$ V taking a line current of $\sqrt{2}\times 10$ A. The resistance $R^s = R^r = 0.1\ \Omega$/phase. If the motor runs at 1710 r/min, calculate (a) the developed torque, (b) the developed power, and (c) the stator and rotor copper losses.

The synchronous speed n_s is obtained from

$$n_s = \frac{120f}{p} = \frac{120\times 60}{4} = 1800 \quad \text{r/min}$$

The slip s is given by

$$s = \frac{n_s - n}{n_s} = \frac{1800-1710}{1800} = 0.05$$

and

$$\omega_m = 2\pi\times\frac{1710}{60} = 57\pi = 179 \text{ rad/s}$$

In order to calculate T_e, we refer to Eq. (7.7) and Fig. 7-2. From Fig. 7-2 we have

$$\begin{aligned} V^s\cos\phi^s - I^sR^s &= \omega^sL^{sr}I^r\cos(\phi^r-90^\circ-\phi^s-\nu\delta) \\ &= \omega^sL^{sr}I^r\sin(\phi^r-\phi^s-\nu\delta) \end{aligned}$$

which when substituted in Eq. (7.7) yields

$$T_e = \frac{1-s}{\omega_m}[V^sI^s\cos\phi^s-(I^s)^2R^s] \tag{7.15}$$

But

$$V^sI^s\cos\phi^s = \text{input power}$$

and

$$(I^s)^2R^s = \text{stator copper loss}$$

From Eq. (7.15), therefore,

$$\text{developed power} = (1-s)\quad(\text{input power} - \text{stator copper loss}) \tag{7.16}$$

Balance of power requires that

input power = developed power + stator copper loss + rotor copper loss (7.17)

From Eqs. (7.15–7.17) we have

$$\text{rotor copper loss} = s[V^s I^s \cos \phi^s - (I^s)^2 R^s] \tag{7.18}$$

The required results, on a per-phase basis, are as follows:

(a) From Eq. (7.15),

$$T_e = \frac{1-0.05}{179}[\sqrt{2}\times 110 \times \sqrt{2}\times 10 \times 0.8 - (\sqrt{2}\times 10)^2 \times 0.1] \quad \text{W·s}$$

$$= 9.24 \quad \text{N·m}$$

(b) Developed power = $9.24 \times 179 = 1655$ W.

(c) Stator copper loss = $(\sqrt{2}\times 10)^2 \times 0.1 = 20$ W; from Eq. (7.18), rotor copper loss = 0.05(1740) = 87 W. Neglecting *other* losses, the efficiency of conversion = 1655/(1655 + 20 + 87) = 94%.

7.1.2 The Equations of Motion in *dq* Variables

In the preceding discussions we considered the operation of the induction machine under steady-state conditions. For dynamic conditions, $p \neq j\omega$, although the equations of motion can be linearized, for constant-speed operation, by using the *dq* transformation (previously used in Example 5-5 and in the analysis of dc machines in Sec. 6.1.1). To make this point clear, we begin with the original equations of motion, Eqs. (7.1a) and (7.1b), with the constants defined by Eqs. (7.2a–c). Now, recalling Eq. (5.39),

$$S_{dq} = \begin{bmatrix} \cos \nu\theta & \sin \nu\theta \\ -\sin \nu\theta & \cos \nu\theta \end{bmatrix} \tag{7.19}$$

and operating rotor quantities with S_{dq} matrix such that

$$v_{ab}^r = S_{dq} v_{dq}^r \tag{7.20a}$$

and

$$i_{ab}^r = S_{dq} i_{dq}^r \tag{7.20b}$$

we have, from Eqs. (7.1a) and (7.19–7.20b),

$$\begin{bmatrix} v_{ab}^s \\ \hline v_{dq}^r \end{bmatrix} = \left[\begin{array}{c|c} R^s + pL^s & (pL^{sr})S_{dq} \\ \hline S_{dq}^{-1}(pL^{rs}) & R^r + S_{dq}^{-1}(pL^r)S_{dq} \end{array}\right] \begin{bmatrix} i_{ab}^s \\ \hline i_{dq}^r \end{bmatrix} \tag{7.21a}$$

In Eq. (7.21a) the subscript *ab*'s denote the original variables and the *dq*'s denote the transformed variables. From Eqs. (7.2a–c), (7.19), and (7.21a) we have, for a speed ω_m of the rotor,

$$\begin{bmatrix} v_a^s \\ v_b^s \\ v_d^r \\ v_q^r \end{bmatrix} = \begin{bmatrix} R^s+pL^s & 0 & pL^{sr} & 0 \\ 0 & R^s+pL^s & 0 & pL^{sr} \\ pL^{sr} & \nu\omega_m L^{sr} & R^r+pL^r & \nu\omega_m L^r \\ -\nu\omega_m L^{sr} & pL^{sr} & -\nu\omega_m L^r & R^r+pL^r \end{bmatrix} \begin{bmatrix} i_a^s \\ i_b^s \\ i_d^r \\ i_q^r \end{bmatrix} \tag{7.21b}$$

When Eqs. (7.2c), (7.19), and (7.20b) are substituted in the torque equation, Eq.(7.1b), it simply reduces to

$$T_e = -\nu L^{sr}(i_b^s i_d^r - i_a^s i_q^r) \tag{7.21c}$$

The above equations, Eqs. (7.21b,c), are linear differential equations with constant coefficients. The transient, or steady-state, performance of the induction machine can be derived from these equations by setting $v_d^r = v_q^r = 0$. Thus, for given stator voltages v_a^s and v_b^s, the instantaneous stator and rotor currents can be obtained from Eq. (7.21b). Knowing the transient currents, the transient torques are found from Eq. (7.21c). We shall discuss the transients in the induction machine in a later section. At present we shall show that the steady-state characteristics of the induction machine are also obtainable from Eqs. (7.21b,c).

From Eqs. (7.19) and (7.20b) we have

$$\begin{aligned} i_a^r &= i_d^r \cos \nu\theta + i_q^r \sin \nu\theta \\ &= \text{Re}\,[(i_d^r + j i_q^r)e^{-j\nu\theta}] \\ &= \sqrt{2}\,\text{Re}\,[\mathbf{I}^r e^{-j\nu\theta}] \end{aligned} \tag{7.22}$$

where

$$\mathbf{I}^r = \frac{1}{\sqrt{2}}(i_d^r + j i_q^r) \tag{7.23a}$$

Since, for steady-state and balanced conditions, the stator currents i_a^s and i_b^s are in quadrature, we may write

$$\mathbf{I}^s = \frac{1}{\sqrt{2}}(i_a^s + j i_b^s) \tag{7.23b}$$

Using Eqs. (7.23a,b), the expression for torque, Eq. (7.21c), can be expressed as

$$T_e = 2\nu L^{sr}\,\text{Im}\,[\mathbf{I}^r \mathbf{I}^{s*}] \tag{7.24}$$

where $\mathbf{I}^{s*}$ is the complex conjugate of $\mathbf{I}^s$. In Eqs. (7.22) and (7.24), Re and Im respectively denote the real and imaginary parts of the quantities in brackets. In phasor notation, using Eqs. (7.23a,b), Eq. (7.21b) can be written as

$$\mathbf{V}^s = (R^s + L^s p)\mathbf{I}^s + L^{sr} p \mathbf{I}^r \tag{7.25a}$$

$$0 = L^{sr}(p + j\nu\omega_m)\mathbf{I}^s + [R^r + L^r(p + j\nu\omega_m)]\mathbf{I}^r \tag{7.25b}$$

From Eqs. (7.24) and (7.25b) it can be shown that the final expression for the torque (see Problem 7-1) takes the form

$$T_e = \frac{(I^r)^2}{\omega_m} \frac{1-s}{s} R^r \tag{7.26}$$

Notice that Eqs. (7.9) and (7.26) are identical.

7.2 The Single-Phase Induction Motor

The single-phase induction motor differs in construction from the 2-phase motor in that the former carries only one exciting winding on the stator, although an auxiliary winding is also provided for starting. We shall show that theory developed for the 2-phase machine (Sec. 7.1.2) is also applicable to the single-phase machine. A single-phase machine is shown schematically in Fig. 7-4. Assuming $\nu = 1$, the equations of motion in dq variables are, from Eq. (7.21b),

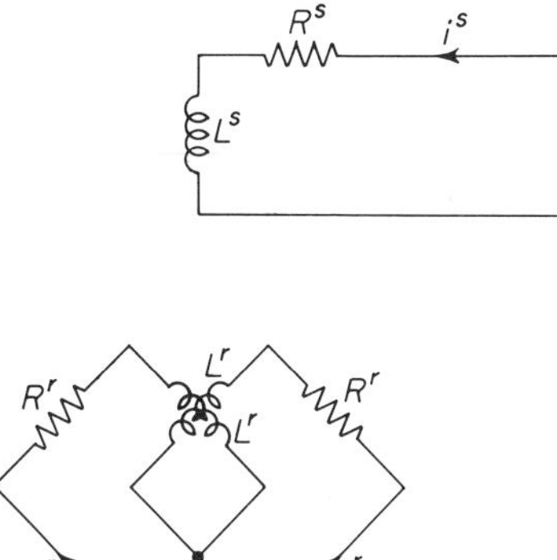

Fig. 7-4 Schematic representation of a 1-phase motor.

$$\begin{bmatrix} v^s \\ 0 \\ 0 \end{bmatrix} = \begin{bmatrix} R^s + L^s p & L^{sr} p & 0 \\ L^{sr} p & R^r + L^r p & L^r \omega_m \\ -L^{sr}\omega_m & -L^r \omega_m & R^r + L^r p \end{bmatrix} \begin{bmatrix} i^s \\ i_d^r \\ i_q^r \end{bmatrix} \tag{7.27}$$

$$T_e = L^{sr} i_q^r i^s \tag{7.28}$$

Solving for i_q^r, from Eq. (7.27), and putting $p = j\omega^s$ for steady-state operation, we have

$$i_q^r = \frac{\omega_m R^r L^{sr} i^s}{[R^r + L^r j(\omega^s + \omega_m)][R^r + L^r j(\omega^s - \omega_m)]} \tag{7.29}$$

Furthermore, the last two equations of Eq. (7.27) yield (with $p = j\omega^s$)

$$0 = L^{sr}j(\omega^s+\omega_m)i^s+[R^r+L^r j(\omega^s+\omega_m)](i_d^r-ji_q^r) \tag{7.30a}$$

$$0 = L^{sr}j(\omega^s-\omega_m)i^s+[R^r+L^r j(\omega^s-\omega_m)](i_d^r+ji_q^r) \tag{7.30b}$$

Now, ω^s and ω_m are related by the slip as

$$s = \frac{\omega^s-\omega_m}{\omega^s}$$

so that

$$\omega^s-\omega_m = s\omega^s$$

and

$$\omega^s+\omega_m = (2-s)\omega^s$$

which when substituted in Eqs. (7.29–7.30b) yield

$$i_q^r = \frac{\omega_m R^r L^{sr} i^s}{[R^r+j(2-s)\omega^s L^r](R^r+js\omega^s L^r)} \tag{7.31a}$$

$$-j(2-s)\omega^s L^{sr} i^s = [R^r+j(2-s)\omega^s L^r]\mathbf{I}^{r*}\sqrt{2} \tag{7.31b}$$

$$-js\omega^s L^{sr} i^s = (R^r+js\omega^s L^r)\mathbf{I}^r\sqrt{2} \tag{7.31c}$$

Equations (7.31b,c) can be substituted in Eq (7.31a) to obtain

$$i_q^r = \frac{2\omega_m R^r|I^r|^2}{(\omega^s)^2 L^{sr} i^s s(2-s)} \tag{7.31d}$$

Equations (7.28) and (7.31d) finally yield

$$\begin{aligned} T_e &= \frac{2\omega_m R^r|I^r|^2}{(\omega^s)^2 s(2-s)} = \frac{2R^r|I^r|^2}{\omega_m}\frac{(1-s)^2}{s(2-s)} \\ &= \frac{R^r|I^r|^2}{\omega_m}\left(\frac{1-s}{s}-\frac{1-s}{2-s}\right) \\ &= \frac{|I^r|^2}{\omega_m}\frac{1-s}{s}R^r - \frac{|I^r|^2}{\omega_m}\frac{1-s}{2-s}R^r \end{aligned} \tag{7.32}$$

Comparing Eqs. (7.26) and (7.32) we observe that the single-phase motor develops a torque which can be resolved into two components, as given by the two terms in Eq. (7.32), opposing each other. As a consequence, we may draw an analogy with the 2-phase motor and assume that the single-phase motor is equivalent to two 2-phase motors with a common shaft, but with stators connected to produce counter-rotating fields. This equivalence is illustrated in Fig. 7-5, and it suggests that the pulsating single-phase field can be resolved into two revolving fields—one a forward rotating field having a velocity ω^s, and the other a backward rotating field having a velocity $-\omega^s$.

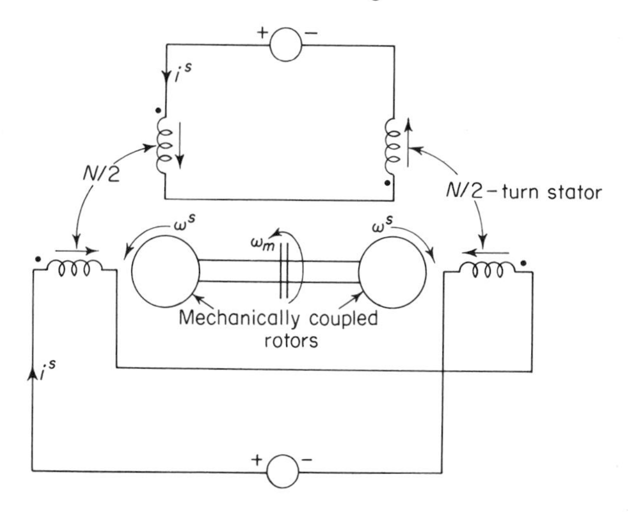

Fig. 7-5 Equivalent two-phase motor for a single-phase motor: N = no. of turns on the stator of the single-phase motor, and i^s = stator current.

This theory is known as the *revolving-field theory*; the equivalent circuit of the induction motor based on this theory is shown in Fig. 7-6. The torque–slip characteristics derived from Eq. (7.32), or from Fig. 7-6, are shown in Fig. 7-7, from which we conclude that (1) the single-phase motor is not self-starting, and (2) that the single-phase motor will continue to run in the positive or negative direction, if given a starting torque in that direction. Notice that the

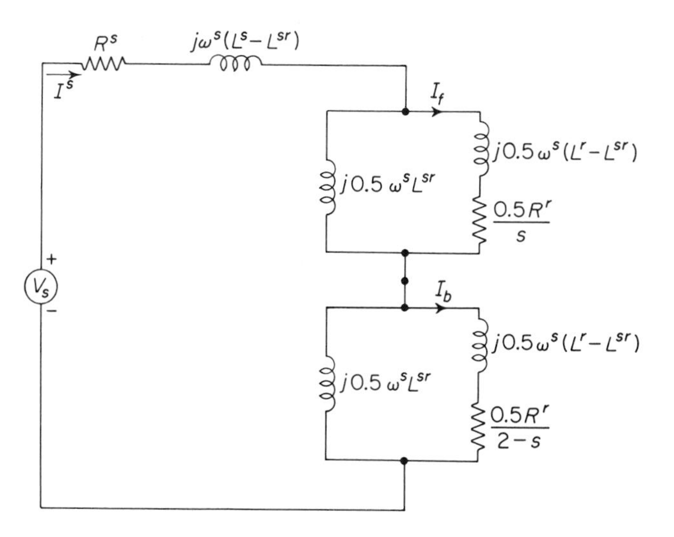

Fig. 7-6 Equivalent circuit of a single-phase induction motor based on revolving-field theory.

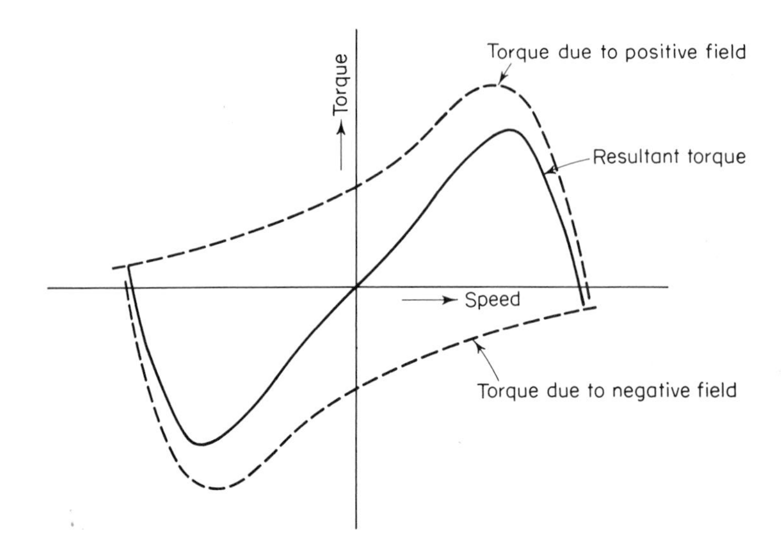

Fig. 7-7 Torque-speed characteristic of a single-phase induction motor based on revolving-field theory.

backward field travels with a slip $(2-s)$ with respect to ω_m, the rotor angular velocity. This fact is reflected in the equivalent circuit (Fig. 7-6) by the resistance $0.5R^r/(2-s)$, whereas the rotor resistance for the forward field is simply $0.5R^r/s$. As compared with the equivalent circuit for the 2-phase machine (Fig. 7-1), the rotor quantities for the single-phase machine (at a standstill) have been equally divided for the forward and backward fields. Once the equivalent circuit is known, the machine performance can be calculated, as illustrated by the following example. Note that in the circuits developed so far we have not included the effects of iron losses. It is, however, a fairly straightforward matter to represent iron losses in equivalent circuits and we shall consider this in a later section.

EXAMPLE 7–2[3]

With reference to Fig. 7-6, the constants of a $\frac{1}{4}$-hp, 230-V, 4-pole, 60-cycle, single-phase induction motor are; $R^s = 10.0\ \Omega$; $R^r = 11.65\ \Omega$; $\omega^s(L^s - L^{sr}) = \omega^s(L^r - L^{sr}) = 12.8\ \Omega$; $\omega^s L^{sr} = 258.0\ \Omega$. For an applied voltage of 210 V determine (a) input current, (b) power factor, (c) developed power, (d) shaft power (if rotational losses are 7 W), and (e) efficiency (if iron losses at 210 V are 35.5 W), at 3% slip.

For the circuit shown in Fig. 7-6, we have

$$\frac{0.5\,R^r}{s} = \frac{11.65}{2 \times 0.03} = 194.1\quad \Omega$$

and

$$\frac{0.5\ R^r}{2-s} = \frac{11.65}{2\times 1.97} = 2.96 \quad \Omega$$

$$j\times 0.5\times \omega^s L^{sr} = j\times 129.0 \quad \Omega$$

and

$$j\times 0.5\times \omega^s(L^s - L^{sr}) = j\times 0.5\times \omega^s(L^r - L^{sr}) = j6.4 \quad \Omega$$

For the forward-field circuit,

$$Z_f = \frac{194.1\times j129}{194.1+j129} = 59.2+j89$$

and for the backward-field circuit,

$$Z_b = \frac{2.96\times j129}{2.96+j129} \simeq 2.96$$

For the total series impedance Z_e,

$$\begin{aligned} Z_e &= Z^s + Z_f + Z_b \\ &= (10+j12.8)+(59.2+j89)+2.96 \\ &= 72.16+j101.8 = 124\ \angle 55^\circ \end{aligned}$$

(a) Input current:

$$I = \frac{V}{Z_e} = \frac{210}{124\angle 55^\circ} = 1.7\ \angle -55^\circ \quad \text{A}$$

(b) Power factor $= \cos 55^\circ = 0.573$ lagging

(c) Developed power:

$$\begin{aligned} P_d &= \left(\frac{0.5\ R^r}{s} I_f^2 - \frac{0.5\ R^r}{2-s} I_b^2\right)(1-s) \\ &= \left(\frac{V_f^2}{0.5R^r/s} - \frac{V_b^2}{0.5R^r/(2-s)}\right)(1-s), \text{ since } s = 0.03 \text{ (small)} \end{aligned}$$

but

$$|V_f| = IZ_f = 1.7(59.2+j89) = 182 \quad \text{V}$$

and

$$|V_b| = IZ_b = 1.7(2.96) = 5.04 \quad \text{V}$$

Or,

$$P_d = \left(\frac{182^2}{194} - \frac{5.04^2}{2.96}\right)(1-0.03) = 156 \quad \text{W}$$

(d) Shaft power $P_s = P_d - P_{\text{rot}} = 156 - 7 = 149$ W

(e) Input power $= VI\cos\theta = 210\times 1.7\times 0.573 = 204$ W

Output power $= P_s - P_{\text{iron}} = 149 - 35.5 = 113.5$

$$\text{Efficiency} = \frac{113.5}{204} = 55.6\%$$

7.3 An Alternate Formulation of the Equations of Motion[1]

In Sec. 7.1.2 we obtained the equations of motion for a 2-phase induction machine in terms of the *dq* variables and showed that the equations are linear for constant-speed operation. We derived the steady-state volt-ampere and torque characteristics using the *dq* variables and demonstrated that the results are consistent with those obtained in Sec. 7.1.1, using the frequency constraints. Finally, we analyzed the single-phase induction machine and developed the double-revolving-field theory of the single-phase machine. We thus observe that analysis in terms of *dq* variables has many applications. (*Note*: We analyzed the synchronous machine in Example 5-5.)

We now consider the alternate formulation of the equations of motion of the induction machine in terms of the $\pm$ and fb components. These variables were introduced in Example 5-6. For the induction machine, $v_f^r = v_b^r = 0$ and the equations of Example 5-6 become

$$\begin{bmatrix} v_+^s \\ v_-^s \\ 0 \\ 0 \end{bmatrix} = \begin{bmatrix} R^s + L^s p & 0 & L^{sr} p & 0 \\ 0 & R^s + L^s p & 0 & L^{sr} p \\ L^{sr}(p - jv\omega_m) & 0 & R^r + L^r(p - jv\omega_m) & 0 \\ 0 & L^{sr}(p + jv\omega_m) & 0 & R^r + L^r(p + jv\omega_m) \end{bmatrix} \begin{bmatrix} i_+^s \\ i_-^s \\ i_f^r \\ i_b^r \end{bmatrix} \tag{7.33a}$$

and

$$T_e = jvL^{sr}(i_+^{s*} i_f^r - i_+^s i_b^r) = jvL^{sr}(i_+^{s*} i_f^r - i_+^s i_f^{r*}) \tag{7.33b}$$

To show that Eq. (7.13) or Eq. (7.26) follows from Eqs. (7.33a,b), we assume balanced 2-phase steady-state excitations as given by Eq. (7.2d). Now, using

$$v_{+-} = S_{+-}^{-1} v_{ab} \tag{7.34a}$$

where

$$S_{+-}^{-1} = \frac{1}{\sqrt{2}} \begin{bmatrix} 1 & j \\ 1 & -j \end{bmatrix} \tag{7.34b}$$

we have

$$v_+^s = \mathbf{V}_+^s e^{j\omega^s t} \tag{7.35a}$$

$$v_-^s = \mathbf{V}_+^{s*} e^{-j\omega^s t} \tag{7.35b}$$

where

$$\mathbf{V}_+^s = \frac{1}{\sqrt{2}} V^s$$

and

$$\mathbf{V}_-^s = \frac{1}{\sqrt{2}} (\mathbf{V}_a^s - j\mathbf{V}_b^s) = 0$$

Similarly $$\mathbf{I}^s_+ = \frac{1}{\sqrt{2}} I^s$$

and $$\mathbf{I}^s_- = 0$$

Also $$\mathbf{I}^r_f = \frac{1}{\sqrt{2}} I^r_f$$

and $$\mathbf{I}^r_b = 0$$

Substituting these voltages and currents in Eq. (7.33a) we have, for *steady-state* and *balanced* conditions,

$$\mathbf{V}^s_+ = (R^s + j\omega^s L^s)\mathbf{I}^s_+ + j\omega^s L^{sr}\mathbf{I}^r_f$$

$$0 = j(\omega^s - \nu\omega_m)L^{sr}\mathbf{I}^s_+ + [R^r + j(\omega^s - \nu\omega_m)L^r]\mathbf{I}^r_f$$

Or, in terms of slip, $s = (\omega^s - \nu\omega_m)/\omega^s$, these equations reduce to

$$\mathbf{V}^s_+ = (R^s + j\omega^s L^s)\mathbf{I}^s_+ + j\omega^s L^{sr}\mathbf{I}^r_f \tag{7.36a}$$

$$0 = j\omega^s L^{sr}\mathbf{I}^s_+ + \left(\frac{R^r}{s} + j\omega^s L^r\right)\mathbf{I}^r_f \tag{7.36b}$$

The equivalent circuit representing Eqs. (7.36a,b) is shown in Fig. 7-8. Notice that this circuit is identical to the circuit shown in Fig. 7-1. Thus

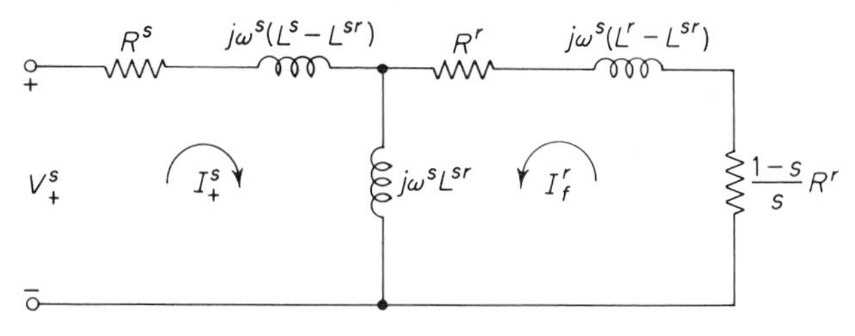

Fig. 7-8 Equivalent circuit of a balanced induction machine.

applying (1) the frequency constraints (Sec. 7.1), (2) the *dq* transformation (Sec. 7.1.2), and (3) the $+-$ *fb* transformation just discussed, we obtain identical results, which should not be surprising. However, the real advantage of the transformations is in the analysis of the machine under abnormal operating conditions, such as unbalanced operation or under transient conditions.

†7.3.1 *The Unbalanced Machine*

Consider the 2-phase induction machine of Example 5-6, for which the equations of motion are rewritten as Eqs. (7.33a,b). By unbalanced operating conditions we mean

$$v_a^s = V_a^s \cos \omega t \tag{7.37}$$
$$v_b^s = V_b^s \sin (\omega t + \phi)$$

and

$$v_a^r = v_b^r = v_f^r = v_b^r = 0 \tag{7.38}$$

From Eqs. (7.34a,b) and (7.37) we have

$$\begin{aligned} v_+^s &= \frac{1}{2}\left[\frac{1}{\sqrt{2}}(\mathbf{V}_a^s + j\mathbf{V}_b^s)e^{j\omega t} + \frac{1}{\sqrt{2}}(\mathbf{V}_a^{s*} + j\mathbf{V}_b^{s*})e^{-j\omega t}\right] \\ v_-^s &= \frac{1}{2}\left[\frac{1}{\sqrt{2}}(\mathbf{V}_a^s - j\mathbf{V}_b^s)e^{j\omega t} + \frac{1}{\sqrt{2}}(\mathbf{V}_a^{s*} - j\mathbf{V}_b^{s*})e^{-j\omega t}\right] \end{aligned} \tag{7.39}$$

where

$$\mathbf{V}_a^s = V_a^s$$
$$\mathbf{V}_b^s = -jV_b^s e^{j\phi} \tag{7.40}$$

and the superscript * denotes the complex conjugate. Under steady-state conditions

$$\begin{aligned} \mathbf{V}_+^s &= \frac{1}{\sqrt{2}}(\mathbf{V}_a^s + j\mathbf{V}_b^s) \\ \mathbf{V}_-^s &= \frac{1}{\sqrt{2}}(\mathbf{V}_a^s - j\mathbf{V}_b^s) \end{aligned} \tag{7.41}$$

Using Eq. (7.41), Eq. (7.39) can be expressed as

$$\begin{aligned} v_+^s &= \tfrac{1}{2}(\mathbf{V}_+^s e^{j\omega t} + \mathbf{V}_-^{s*} e^{-j\omega t}) \\ v_-^s &= (v_+^s)^* \end{aligned} \tag{7.42}$$

Similarly, for the steady-state currents we can write

$$\begin{aligned} i_+^s &= \tfrac{1}{2}(\mathbf{I}_+^s e^{j\omega t} + \mathbf{I}_-^{s*} e^{-j\omega t}) \\ i_-^s &= (i_+^s)^* \\ i_f^r &= \tfrac{1}{2}(\mathbf{I}_f^r e^{j\omega t} + \mathbf{I}_b^{r*} e^{-j\omega t}) \\ i_b^r &= (i_f^r)^* \end{aligned} \tag{7.43}$$

We observe, from Eq. (7.39), that the voltages can be resolved into two components: $\mathbf{V}_+^s e^{j\omega t}$ and $\mathbf{V}_-^{s*} e^{-j\omega t}$. Consequently, Eqs. (7.39–7.43), when substituted in Eq. (7.33a), yields, for a constant rotor speed ω_m, for an excitation $\mathbf{V}_+^s e^{j\omega t}$,

$$\begin{aligned} \mathbf{V}_+^s &= (R^s + j\omega L^s)\mathbf{I}_+^s + j\omega L^{sr}\mathbf{I}_f^r \\ 0 &= j(\omega - \nu\omega_m)L^{sr}\mathbf{I}_+^s + [R^r + j(\omega - \nu\omega_m)L^r]\mathbf{I}_f^r \end{aligned} \tag{7.44}$$

for an excitation $\mathbf{V}_-^{s*} e^{-j\omega t}$,

$$\begin{aligned} \mathbf{V}_-^{s*} &= (R^s - j\omega L^s)\mathbf{I}_-^{s*} - j\omega L^{sr}\mathbf{I}_b^{r*} \\ 0 &= -j(\omega + \nu\omega_m)L^{sr}\mathbf{I}_-^{s*} + [R^r - j(\omega + \nu\omega_m)L^r]\mathbf{I}_b^{r*} \end{aligned} \tag{7.45}$$

But $(\omega-\nu\omega_m) = s\omega$ and $(\omega+\nu\omega_m) = (2-s)\omega$, where s = slip. For an excitation $\mathbf{V}^s_+$, Eqs. (7.44) and (7.45) can be expressed in the final form in terms of unconjugated voltages and current as follows:

$$\begin{aligned} \mathbf{V}^s_+ &= (R^s+j\omega L^s)\mathbf{I}^s_+ + j\omega L^{sr}\mathbf{I}^r_f \\ 0 &= j\omega L^{sr}\mathbf{I}^s_+ + \left(\frac{R^r}{s}+j\omega L^r\right)\mathbf{I}^r_f \end{aligned} \tag{7.46}$$

and for an excitation $\mathbf{V}^s_-$,

$$\begin{aligned} \mathbf{V}^s_- &= (R^s+j\omega L^s)\mathbf{I}^s_- + j\omega L^{sr}\mathbf{I}^r_b \\ 0 &= j\omega L^{sr}\mathbf{I}^s_- + \left(\frac{R^r}{2-s}+j\omega L^r\right)\mathbf{I}^r_b \end{aligned} \tag{7.47}$$

The equivalent circuit representing Eqs. (7.46) and (7.47) is shown in Fig. 7-9, from which the performance of the machine can be determined.

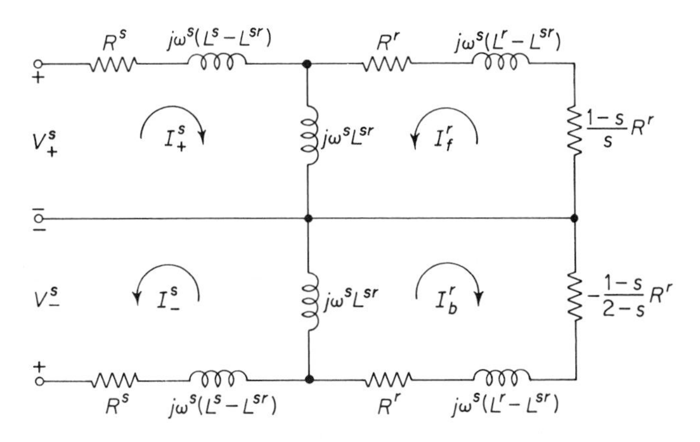

Fig. 7-9 Steady-state equivalent circuit of an unbalanced induction machine.

Having obtained the volt-ampere characteristic, we now turn to the torque equation, Eq. (7.33b). Substituting Eq. (7.43) in Eq. (7.33b), we obtain

$$\begin{aligned} T_e = \tfrac{1}{4} j\nu L^{sr}[&(\mathbf{I}^{s*}_+\mathbf{I}^r_f - \mathbf{I}^s_+\mathbf{I}^{r*}_f) + (\mathbf{I}^s_-\mathbf{I}^{r*}_b - \mathbf{I}^{s*}_-\mathbf{I}^r_b) \\ &+ (\mathbf{I}^s_-\mathbf{I}^r_f - \mathbf{I}^s_+\mathbf{I}^r_b)e^{j2\omega t} + (\mathbf{I}^{s*}_+\mathbf{I}^{r*}_b - \mathbf{I}^{s*}_-\mathbf{I}^{r*}_f)e^{-j2\omega t}] \end{aligned} \tag{7.48}$$

Solving for the currents from Eqs. (7.46) and (7.47), substituting in Eq. (7.48), and noting that the average value of $e^{\pm j2\omega t}$ is zero, we obtain the expression for the average torque $\langle T_e \rangle$ as

$$\langle T_e \rangle = \frac{\frac{1}{2}\nu\omega(L^{sr})^2 \frac{R^r}{s}(V^s_+)^2}{\left[\frac{R^rR^s}{s} - \omega^2[L^rL^s-(L^{sr})^2]\right]^2 + \omega^2\left(R^sL^r + \frac{R^r}{s}L^s\right)^2}$$

$$-\frac{\frac{1}{2}\nu\omega(L^{sr})^2\frac{R^r}{2-s}(V^s_-)^2}{\left[\frac{R^rR^s}{2-s}-\omega^2[L^rL^s-(L^{sr})^2]\right]^2+\omega^2\left(R^sL^r+\frac{R^r}{2-s}L^s\right)^2} \tag{7.49}$$

We can readily verify that Eq. (7.49) reduces to Eq. (7.13) for balanced operation, when $V^s_- = 0$ and $I^s_- = 0$.

7.3.2 *The Single-Phase Machine*

The equivalent circuit for the single-phase machine follows directly from the preceding analysis of the unbalanced machine. In this case, $\mathbf{V}^s_a = V^s$ and $\mathbf{V}^s_b = 0$, so that $\mathbf{V}^s_+ = (1/\sqrt{2})V^s = \mathbf{V}^s_-$. Similarly, $\mathbf{I}^s_+ = (1/\sqrt{2})I^s = \mathbf{I}^s_-$, which implies that the current in the middle lead of the circuit shown in Fig. 7-9 is zero. Consequently, this lead can be removed, and the circuit reduces to that shown in Fig. 7-10. Comparing this with Fig. 7-6, we notice

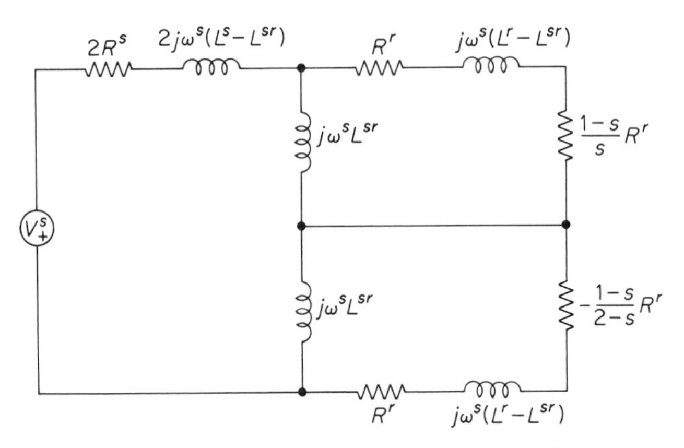

Fig. 7-10 Equivalent circuit of a single-phase induction motor.

that the circuit of Fig. 7-10 is very similar to the circuit shown in Fig. 7-6. The values of the parameters of the circuit shown in Fig. 7-10 can be adjusted so that an exact equivalence between the circuits of Figs. 7-6 and 7-10 can be derived (see Problem 7-2). To demonstrate further the usefulness of the method of analysis of the unbalanced machine, we can also obtain the speed–torque characteristics of 2-phase servomotors directly from Eq. (7.49) (see Problem 7-3). In order to estimate the effects of unbalance in a 2-phase machine, we now consider the following example.

EXAMPLE 7–3

The voltages applied to a 60-cycle, 2-pole, 2-phase induction motor are $100\sqrt{2}$

$\angle 0°$ and $80\sqrt{2}\ \angle -90°$. The motor runs with a slip of 5%. Calculate the electrical torque and the line currents if $R^s = 1\ \Omega$, $R^r = 5\ \Omega$, $\omega L^{sr} = 4\ \Omega$, $\omega L^s = 10\ \Omega$, and $\omega L^r = 6\ \Omega$. Compare the results if the applied voltages are perfectly balanced.

From Eq. (7.41) we have

$$\mathbf{V}^s_+ = \frac{1}{\sqrt{2}}[\sqrt{2}100 + j\sqrt{2}(-j80)] = 180 \quad \text{V}$$

$$\mathbf{V}^s_- = \frac{1}{\sqrt{2}}[\sqrt{2}100 - j\sqrt{2}(-j80)] = 20 \quad \text{V}$$

Substituting these values in Eq. (7.49) we have

$$\langle T_e \rangle = \frac{4 \times \dfrac{4}{377} \times \dfrac{5}{0.05} \times 180^2}{\left[\dfrac{5}{0.05} - (60-16)\right]^2 + (6+1000)^2} - \frac{4 \times \dfrac{4}{377} \times \dfrac{5}{1.95} \times 20^2}{\left[\dfrac{5}{1.95} - (60-16)\right]^2 + \left(6 + \dfrac{50}{1.95}\right)^2}$$

$$= 0.13 - 0.0157 = 0.1143 \quad \text{N·m}$$

If the applied voltage is balanced, the negative-sequence voltage V_- is zero, the second term in the above torque expression is zero, and $V_+ = 200$ V. Thus, for balanced operation, from Eq. (7.13),

$$\langle T_e \rangle = 0.158 \text{ N·m}$$

To calculate the currents, we use Eqs. (7.46) and (7.47), from which

$$\mathbf{I}^s_+ = \frac{\left(\dfrac{R^r}{s} + j\omega L^r\right)\mathbf{V}^s_+}{\left(\dfrac{R^r}{s} + j\omega L^r\right)(R^s + j\omega L^s) + (\omega L^{sr})^2}$$

$$= \frac{(100 + j6)\ 180}{(100 + j6)(1 + j10) + 16} \simeq 18\ \angle -90° = -j18 \quad \text{A}$$

$$\mathbf{I}^s_- = \frac{\left(\dfrac{R^r}{2-s} + j\omega L^r\right)\mathbf{V}^s_-}{\left(\dfrac{R^r}{2-s} + j\omega L^r\right)(R^s + j\omega L^s) + (\omega L^{sr})^2}$$

$$= \frac{(2.56 + j6)\ 20}{(2.56 + j6)\ (1 + j10) + 16} \simeq 0.61 - j2.44 \quad \text{A}$$

Recalling

$$\begin{bmatrix} \mathbf{I}_a \\ \mathbf{I}_b \end{bmatrix} = \frac{1}{\sqrt{2}}\begin{bmatrix} 1 & 1 \\ -j & j \end{bmatrix}\begin{bmatrix} \mathbf{I}_+ \\ \mathbf{I}_- \end{bmatrix}$$

we have

$$I_a = 14.4 \quad \text{A}$$

and

$$I_b = 11.6 \quad \text{A}$$

7.4 The 3-Phase Machine

So far we have considered only the 2-phase induction machine, and showed that the equations of motion of the single-phase induction machine emerge from the equations pertaining to the 2-phase machine by appropriate constraints. In practice, however, the 3-phase induction machine (operating as a motor) is more commonly encountered than the 2-phase machine. A typical, large 3-phase induction motor is shown in Fig. 7-11. Three-phase alternating

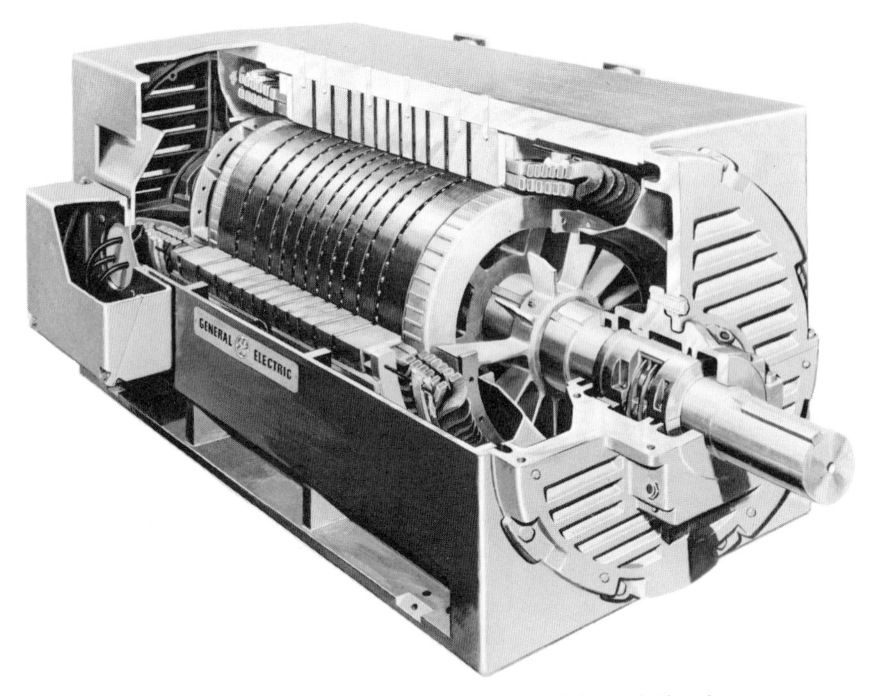

Fig. 7-11 An induction motor. (Courtesy of General Electric Company.)

current is supplied to the stator; the rotating magnetic field is produced in the airgap. The mechanism of production of the traveling field was discussed in Sec. 4.3, and it can be further demonstrated from Fig. 7-12 that a 3-phase balanced excitation results in the production of a rotating field. Normally (except in a doubly fed machine) the rotor is not connected to an external source. Currents are "induced" in the rotor circuit, and energy conversion results from the interaction of the airgap fields and the induced rotor currents. We thus see that, qualitatively, the 3-phase induction machine is not very

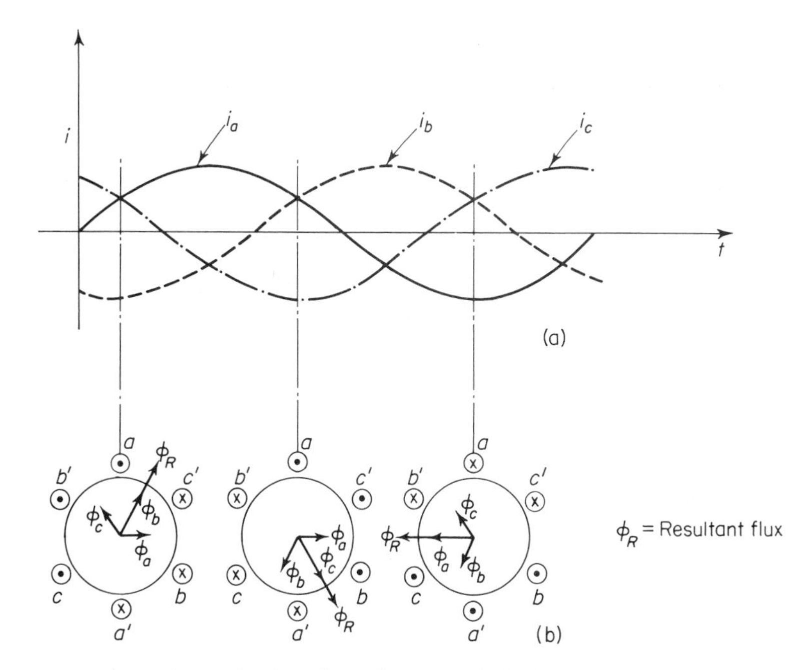

Fig. 7-12 Production of rotating magnetic field: (a) time diagram; (b) space diagram.

much different from the 2-phase machine. Using theorems 3 and 4 of Sec. 5.7.1, we can show that, quantitatively, a 3-phase machine (or an n-phase machine for that matter) is in fact equivalent to a 2-phase machine. The restrictions on the stator and rotor impedance matrices are as given in theorems 3 and 4 (Sec. 5.7.1). Using the $+-$, or *symmetrical-component*, transformation, we shall first obtain the equations of motion for the 3-phase machine. Subsequently, we shall show that the 3-phase machine can also be analyzed by the dq transformation.

7.4.1 Symmetrical-Component Transformation of a 3-Phase Machine

We may now consider the general equations of motion in terms of machine variables. These are given as Eqs. (7.1a,b). We assume that the machine has a 3-phase winding on the stator and rotor; and furthermore, that these windings are balanced (that is, each winding on the stator and on the rotor has the same number of turns), symmetric (that is, the windings are displaced from each other by 120°), and sinusoidally distributed.

Because we have assumed that the windings are balanced and symmetric, the resulting Z matrix is cyclic-symmetric (definition 1, Sec. 5.7.1) and we can choose the S matrix given in theorem 3, Sec. 5.7.1. The assumption that the rotor and stator windings are sinusoidally distributed implies that a typical element of the L^{sr} matrix is of the form

$$l_{pq}^{sr} = L^{sr} \cos v \left[\theta + (q-1)\frac{\pi}{3} - (p-1)\frac{\pi}{3} \right]$$

which follows from Eq. (5.21c). We observe that the resulting Z matrix is of the same form as given in theorem 4, Sec. 5.7.1.

Now, proceeding as in Sec. 5.7.2, we define

$$v = Sv'$$

and

$$i = Si' \tag{7.50}$$

where the unprimed quantities are the original machine variables and the primed quantities are transformed variables and the transformation matrix S is given by

$$S = \frac{1}{\sqrt{3}} \begin{bmatrix} 1 & 1 & 1 \\ 1 & s^2 & s \\ 1 & s & s^2 \end{bmatrix} \tag{7.51}$$

$$s = e^{j2\pi/3} = -0.5 + j0.866$$

$$s^2 = e^{j4\pi/3} = -0.5 - j0.866$$

and

$$S^{-1} = \tilde{S}^*$$

From Eqs. (7.50), (7.51), and (7.1a) we have

$$\begin{bmatrix} v^{s'} \\ \hline v^{r'} \end{bmatrix} = \left[\begin{array}{c|c} S^{-1}R^sS + pS^{-1}L^sS & pS^{-1}L^{sr}S \\ \hline pS^{-1}L^{rs}S & S^{-1}R^rS + pS^{-1}L^rS \end{array} \right] \begin{bmatrix} i^{s'} \\ \hline i^{r'} \end{bmatrix} \tag{7.52}$$

If we denote v^s as

$$v^s = \begin{bmatrix} v_a^s \\ v_b^s \\ v_c^s \end{bmatrix}$$

then

$$v^{s'} = S^{-1}v^s = \frac{1}{\sqrt{3}} \begin{bmatrix} v_a^s + v_b^s + v_c^s \\ v_a^s + sv_b^s + s^2v_c^s \\ v_a^s + s^2v_b^s + sv_c^s \end{bmatrix} = \begin{bmatrix} v_0^s \\ v_+^s \\ v_-^s \end{bmatrix} \tag{7.53a}$$

Similar expressions hold for the stator currents and rotor voltages and currents. In Eq. (7.53a), v_0^s, v_+^s, and v_-^s are respectively known as the *zero-sequence*, *positive-sequence*, and *negative-sequence* voltages. Expanding the various submatrices in Eq. (7.52) we find that $R^{s\prime} = R^s$ and $R^{r\prime} = R^r$. The transformed inductances are given by

$$L^{s\prime} = S^{-1}L^sS = \begin{bmatrix} L_0^s & 0 & 0 \\ 0 & L_+^s & 0 \\ 0 & 0 & L_-^s \end{bmatrix} \tag{7.53b}$$

where

$$\begin{aligned} L_0^s &= L_{aa}^{ss} + L_{ab}^{ss} + L_{ac}^{ss} \\ L_+^s &= L_{aa}^{ss} + s^2L_{ab}^{ss} + sL_{ac}^{ss} \\ L_-^s &= L_{aa}^{ss} + sL_{ab}^{ss} + s^2L_{ac}^{ss} \end{aligned} \tag{7.53c}$$

As for voltages and currents, the rotor self-inductance submatrix transforms in the same manner as L^s, as given by Eq. (7.53b). Because the stator-to-rotor submatrix L^{sr} is of the form given in theorem 4 (Sec. 5.7.1), for the corresponding transformed submatrix we have

$$L^{sr\prime} = S^{-1}L^{sr}S = \tfrac{3}{2}L^{sr}\begin{bmatrix} 0 & 0 & 0 \\ 0 & e^{j\nu\theta} & 0 \\ 0 & 0 & e^{-j\nu\theta} \end{bmatrix} \tag{7.53d}$$

and

$$L^{rs\prime} = \tilde{L}^{sr\prime *} \tag{7.53e}$$

When Eqs. (7.53a–e) are substituted in Eq. (7.52), with $v^r = v^{r\prime} = 0$, the volt-ampere equations for the induction machine become

$$\begin{bmatrix} v_0^s \\ v_+^s \\ v_-^s \\ \hline 0 \\ 0 \\ 0 \end{bmatrix} = \left[\begin{array}{ccc|ccc} R^s+pL_0^s & 0 & 0 & 0 & 0 & 0 \\ 0 & R^s+pL_+^s & 0 & 0 & \tfrac{3}{2}pL^{sr}e^{j\nu\theta} & 0 \\ 0 & 0 & R^s+pL_-^s & 0 & 0 & \tfrac{3}{2}pL^{sr}e^{-j\nu\theta} \\ \hline 0 & 0 & 0 & R^r+pL_0^r & 0 & 0 \\ 0 & \tfrac{3}{2}pL^{sr}e^{-j\nu\theta} & 0 & 0 & R^r+pL_+^r & 0 \\ 0 & 0 & \tfrac{3}{2}pL^{sr}e^{j\nu\theta} & 0 & 0 & R^r+pL_-^r \end{array}\right] \begin{bmatrix} i_0^s \\ i_+^s \\ i_-^s \\ \hline i_0^r \\ i_+^r \\ i_-^r \end{bmatrix} \tag{7.54}$$

In Eq. (7.54) notice that there is no coupling between the rotor and stator for zero-sequence quantities. For instance, $v_0^s = (R^s+pL_0^s)i^s$ is unaffected by the rotor currents, but $v_+^s = (R^s+pL_+^s)i_+^s + \frac{3}{2}\,pL^{sr}e^{j\nu\theta}i_+^r$ depends upon the positive-sequence rotor current i_+^r.

When expressions for the transformed currents and stator-to-rotor mutual inductances, Eqs. (7.53d,e) are substituted in Eq. (7.1b) we obtain, after algebraic simplification (see Problem 7-4),

$$T_e = \frac{\nu}{2}\frac{3}{2} L^{sr} j[(i_+^{s*} i_+^r)e^{j\nu\theta} - (i_+^s i_+^{r*})e^{-j\nu\theta}] \tag{7.55}$$

In Eq. (7.55) too, we find that the zero-sequence quantities are absent. Using Eq. (5.46) (for currents), the network representation for Eq. (7.54) for sinusoidal steady-state operation, in terms of the $+-$ and fb components, is shown in Fig. 7-13. We thus conclude that the 3-phase machine

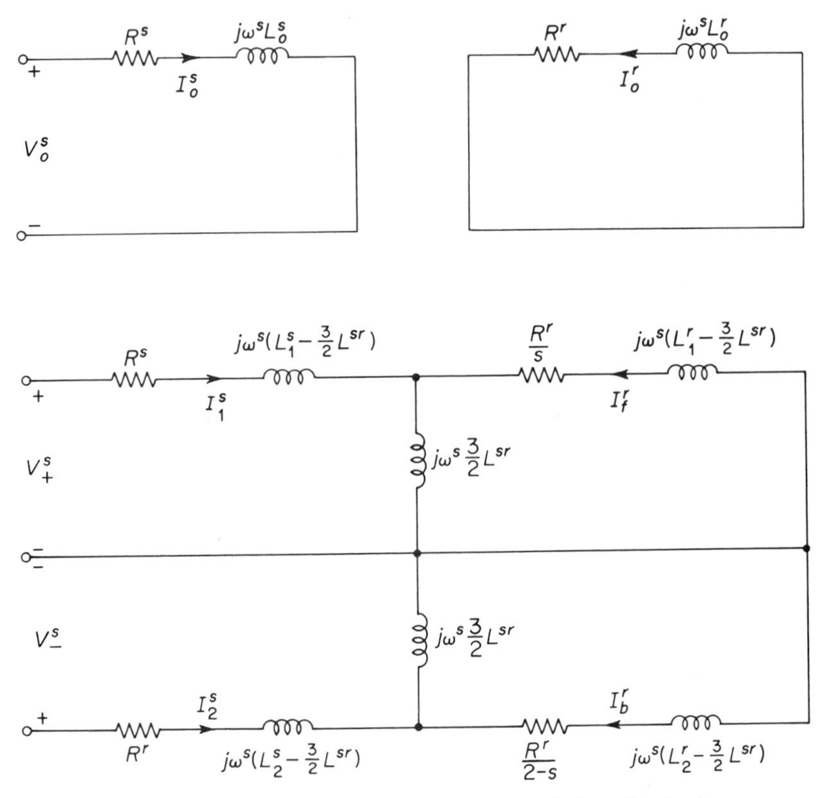

Fig. 7-13 Equivalent circuit of a three-phase unbalanced induction machine.

is equivalent to a 2-phase machine so far as energy-conversion properties are concerned. Therefore, the theory developed in Secs. 7.1 and 7.3 are equally applicable to the 2-phase machine.

7.4.2 Three-Phase to dq Transformation

Using the symmetrical-component transformation, we showed in the last section that a 3-phase machine is equivalent to a 2-phase machine. Because we have earlier used the dq transformation in the analysis of the 2-phase machine and because the dq transformation is useful in certain transient studies of the induction machine, we shall now show that the 3-phase machine too can be analyzed in terms of dq variables. In other words, we shall derive a transformation which relates the 3-phase variables to the dq variables.

Assume a 3-phase, sinusoidally distributed, balanced winding carrying currents i_a, i_b, and i_c as shown in Fig. 7-14(a). The MMF's due to these windings are given by

$$\begin{aligned} F_a &= Ni_a \sin \nu\theta \\ F_b &= Ni_b \sin\left(\nu\theta+\frac{2\pi}{3}\right) \\ F_c &= Ni_c \sin\left(\nu\theta+\frac{4\pi}{3}\right) \end{aligned} \tag{7.56a}$$

and the total MMF F is given by

$$F = F_a+F_b+F_c \tag{7.56b}$$

The MMF F' due to the equivalent dq windings, shown in Fig. 7-14(b), is given by

$$F' = F_d+F_q \tag{7.57}$$

where

$$F_d = N'i_d \sin \nu\theta$$

and

$$F_q = N'i_q \sin\left(\nu\theta-\frac{\pi}{2}\right) \tag{7.58}$$

From Eqs. (7.56a,b) we have

$$F = N\left[\left(i_a+\cos\frac{2\pi}{3}\,i_b+\cos\frac{2\pi}{3}\,i_c\right)\sin\nu\theta-\left(0-\sin\frac{2\pi}{3}\,i_b+\sin\frac{2\pi}{3}\,i_c\right)\cos\nu\theta\right] \tag{7.59}$$

Similarly, from Eqs. (7.57a,b) we have

$$F' = N'[i_d \sin \nu\theta - i_q \cos \nu\theta] \tag{7.60}$$

For the two systems to be equivalent, we must have $F = F'$. Therefore, equating the coefficients of $\sin \nu\theta$ and $\cos \nu\theta$ in Eqs. (7.59) and (7.60) we obtain

$$\begin{bmatrix} i_d \\ i_q \end{bmatrix} = \frac{2}{3}\begin{bmatrix} 1 & -\frac{1}{2} & -\frac{1}{2} \\ 0 & -\frac{\sqrt{3}}{2} & \frac{\sqrt{3}}{2} \end{bmatrix}\begin{bmatrix} i_a \\ i_b \\ i_c \end{bmatrix} \tag{7.61a}$$

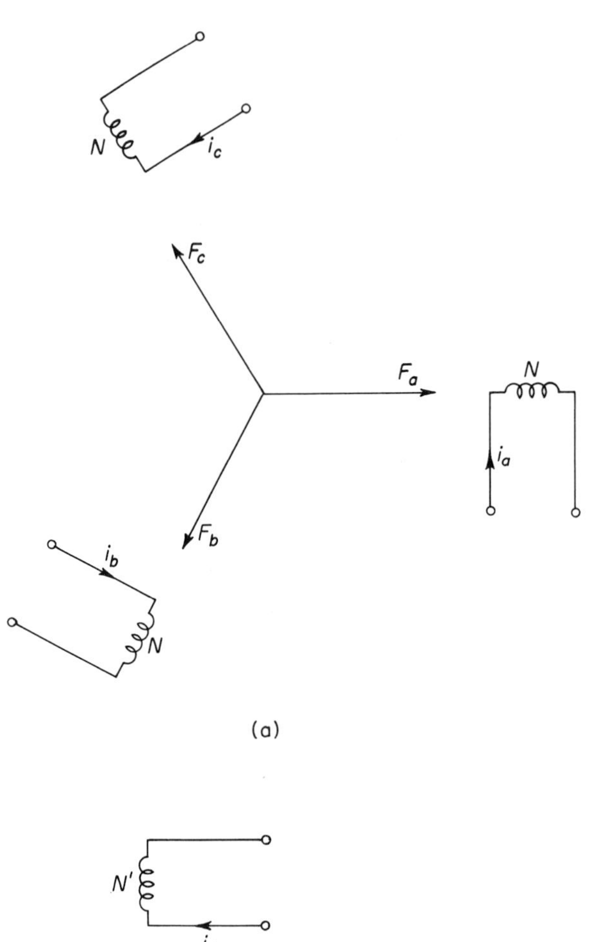

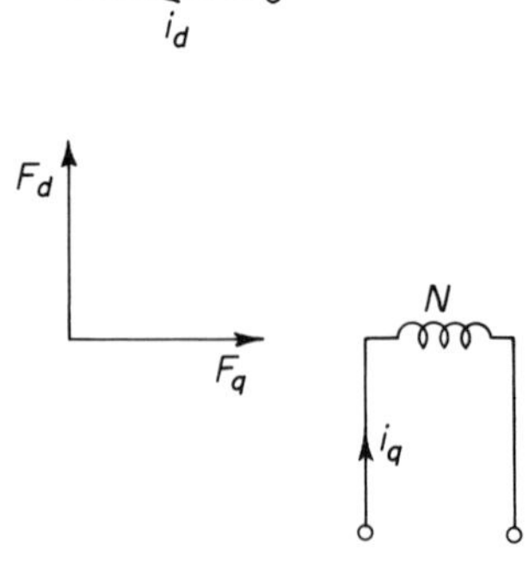

Fig. 7-14 Three-phase to two-phase transformation: (a) a three-phase MMF system; (b) an equivalent two-phase MMF system.

where $2N = 3N'$ for equivalence, and

$$\begin{bmatrix} i_a \\ i_b \\ i_c \end{bmatrix} = \begin{bmatrix} 1 & 0 \\ -\frac{1}{2} & -\frac{\sqrt{3}}{2} \\ -\frac{1}{2} & \frac{\sqrt{3}}{2} \end{bmatrix} \begin{bmatrix} i_d \\ i_q \end{bmatrix} \tag{7.61b}$$

Thus, in terms of dq variables also, the 3-phase machine can be reduced to a 2-phase machine.

7.4.3 *The Per-Phase Equivalent Circuit*

For the 2-phase machine recall that, for balanced conditions, the machine can be represented by the circuit shown in Fig. 7-1. Similarly, the steady-state performance of the 3-phase balanced induction machine can quite easily be calculated from its equivalent circuit drawn on a per-phase basis, as shown in Fig. 7-15. The only difference between this circuit and that shown

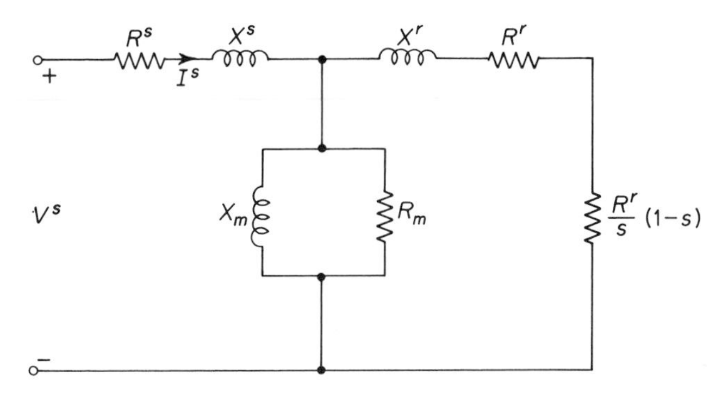

Fig. 7-15 Per-phase equivalent circuit of a balanced three-phase induction motor.

in Fig. 7-1 is that in Fig. 7-15 we have taken the hysteresis and eddy-current losses into account by means of the shunt resistance R_m. As compared to Fig. 7-1, the nomenclature in Fig. 7-15 is $X^s = j\omega^s(L^s - L^{sr})$, $X_m = j\omega L^{sr}$, and $X^r = j\omega^s(L^r - L^{sr})$. To show the usefulness of this equivalent circuit, we consider the following example.

EXAMPLE 7–4

A 3-phase, Y-connected, 220-V, 10-hp, 60-cycle, 4-pole induction motor has the following constants in ohms per phase:

$$R^s = 0.2 \qquad R^r = 0.1$$

$$X^s = 0.5 \qquad X^r = 0.2$$

$$X_m = 20.0$$

The total iron and rotational losses are 350 W. For a slip of 2.5%, calculate (a) input current, (b) output power, (c) output torque, and (d) efficiency.

Because the iron losses are known (350 W), we make an approximation by neglecting the resistance R_m. Thus, from Fig. 7-15, the total impedance is

$$Z_T = R^s + jX^s + \frac{jX_m\left(\dfrac{R^r}{s} + jX^r\right)}{\dfrac{R^r}{s} + j(X_m + X^r)}$$

$$= 0.2 + j \times 0.5 + \frac{j20(4 + j \times 0.2)}{4 + j(20 + 0.2)}$$

$$= (0.2 + j \times 0.5) + (3.8 + j \times 0.81) = 4.21 \angle 18^\circ$$

$$\text{Phase voltage} = 220/\sqrt{3} = 127 \text{ V.}$$

$$\text{Input current} = 127/4.21 = 30 \text{ A.}$$

$$\text{Power factor} = \cos 18^\circ = 0.95.$$

$$\text{Input power} = \sqrt{3} \times 220 \times 30 \times 0.95 = 10.85 \text{ kW.}$$

$$\text{Power across the airgap} = 3 \times 30^2 \times 3.8 = 10.25 \text{ kW.}$$

$$\text{Power developed} = 0.975 \times 10.25 = 10.0 \text{ kW.}$$

$$\text{Output power} = 10 - 0.35 = 9.65 \text{ kW.}$$

$$\text{Output torque} = \text{output power}/\omega_m = (9.65/184) \times 1000 = 52.4 \text{ N·m.}$$

$$\text{where } \omega_m = 0.975 \times 60 \times \pi = 184 \text{ rad/s.}$$

$$\text{Efficiency} = \frac{9.65}{10.85} = 89\%.$$

7.4.4 The Equivalent Circuit from Test Data[4]

The above example illustrates the usefulness of the equivalent circuit. However, we did not actually use the circuit shown in Fig. 7-15. Rather, to simplify the calculations we neglected the shunt-branch resistance R_m. In many calculations, for practical purposes the induction machine is represented by the approximate equivalent circuit shown in Fig. 7-16. In order to calculate the performance of the machine, its parameter must be known. The parameters of the circuit shown in Fig. 7-16 can be obtained from the following two tests.

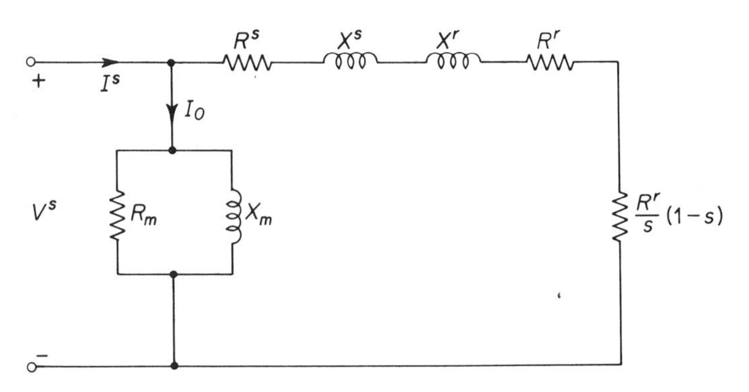

Fig. 7-16 Approximate equivalent circuit of an induction motor.

a. *No-load test.* In this test, rated voltage is applied to the machine and it is allowed to run on no-load. Input power, voltage, and current are measured. These are reduced to per-phase values and respectively denoted by W_0, V_0, and I_0. When the machine runs on no-load, the slip is close to zero and, approximately, the circuit to the right of the shunt branch is taken to be an open circuit. Thus, the parameters R_m and X_m are found from the following equations:

$$R_m = \frac{V_0^2}{W_0} \tag{7.62a}$$

$$X_m = \frac{V_0}{I_0 \sin \phi_0} \tag{7.62b}$$

where

$$\phi_0 = \cos^{-1} \frac{W_0}{V_0 I_0} \tag{7.62c}$$

b. *Blocked-rotor test.* In this test, the rotor of the machine is blocked ($s = 1$) and a reduced voltage is applied to the machine so that the rated current flows through the stator windings. The input power, voltage, and current are recorded and reduced to per-phase values. These are respectively denoted by W_s, V_s, and I_s. In this test, the iron losses are assumed to be negligible and the shunt branch of the circuit shown in Fig. 7-16 is absent. The parameters are thus found from

$$R_e = R^s + a^2 R^r = \frac{W_s}{I_s^2} \tag{7.63a}$$

$$X_e = X^s + a^2 X^r = \frac{V_s \sin \phi_s}{I_s} \tag{7.63b}$$

where

$$\phi_s = \cos^{-1} \frac{W_s}{V_s I_s}$$

In Eqs. (7.63a,b) a is a constant and is analogous to the transformation ratio of a transformer. It takes into account the effect of rotor resistance and reactance as referred to the stator. The tests described here are approximate. Refinements can, however, be made and details are available in References 3, 4, and 5. We now consider an example to illustrate the calculations involved in the determination of the machine constants from test data.

EXAMPLE 7–5

The results of the no-load and blocked-rotor tests on a 3-phase, Y-connected induction motor are as follows:

No-load test—line-to-line voltage = 220 V
total input power = 1000 W
line current = 20 A
friction and windage loss = 400 W

Blocked-rotor test—line-to-line voltage = 30 V
total input power = 1500 W
line current = 50 A

Calculate the parameters of the approximate equivalent circuit shown in Fig. 7-16.

$$V_0 = \frac{220}{\sqrt{3}} = 127 \quad \text{V}$$

$$I_0 = 20 \quad \text{A}$$

$$W_0 = \tfrac{1}{3}(1000 - 400) = 200 \quad \text{W}$$

Thus, from Eqs. (7.62),

$$R_m = \frac{127^2}{200} = 80.5 \quad \Omega$$

$$\phi_0 = \cos^{-1}\frac{200}{20 \times 127} = 86^\circ$$

$$X_m = \frac{127}{20 \times 0.99} = 6.4 \quad \Omega$$

Now,

$$V_s = \frac{30}{\sqrt{3}} = 17.32 \quad \text{V}$$

$$I_s = 50 \quad \text{A}$$

$$W_s = \frac{1500}{3} = 500 \quad \text{W}$$

Thus, from Eqs. (7.63),

$$R_e = \frac{500}{50^2} = 0.2 \quad \Omega$$

$$\phi_s = \cos^{-1}\frac{500}{17.32\times 50} = 54^\circ$$

$$X_e = 17.32\times\frac{0.8}{50} = 0.277 \quad \Omega$$

Knowing the circuit constants, we can calculate the machine performance, as in Example 7-4.

7.4.5 *The Circle Diagram*

We have seen in the previous examples that the performance of the induction machine can be calculated from its equivalent circuit. However, the set of calculations has to be repeated for discrete values of slip ranging from 0 to 1. The complete performance of the machine can be determined quickly, but approximately, by means of a graphical method. This method is based on the fact that the locus of the current vector in a reactive circuit $R+jX$ is a circle for $-\infty \leqslant R \leqslant \infty$ at constant applied voltage; that is,

$$\mathbf{I} = \frac{\mathbf{V}}{R+jX} \tag{7.64}$$

traces a circle as R is varied. In Eq. (7.64) the denominator represents a series circuit. If we recall the approximate equivalent circuit of the induction machine (Fig. 7-16), we see that here too we have a series circuit consisting of a variable resistance R_2/s. Thus, according to Eq. (7.64), the current I^s traces a circle as the load on the machine is varied. To find the total current we have, from Fig. 7-16,

$$\mathbf{I} = \mathbf{I}_0 + \mathbf{I}^s \tag{7.65}$$

Recalling the results of the no-load and blocked-rotor tests, we have

$$I_0 \cos\phi_0 = \frac{V_0}{R_m}$$

$$I_s \cos\phi_s = \frac{W_s}{V_s}$$

Using these test data, we may construct the circle diagram shown in Fig. 7-17. The point P is a typical operating point. It moves to N for no-load conditions and to B for blocked-rotor conditions. The line BC is divided so that $BD/DC = R^r/R^s$. The per-phase results obtained from Fig. 7-17 are as follows:

input current $= OP$ power factor angle $= \theta$

input power $= PQ$, where $OV = \dfrac{\text{applied voltage}}{\text{phase}}$

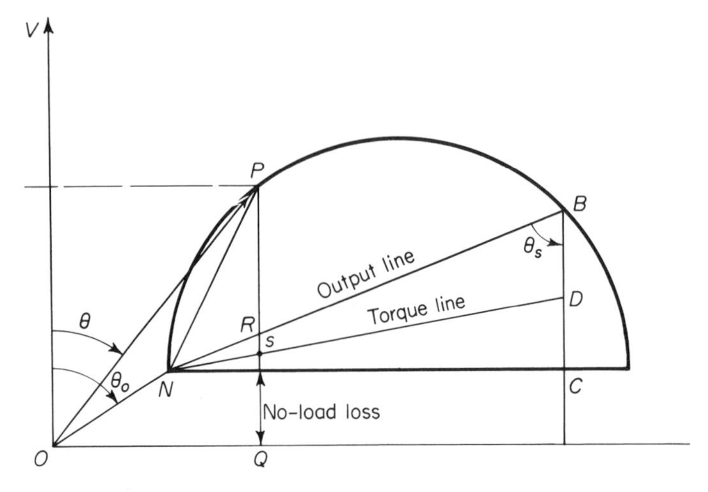

Fig. 7-17 Circle diagram of an induction motor.

$$\text{slip} = \frac{RS}{PS} \qquad 1-s = \frac{PR}{PS}$$

$$\text{torque} = OV(PS)\frac{33{,}000}{2\pi n \times 746}, \quad \text{where } n = \text{r/min of the rotor}$$

$$\text{efficiency} = \frac{PR}{PQ}$$

EXAMPLE 7–6

The results of the no-load and blocked-rotor tests on a 3-phase, 4-pole, 220-V, Y-connected induction motor are as follows:

No-load test—input voltage = 220 V
input power = 900 W
input current = 13.5 A
friction and windage loss = 400 W

Blocked-rotor test—input voltage = 40 V
input power = 1800 W
input current = 40 A

The per-phase stator resistance is 0.187 Ω. Draw the circle diagram and determine the performance of the motor if it takes a 60-A current.

First of all, we reduce various quantities to per-phase values, so that

$$V_0 = \frac{220}{\sqrt{3}} = 127 \quad \text{V}$$

$$I_0 = 13.5 \quad \text{A}$$

$$W_0 = \frac{900}{3} = 300 \quad \text{W}$$

$$\phi_0 = \cos^{-1} \frac{300}{13.5 \times 127} = 80^\circ$$

$$V_s = \frac{40}{\sqrt{3}} = 23 \quad \text{V}$$

$$I_s = 40 \quad \text{A}$$

$$W_s = \frac{1800}{3} = 600 \quad \text{W}$$

$$\phi_s = \cos^{-1} \frac{600}{23 \times 40} = 50^\circ$$

Because the rated voltage is 220 V and the blocked-rotor test is performed at 40 V, the current with the rotor blocked and 220 V applied to the stator is $40 \times 5.5 =$ 220 A (assuming linear dependence of current on the applied voltage). To find the point D we proceed as follows. From the blocked-rotor test the total effective

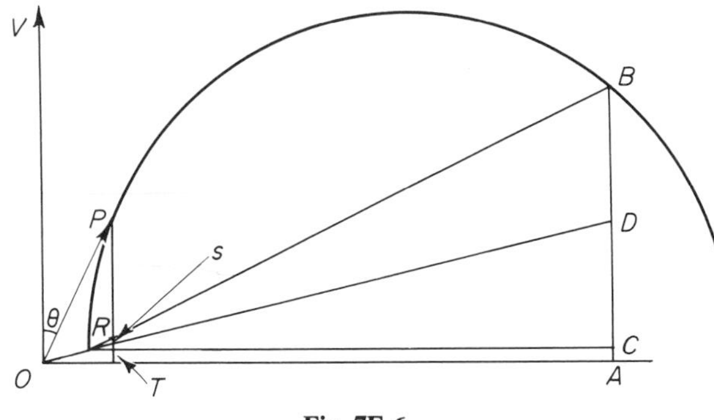

Fig. 7E-6

resistance of the series circuit (of Fig. 7-16) is $R_e = W_s/I_s^2$. Or $R_e = 600/1600$ $= 0.374\ \Omega$. But the stator resistance is 0.187 Ω, which is exactly $\frac{1}{2}\ R_e$, so that D is midway between B and C. We thus have complete data to construct the circle diagram shown in Fig. 7E-6. From this diagram, for an input current of 60 A, shown by OP, we find

$$\text{pf angle} = 24^\circ \qquad \text{power factor} = \cos 24^\circ = 0.914$$

$$\text{total input power} = 3 \times PQ = 3 \times 6600 = 19{,}800 \quad \text{W}$$

(*Note:* Observe that $TQ = W_0 = 300$ W)

$$\text{total output power} = 3 \times PR = 3 \times 5300 = 15{,}900 \quad \text{W}$$

$$\text{efficiency} = \frac{159}{198} = 80\%$$

$$\text{slip} = \frac{5}{48} = 0.104$$

Any other desired result can thus be obtained in a straightforward manner.

7.5 Speed Control of Induction Motors

The induction motor, because of its simplicity and ruggedness, finds numerous applications. However, it suffers from the drawback that, in contrast to dc motors, its speed cannot easily and efficiently be varied continuously over a wide range of operation. We shall briefly review the various possible methods by means of which the speed of the induction motor can be varied either continuously or in discrete steps. It is beyond the scope of this book to consider all these methods in detail and the interested reader should consult References 5–10.

The speed of the induction motor can be varied by either (1) varying the synchronous speed of the traveling field or (2) varying the slip. Because the efficiency of the induction motor is proportional to $(1-s)$, any method of speed control which depends on the variation of slip is inherently inefficient. On the other hand, if the supply frequency is constant, varying the speed by changing the synchronous speed results only in discrete changes in the speed of the motor. We shall now consider these methods of speed control in some detail.

7.5.1 Speed Control by Changing the Synchronous Speed

Recall that the synchronous speed n_s of the traveling field in a rotating induction machine is given by

$$n_s = 120\frac{f}{p} \tag{7.66}$$

where p = number of poles and f = supply frequency. Equation (7.66) indicates that n_s can be varied by either (1) changing the number of poles p, or (2) changing the frequency f. Both of these methods have found applications and we consider here the pertinent qualitative details.

(1) *The pole-changing method.* In this method, the stator winding of the motor is so designed that by changing the connections of the various coils (the terminals of which are brought out), the number of poles of the winding can be changed in the ratio 2 to 1. Accordingly, two synchronous speeds result. We observe that only two speeds of operation are possible. If more (say two) independent windings are provided, each arranged for pole chang-

ing, more (say four) synchronous speeds can be obtained. However, the fact remains that only discrete changes in the speed of the motor can be obtained by this method. The method has the advantage of being efficient and reliable (because the motor has a squirrel-cage rotor and no brushes).

Another method of pole-changing is by means of pole-amplitude modulation, and single-winding squirrel-cage motors are reported to have been developed which yield three operating speeds.[8] Another method, based on pole-changing, which produces three or five speeds, has been termed "phase-modulated pole-changing."[6] Like the simplest pole-changing method, the pole-amplitude modulation and the phase-modulated pole-change method give discrete variation in the synchronous speed of the motor.

(2) *The variable-frequency method.* Notice from Eq. (7.66) that the synchronous speed is directly proportional to the frequency. Therefore, if it is practicable to vary the supply frequency, the synchronous speed of the motor can also be varied. The variation in speed is continuous or discrete according to continuous or discrete variation of the supply frequency. However, the maximum torque developed by the motor is inversely proportional to the synchronous speed. Therefore, if we desire a constant maximum torque, the supply voltage should be increased with an increase in the supply frequency if we wish to increase the synchronous speed of the motor. The inherent difficulty in the application of this method is that the supply frequency commonly available is fixed; thus the method is applicable only if a variable-frequency supply is available. Various schemes have been proposed to obtain a variable-frequency supply. With the advent of solid-state devices with comparatively large power ratings it is now possible to use static inverters to drive the induction motor. One such inverter-driven single-phase induction motor is shown in Fig. 7-18. The output frequency as well as

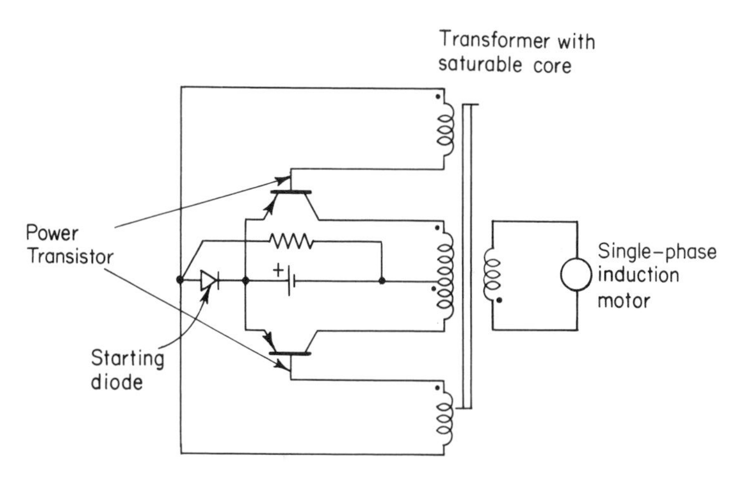

Fig. 7-18 Inverter-driven one-phase induction motor.

the voltage of the inverter vary simultaneously with the input dc voltage. Therefore, this method does not provide independent torque control on the motor. In Fig. 7-18, only the single-phase inverter circuit is shown. But this can be modified to yield a 2- or 3-phase output by using locked oscillators or some other modifications.[11]

(3) *The variable pole-pitch method.*[6,7] Using Eq. (7.66), we have outlined two methods of changing the synchronous speed of an induction motor. A third method of varying the synchronous speed follows from a consideration of the linear-induction machine, such as the one discussed in Example 4-1. We observe from Fig. 4E-1 that the linear synchronous speed u_s of the traveling field is given by

$$u_s = \lambda f \tag{7.67}$$

where $\lambda/2$ = pole pitch. Evidently, from Eq. (7.67), the synchronous speed can be varied by varying the pole pitch. Although Eq. (7.67) was written for the linear machine, it can be applied to the rotating machine by assuming that the rotating machine is made by "rolling up" the linear machine with a discontinuity where the two edges of the linear machine join. It must be pointed out that varying the pole pitch in a machine is an extremely difficult task. The method is simple and efficient in principle; in practice, however, only one method of varying the pole pitch has been reported[7] and, because of the special phase shifter required for changing the pole pitch, the overall efficiency and power/weight ratio of the machine are low. In short, the variable pole pitch method for rotating induction motors has not yet found practical applications.

7.5.2 Speed Control—Changing the Slip

The method of controlling the speed of an induction motor by changing its slip is best understood by referring to Fig. 7-19. The dotted curve shows the speed–torque characteristic of the load; the curves with solid lines are the speed–torque characteristics of the induction motor under various conditions (such as different rotor resistances—R_1, R_2, R_3; or different stator voltages—V_1, V_2). We have four different torque–speed curves and, therefore, the motor can run at any one of four speeds—N_1, N_2, N_3, or N_4—for the given load. Note that to the right of the peak torque is the stable operating region of the motor. In practice, the slip of the motor can be changed by one of the following methods.

(1) *Variable stator voltage method.* Since the electromagnetic torque developed by the machine is proportional to the square of the applied voltage, we obtain different torque–speed curves for different voltages applied to the motor. For a given rotor resistance R_2, two such curves are shown in Fig. 7-19 for two applied voltages V_1 and V_2. Thus, the motor can run at speeds

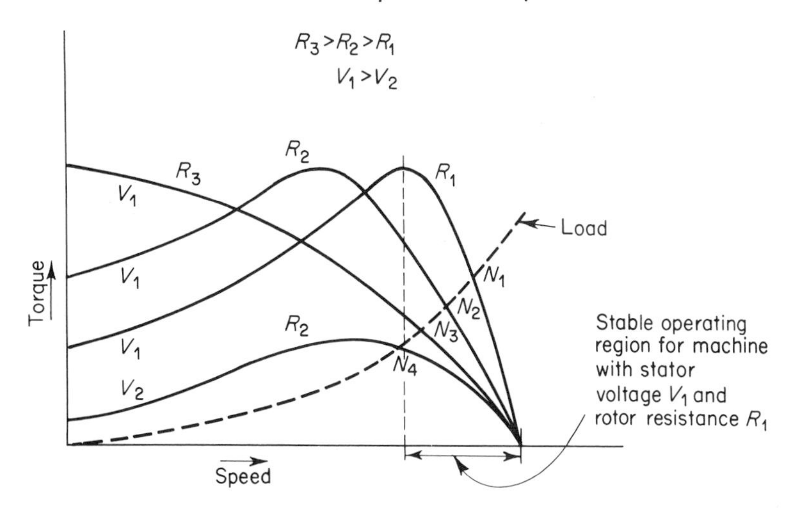

Fig. 7-19 Speed control by changing the slip.

N_2 or N_4. If the voltage can be varied continuously from V_1 to V_2, the speed of the motor can also be varied continuously between N_2 and N_4 for the given load. This method is applicable to the cage-type as well as the wound-rotor–type induction motors.

(2) *Variable rotor-resistance method.* This method is applicable only to the wound-rotor motor. The effect on the speed–torque curves of inserting external resistances in the rotor circuit is shown in Fig. 7-19 for three different rotor resistances R_1, R_2, and R_3. For the given load, three speeds of operation are possible. Of course, by continuous variation of the rotor resistance, continuous variation of the speed is possible.

(3) *Control by solid-state switching.* Other than the inverter-driven motor shown in Fig. 7-18, the speed of the wound-rotor motor can be controlled by inserting the inverter in the rotor circuit,[9,10] or by controlling the stator voltage by means of solid-state switching devices such as silicon-controlled rectifiers (also known as SCR's or thyristors). The output from the SCR feeding the motor is controlled by adjusting its firing angle; in this respect the method is similar to the variable-voltage method outlined earlier. However, it has been found that control by an SCR gives a wider range of operation and is more efficient than other slip-control methods.

(4) *Speed control by auxiliary machines.* There are numerous other schemes available for controlling the speed of the induction motor. These include concatenation, the Schrage motor, the Kramer control, the Scherbius control, etc., described in some detail in References 4 and 5.

In summary, it should be pointed out that the method controlling the speed of an induction motor by controlling its slip is basically inefficient and may have other disadvantages also. The method based on controlling synchronous speed is efficient but difficult to achieve in practice, especially with brushless (or cage-type) machines for continuous-speed variation. A satisfactory method of speed control of the brushless motor is, it seems, yet to be developed.

7.6 Transients in Induction Machines [12–16]

Transient conditions in induction machines occur under three circumstances: (1) at standstill, e.g., at the time of starting the machine; (2) at constant speed; and (3) at variable speed. These transients take place because of switching, plugging, overspeeding, reversing, and sudden applications of load on the machine. Induction machines supplied with a variable-frequency source also undergo transients. In addition to these causes, of course, there are other possible reasons for transient conditions. All in all, an understanding of the transients is quite important, and we shall therefore develop some quantitative analysis to obtain the transient performance of the machine. Unfortunately, no general method is available which can conveniently be used for all transient cases. But considerable research has been done on this topic, as reflected in References 12–16.

In Sec. 7.1.2 we mentioned that using the *dq* transformation the transient torques and currents can be obtained. In fact, most of the recent work on transients in induction machines is based on the use of *dq* transformation,[13,15,17] instantaneous symmetrical-component transformation[9,12] or by transformation to variable-speed axes.[14] Invariably, solutions to the equations of motion thus obtained require use of either an analog or digital computer, and we shall not consider these here. Rather, we shall take a somewhat simplified view and consider only certain special cases to justify certain approximations.

7.6.1 Constant-Speed Equations[1]

First of all, let us consider the operation of an induction motor running at a speed ω_m for a given load. Note from Fig. 7-19 that the normal operating region is to the right of the peak torque. If balanced voltages are applied to the machine, we know, from Secs. 7.1–7.4, that the electromagnetic torque developed by the machine is given by

$$T_e = \frac{\dfrac{\nu}{2}\omega(L^{sr})^2 \dfrac{R^r}{s}(V^s)^2}{\left\{\dfrac{R^r R^s}{s} - \omega^2[L^r L^s - (L^{sr})^2]\right\}^2 + \omega^2\left(L^r R^s + L^s \dfrac{R^r}{s}\right)^2} \tag{7.68}$$

For normal operating conditions, the slip is very small (of the order of 0.05 or less); that is, $s \simeq 0$, so that Eq. (7.68) can be approximately expressed as

$$T_e \simeq \frac{\nu}{2} \frac{\omega(L^{sr})^2(V^s)^2 s}{R^r[(R^s)^2+(\omega L^s)^2]} = ks \tag{7.69}$$

where

$$k = \frac{\nu}{2} \frac{\omega(L^{sr}V^s)^2}{R^r[(R^s)^2+(\omega L^s)^2]}$$

Because the electrical time constants are much smaller compared with the mechanical time constant, the electrical transients die away very quickly and do not affect the mechanical transients. The term k in Eq. (7.69) is a constant for specified V^s and ω, and the torque–slip characteristic is approximately linear. The linearized mechanical equation of motion is, for example,

$$J\dot{\omega}_m + b\omega_m + T_L = T_e = ks \tag{7.70}$$

where T_L is the externally applied torque (assumed to act in a direction opposite to the electrical torque T_e). But the slip is related to the rotor velocity by

$$s = 1 - \frac{\nu\omega_m}{\omega} \tag{7.71}$$

which when substituted in Eq. (7.70) yields

$$J\dot{\omega}_m + (b+b')\omega_m = k - T_L \tag{7.72}$$

where $b' = k\nu/\omega$. Equation (7.72) is a linear differential equation valid for small-slip operating region. The dynamic behavior around this quiescent operating point can be obtained by the method developed in Chapter 2 (Sec. 2.57). We assume

$$\left.\begin{aligned} \omega_m &= \Omega_0 + \omega_1 \\ T_L &= T_0 + \tau_1 \\ s &= S_0 + s_1 \end{aligned}\right\} \tag{7.73}$$

where (Ω_0, T_0, S_0) are steady time-independent values and (ω_1, τ_1, s_1) are small perturbations around (Ω_0, T_0, S_0). Substituting Eq. (7.73) in Eq. (7.72), we obtain

$$J\dot{\omega}_1 + (b+b')\omega_1 = -\tau_1 \tag{7.74a}$$

$$(b+b')\Omega_0 = k - T_0 \tag{7.74b}$$

The effect of the application of a load torque τ_1 is given by Eq. (7.74a), whereas Eq. (7.74b) gives the steady-state behavior. For the sake of illustration, we may let $\tau_1 = -U(t)$; that is, the load torque is represented by a unit step. The variation of the motor speed is given by

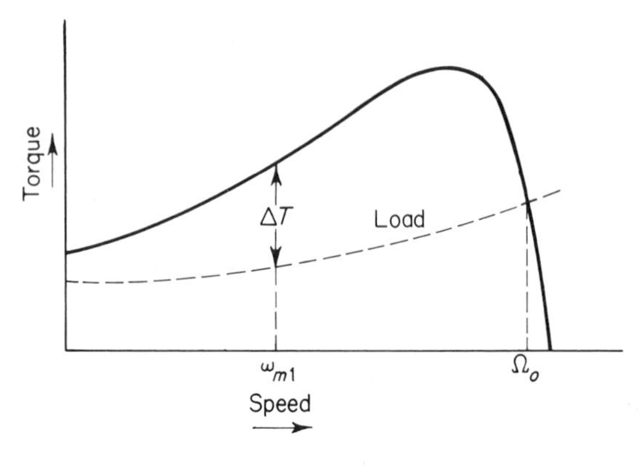

(a)

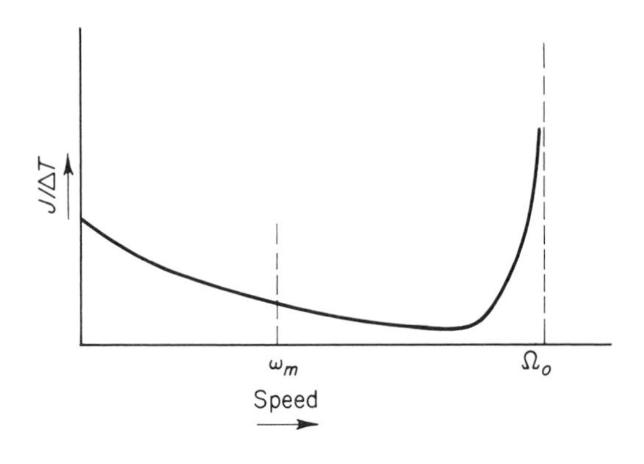

(b)

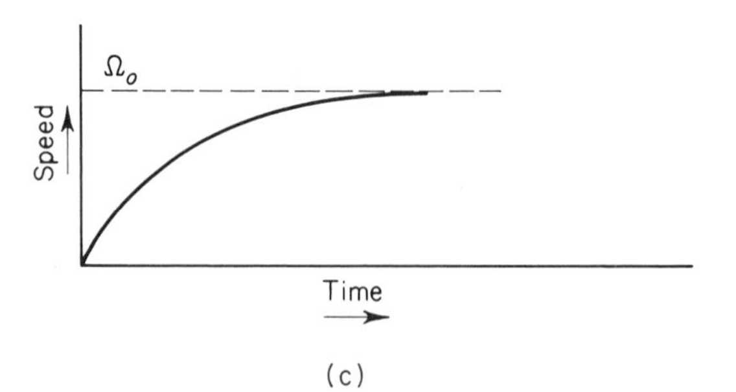

(c)

Fig. 7-20 Speed build-up in an induction motor.

$$\omega_1 = \frac{1 - e^{-[(b+b')/J]t}}{(b+b')} \tag{7.75}$$

Knowing ω_1, s_1 can be found from Eq. (7.71) and hence the variations in the input current due to a sudden application of load can be determined.

7.6.2 Speed Build-Up[1,5]

The speed–build-up characteristics of an induction motor can be very simply obtained graphically if we assume that the torque developed by the motor is given by a steady-state torque–speed curve such as shown in Fig. 7-20(a). The load torque is also shown in Fig. 7-20(a). Note that ΔT is the accelerating torque, so that the speed differential equation is

$$\Delta T = J\dot{\omega}_m \tag{7.76a}$$

and the speed–build-up time is given by

$$t = \int_0^{\omega_m} \frac{J}{\Delta T} d\omega_m \tag{7.76b}$$

Knowing J, we plot $J/\Delta T$ in Fig. 7-20(b) where ΔT is obtained from Fig. 7-20(a). Performing the graphical integration of the $[(J/\Delta T)–\omega_m]$ curve of Fig. 7-20(b) we obtain the curve shown in Fig. 7-20(c), which shows the speed–build-up characteristic of the motor. It should be pointed out that the above analysis is quite approximate, and often the departure from the actual speed–torque characteristic is quite substantial. A typical comparison between the curves obtained from steady-state and from transient analysis is shown in Fig. 7-21.

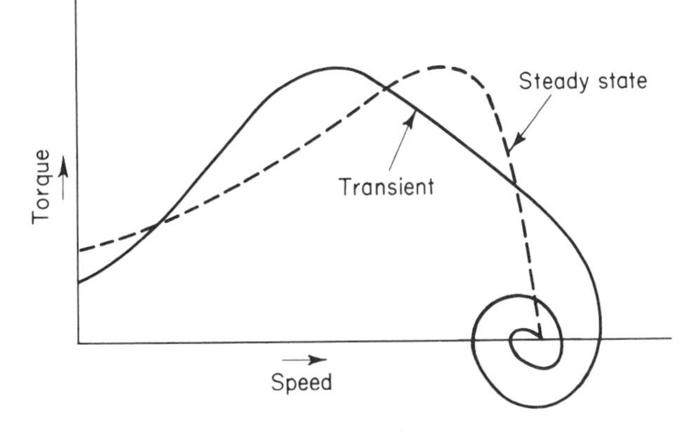

Fig. 7-21 Typical transient torque–speed curve, showing over-speeding.

†7.6.3 Variable-Speed Operation[16]

Using the theory developed in Chapters 4 and 5, we can obtain the expressions for transient torques in an induction motor as follows. Let us consider a 2-phase rotor in the magnetic field of the stator. If $\omega = \omega^s - \omega_m$ is the relative angular velocity between the traveling field and the rotor, then from Eq. (c), Example 5-1, the voltages induced in the two phases of the rotor are

$$v_a^r = \omega N\phi \sin \omega t \tag{7.77a}$$

$$v_b^r = \omega N\phi \sin\left(\omega t - \frac{\pi}{2}\right) \tag{7.77b}$$

Following the approach of Example 4-1, the currents in the rotor are given by

$$v_a^r = (R^r + L^r p) i_a^r \tag{7.78a}$$

$$v_b^r = (R^r + L^r p) i_b^r \tag{7.78b}$$

Here we have assumed balanced conditions. Now consider phase "a" only. From Eqs. (7.77a) and (7.78a) we have

$$i_a^r = \frac{N\phi}{L^r} e^{-t/\tau} \int_{-\infty}^{t} e^{t/\tau} \omega \sin \omega t \, dt \tag{7.79}$$

where $\tau = L^r/R^r$. We note that ω is time-dependent under transient conditions. Knowing the current, the torque is found from

$$T_a = 2rli_a^r B$$

where

$$\left.\begin{aligned} B &= B_m \sin \theta \\ \phi &= 2B_m rl \\ \omega &= \dot{\theta} \end{aligned}\right\} \tag{7.80}$$

From Eqs. (7.79) and (7.80) we obtain

$$T_a = \frac{N\phi^2}{L^r} e^{-t/\tau} \sin \theta \int_{-\infty}^{t} e^{t/\tau} \dot{\theta} \sin \theta \, dt \tag{7.81}$$

Similarly, for phase "b"

$$T_b = \frac{N\phi^2}{L^r} e^{-t/\tau} \cos \theta \int_{-\infty}^{t} e^{t/\tau} \dot{\theta} \cos \theta \, dt \tag{7.82}$$

The resultant instantaneous torque is, therefore,

$$T_e = T_a + T_b = \frac{N\phi^2}{L^r} \int_{-\infty}^{t} e^{(\omega - t)/\tau} \cos(\delta - \theta)\, d\omega \tag{7.83}$$

where the variable of integration has been changed from t to ω such that $\theta(t) = \delta(\omega)$ and $\dot{\theta} = \dot{\delta}$. If the variation in velocity is known, the instantaneous torque can be found by integrating Eqs. (7.81) and (7.82). It is, however, more convenient to express the torque in the form of a differential equation. After carrying through somewhat involved manipulations,[16] the torque differential equation takes the form

$$\frac{d^2T_e}{dt^2} + \left(\frac{2}{\tau} - \frac{\ddot{\theta}}{\dot{\theta}}\right)\frac{dT_e}{dt} + \left(\dot{\theta}^2 + \frac{1}{\tau^2} - \frac{\ddot{\theta}}{\tau\dot{\theta}}\right)T_e = \frac{a\dot{\theta}}{\tau} \tag{7.84}$$

where $a = N\phi^2/L^r$ and $\tau = L^r/R^r$. This equation for the torque can be solved on an analog computer, but the objection to the analysis is that it is difficult to determine ϕ and that it is based on the assumption that the motor is current-excited rather than voltage-excited, which is somewhat unrealistic.

†7.7 A Note on Analog-Computer Representation

We suggested in the last section that certain transients in the induction machine can be studied on an analog computer. We shall now briefly consider a particular analog-computer representation. First let us recall the equations of motion of the 2-phase machine in terms of dq variables (Sec. 7.1.2):

$$\begin{bmatrix} v_a^s \\ v_b^s \\ \hdashline 0 \\ 0 \end{bmatrix} = \left[\begin{array}{cc:cc} R^s + pL^s & 0 & pL^{sr} & 0 \\ 0 & R^s + pL^s & 0 & pL^{sr} \\ \hdashline pL^{sr} & v\dot{\theta}L^{sr} & R^r + pL^r & v\dot{\theta}L^r \\ -v\dot{\theta}L^{sr} & pL^{sr} & -v\dot{\theta}L^r & R^r + pL^r \end{array}\right] \begin{bmatrix} i_a^s \\ i_b^s \\ \hdashline i_d^r \\ i_q^r \end{bmatrix} \tag{7.85a}$$

$$T_e = -vL^{sr}(i_b^s i_d^r - i_a^s i_q^r) \tag{7.86b}$$

Substituting

$$i_d^r = \frac{L^{sr}}{R^r} i_d^{r\prime}$$

and (7.86c)

$$i_q^r = \frac{L^{sr}}{R^r} i_q^{r\prime}$$

in Eqs. (7.86a,b), we obtain

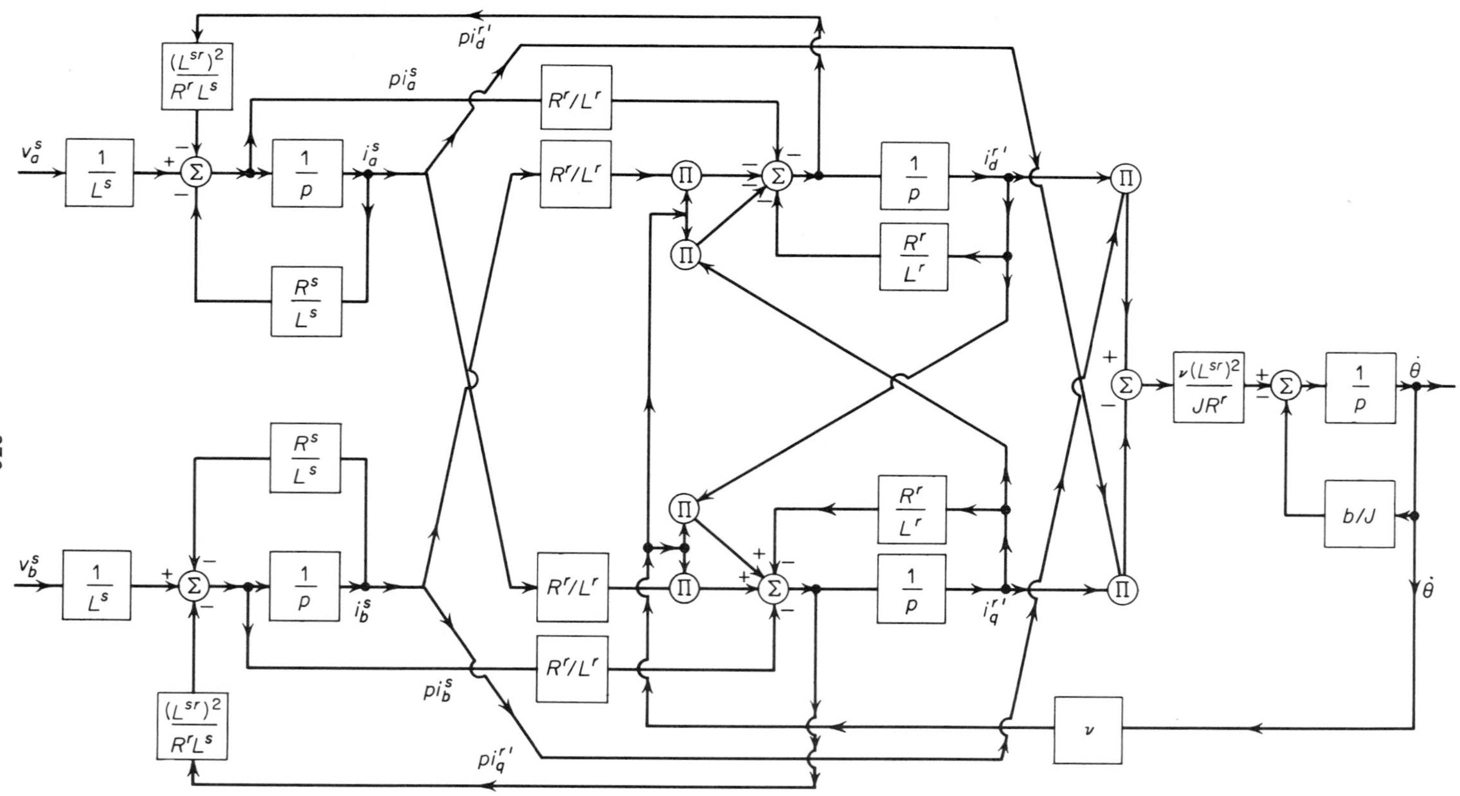

Fig. 7-22 Analog computer representation of an induction motor.

$$\begin{bmatrix} v_a^s \\ v_b^s \\ \hdashline 0 \\ 0 \end{bmatrix} = \left[\begin{array}{cc:cc} R^s + pL^s & 0 & p\dfrac{(L^{sr})^2}{R^r} & 0 \\ 0 & R^s + pL^s & 0 & p\dfrac{(L^{sr})^2}{R^r} \\ \hdashline p & \nu\dot{\theta} & 1 + p\dfrac{L^r}{R^r} & \nu\dot{\theta}\dfrac{L^r}{R^r} \\ -\nu\dot{\theta} & p & -\nu\dot{\theta}\dfrac{L^r}{R^r} & 1 + p\dfrac{L^r}{R^r} \end{array}\right] \begin{bmatrix} i_a^s \\ i_b^s \\ \hdashline i_d^{r\prime} \\ i_q^{r\prime} \end{bmatrix} \tag{7.87a}$$

$$T_e = -\nu \frac{(L^{sr})^2}{R^r} (i_b^s i_d^{r\prime} - i_a^s i_q^{r\prime}) \tag{7.87b}$$

If the mechanical load on the machine is of the form $(J\ddot{\theta} + b\dot{\theta})$, then

$$J\ddot{\theta} + b\dot{\theta} - T_e = 0 \tag{7.87c}$$

An analog-computer representation of Eqs. (7.87a–c) is shown in Fig. 7-22. From this representation the dynamic characteristic of the machine can be obtained.

7.8 Summary

In this chapter we studied the 2-phase induction machine using the *dq* and symmetrical-component transformations. We showed the equivalence of the two transformations and derived the single-phase machine from the 2-phase machine. We also showed that the analysis of the 2-phase machine is also applicable to the 3-phase machine. We derived a number of equivalent circuits for the induction machine and obtained its characteristics from equivalent circuits as well as from test data and circle diagrams. We discussed qualitatively the possible methods of speed control of induction motors, considered transients in the induction machine, and, finally, outlined an analog-computer representation for the 2-phase induction motor.

PROBLEMS

7–1. Derive Eq. (7.26) from Eqs. (7.24) and (7.25b).

7–2. Show that the circuit of Fig. 7-10 is exactly equivalent to the circuit shown in Fig. 7-6.

7–3. The voltages applied to the stator of a 2-phase servomotor are known as the *reference voltage* and *control voltage*. For a given machine, these voltages are respectively $v_a^s = v_R = V_R \cos \omega t$ and $v_b^s = v_c = V_c \sin \omega t$. Using Eq. (7.49) and the approximations $s\omega^2[L^r L^s - (L^{sr})^2] \ll R^r R^s$, $sL^r R^s \ll L^s R^r$, $(2-s)\omega^2 \cdot [L^r L^s - (L^{sr})^2] \ll R^r R^s$, and $(2-s)L^r R^s \ll L^s R^r$, show that the torque equation reduces to[1]

$$\langle T_e \rangle = \frac{(\nu/2)(L^{sr})^2 V_R}{R^r[(R^s)^2 + (\omega L^s)^2]}(2\omega V_c - V_R \nu \omega_m)$$

7–4. Derive Eq. (7.55) from Eq. (7.1a) using Eqs. (7.50), (7.51), (7.53d), and (7.53e).

7–5. In the 3-phase to dq transformations, defined by Eqs. (7.61a,b), verify if (a) the instantaneous power is invariant and (b) the average power is invariant under the transformation. Hence justify the assumption that $2N = 3N'$.

7–6. Using the circuit shown in Fig. 7-2, show that the induction machine can also be operated as a generator. Draw the corresponding phasor diagram.

7–7. From the circuit shown in Fig. 7-16, show that the per-phase torque of the induction motor can be expressed as

$$T = \frac{(V^s)^2(R^{r\prime}/s\omega)}{(R^s + R^{r\prime}/s)^2 + (x^s + x^{r\prime})^2}$$

where $R^{r\prime}$ and $x^{r\prime}$ are respectively the per-phase values of the rotor resistance and rotor reactance referred to the stator. [*Note*: The above expression is more commonly used than expressions such as Eq. (7.68).]

7–8. Derive Eq. (7.84) from Eqs. (7.81) and (7.82). (*Hint*: See Reference 16.)

7–9. A 220-V, 3-phase, 60-cycle, 4-pole, Y-connected induction motor has a per-phase stator resistance of 0.25 Ω. The no-load and blocked-rotor test data on this motor are

No-load test—stator voltage = 220 V
input current = 3.0 A
input power = 600 W
friction and windage loss = 300 W

Blocked-rotor test—stator voltage = 34.6 V
input current = 15.0 A
input power = 720 W

(a) Obtain the approximate equivalent circuit for the machine.
(b) If the machine runs as a motor with 5% slip, calculate the developed power, developed torque, and efficiency.
(c) Determine the slip at which maximum torque occurs and calculate the maximum torque.

7–10. From the test data of Problem 7-9, draw the circle diagram for the motor. Compare the results of part (b) of Problem 7-9 with those obtained from the circle diagram. Determine the maximum torque and the slip at which the maximum torque occurs from the circle diagram and compare with part (c) of Problem 7-9.

7–11. Using Eqs. (7.70–7.73), derive an expression for the slip as a function of time.

7–12. Using the method of Sec. 7.6.2, sketch the speed–build-up characteristic of an induction motor from the following data:
Motor torque–speed characteristic is given as

Speed (r/min)	Torque (lb-ft)	Speed (r/min)	Torque (lb-ft)
0	30	550	70
100	35	600	75
200	40	700	78
300	50	750	76
400	58	800	60
500	66	850	29
		900	0

The load on the machine is pure inertia of 20 lb-ft^2 and requires a constant torque of 15 lb-ft at all speeds.

REFERENCES

1. White, D. C., and H. H. Woodson, *Electromechanical Energy Conversion.* New York: John Wiley & Sons, Inc., 1959.

2. Seely, S., *Electromechanical Energy Conversion.* New York: McGraw-Hill Book Company, 1962.

3. Nasar, S. A., "Validity of No-Load and Locked-Rotor Tests on 1-Phase Motors," *Inter. Jour. E. E. Education.* Oxford: Pergamon Press Ltd., January 1966, p. 11.

4. Puchstein, A. F., T. C. Lloyd, and A. G. Conrad, *Alternating-current Machines,* 3rd ed. New York: John Wiley & Sons, Inc., 1954.

5. Fitzgerald, A. E., and C. Kingsley, Jr., *Electric Machinery,* 2nd ed. New York: McGraw-Hill Book Company, 1961.

6. Laithwaite, E. R., *Induction Machines for Special Purposes.* New York: Chemical Publishing Company, Inc., 1966.

7. Williams, F. C., E. R. Laithwaite, J. F. Eastham, and L. S. Piggott, "The Logmotor—A Cylindrical Brushless Variable Speed Induction Motor," *Proc. IEE* (London), Vol. 108(a), 1961, pp. 91–99.

8. Fong, W., and G. H. Rawcliffe, "Three-Speed, Single-Winding, Squirrel-Cage Induction Motors," *Proc. IEE* (London), Vol. 110, 1963, p. 1649.

9. Erlicki, M. S., J. Ben Uri, and Y. Wallach, "Switching Drive of Induction Motors," *Proc. IEE* (London), Vol. 110, 1963, pp. 1441–1450.

10. Lavi, A., and R. J. Polge, "Induction Motor Speed Control with Static Inverter in the Rotor," *IEEE Trans.*, Vol. PAS-85 ("Power Apparatus and Systems"), 1966, pp. 76–84.

11. Milnes, A. G., "Phase Locking of Switching-Transistor Converters for Polyphase Power Supplies," *AIEE Trans.*, Vol. 74, Part I, 1955, p. 587.

12. Lyon, W. V., *Transient Analysis of Alternating Current Machinery*. New York: John Wiley & Sons, Inc., 1954.

13. Rogers, G. J., "Linearised Analysis of Induction-Motor Transients," *Proc. IEE* (London), Vol. 112, 1965, pp. 1917–1926.

14. Lawrenson, P. J., and J. M. Stephenson, "Note on Induction-Machine Performance with a Variable-Frequency Supply," *Proc. IEE* (London), Vol. 113, 1966, pp. 1617–1623.

15. Ward, E. E., A. Kazi, and R. Farkas, "Time-Domain Analysis of the Inverter-Fed Induction Motor," *Proc. IEE* (London), Vol. 114, 1967, pp. 361–369.

16. West, J. C., B. V. Jayawant, and G. Williams, "Analysis of Dynamic Performance of Induction Motors in Control Systems," *Proc. IEE* (London), Vol. 111, 1964, pp. 1468–1478.

17. Slater, R. D., and W. S. Wood, "Constant-Speed Solutions Applied to the Evaluation of Induction-Motor Transient Torque Peaks," *Proc. IEE* (London), Vol. 114, 1967, pp. 1429–1435.

8

Synchronous Machines

In Chapter 5 (Sec. 5.1.1) we briefly introduced the synchronous machine and identified some of its parts. In Chapter 2 (Example 2-6) we showed that the reluctance machine is a synchronous machine, and in Example 5-5 we used the *dq* transformation in deriving the torque equation of a salient-pole synchronous machine. We are, then, already roughly familiar with the synchronous machine and with certain forms of its equations of motion. In the present chapter, we shall approach the subject more formally. Beginning with the idealized machine introduced in Chapter 5, we shall obtain the equations of motion of the round-rotor and salient-pole synchronous machines. We shall solve the equations of motion to obtain the steady-state and dynamic characteristics of the machine by using certain linear transformations. We shall also consider some of the physical characteristics of these machines and outline a few problems in which the synchronous machine is used as an element of a system. In this chapter we shall restrict our discussions to the idealized machine—that is, we shall ignore saturation, harmonics, and armature reaction—and shall comment on realistic considerations in Chapter 9.

We noted in Chapter 7 (Sec. 7.1.1) that, although the application of frequency constraints (Table 5-1) yields the equations of motion, the algebraic details are cumbersome. We also recall that for the analysis of dynamic operation of the induction machine it was convenient to use linear transformations. For the synchronous machine, therefore, we shall begin our analysis by obtaining the equations of motion in terms of transformed variables. We shall consider the 2-phase machine in detail, because the analysis is applicable to the more conventional 3-phase machine. We shall discuss the salient-pole machine first and obtain the results for the cylindrical-rotor machine by applying simple constraints.

8.1 The Equations of Motion[1, 2]

Recall from Sec. 5.1.1 that a 2-phase synchronous machine has two independent windings, called *armature windings*, on the stator, and one winding, known as the *field winding*, on the rotor. The armature windings carry alternating current whereas the field winding is fed with direct current. In some cases, especially in large machines, the rotors also carry *damper windings*, which come into action during transient conditions but otherwise remain idle during steady-state operation, when the machine is running at synchronous speed. As pointed out earlier, the cylindrical-rotor machine is a special case of the salient-pole machine. Therefore, we need to consider only the salient-pole machine in detail.

As for the induction machine, we repeat the equations of motion, Eqs. (7.1a,b), in terms of machine variables. These equations are

$$\begin{bmatrix} v^s \\ \hline v^r \end{bmatrix} = \left[\begin{array}{c|c} R^s + pL^{ss} & pL^{sr} \\ \hline pL^{rs} & R^r + pL^{rr} \end{array}\right] \begin{bmatrix} i^s \\ \hline i^r \end{bmatrix} \tag{8.1a}$$

$$T_e = -\frac{1}{2}[i^s \quad i^r]\frac{\partial}{\partial\theta}\left[\begin{array}{c|c} L^{ss} & L^{sr} \\ \hline L^{rs} & L^{rr} \end{array}\right] \begin{bmatrix} i^s \\ \hline i^r \end{bmatrix} \tag{8.1b}$$

An idealized 2-phase salient-pole synchronous machine is shown in Fig. 8-1, for which the various inductances are shown in Fig. 8-2. For simplicity we have shown a 2-pole machine, but we shall assume ν pairs of poles for generality. For the machine with ν pairs of poles, the various submatrices to be used in Eqs. (8.1a,b) are

$$R^s = \begin{bmatrix} R^s & 0 \\ 0 & R^s \end{bmatrix} \qquad R^r = R^r \tag{8.2a}$$

$$\left[\begin{array}{c|c} L^{ss} & L^{sr} \\ \hline L^{rs} & L^{rr} \end{array}\right] = \left[\begin{array}{cc|c} L^s + L^s_o \cos 2\nu\theta & -L^s_o \sin 2\nu\theta & L^{sr} \cos \nu\theta \\ -L^s_o \sin 2\nu\theta & L^s - L^s_o \cos 2\nu\theta & -L^{sr} \sin \nu\theta \\ \hline L^{sr} \cos \nu\theta & -L^{sr} \sin \nu\theta & L^{sr} \end{array}\right] \tag{8.2b}$$

$$\begin{bmatrix} i^s \\ i^r \end{bmatrix} = \begin{bmatrix} i^s_a \\ i^s_b \\ -I^r \end{bmatrix} \qquad \begin{bmatrix} v^s \\ v^r \end{bmatrix} = \begin{bmatrix} v^s_a \\ v^s_b \\ V^r \end{bmatrix} \tag{8.2c}$$

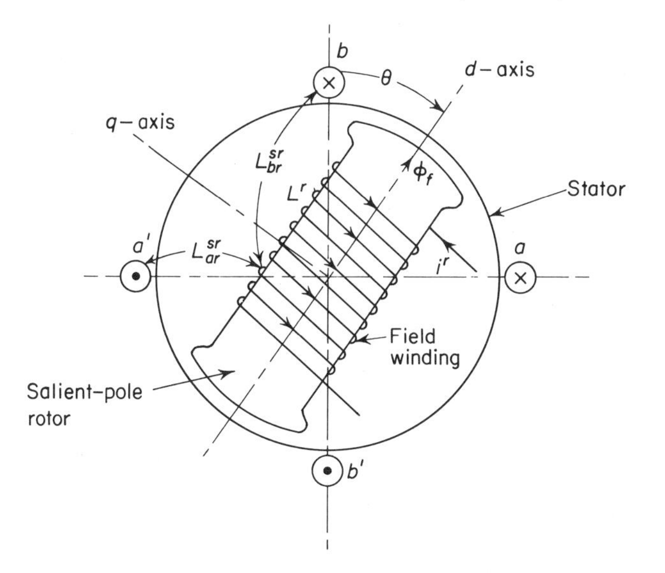

Fig. 8-1 A two-phase, salient-pole synchronous machine.

Note that Eq. (8.2b) was obtained in Example 5-4, Eq. (5.27). At this stage we might recall that we have considered this machine in Example 5-5 to introduce the *dq* transformation. We continue with Example 5-5 and consider the equations of motion in *dq* variables.

8.1.1 Equations of Motion in dq Variables

Slightly modifying the S_{dq} matrix derived in Example 5-5, Eq. (5.39), we may define a transformation matrix S_{dq} by

$$S_{dq} = \begin{bmatrix} \cos \nu\theta & \sin \nu\theta & 0 \\ -\sin \nu\theta & \cos \nu\theta & 0 \\ 0 & 0 & 1 \end{bmatrix} \tag{8.3a}$$

such that

$$i_{ab} = S_{dq}\, i_{dq} \tag{8.3b}$$

$$v_{ab} = S_{dq}\, v_{dq} \tag{8.3c}$$

$$S_{dq}^{-1} = \tilde{S}_{dq} \tag{8.3d}$$

The equations of motion in terms of transformed variables are

$$v_{dq} = Ri_{dq} + S_{dq}^{-1}(pL)S_{dq}i_{dq} \tag{8.4a}$$

$$T_e = -\frac{1}{2}\, \tilde{i}_{dq}\tilde{S}_{dq} \frac{\partial L}{\partial \theta} S_{dq}i_{dq} \tag{8.4b}$$

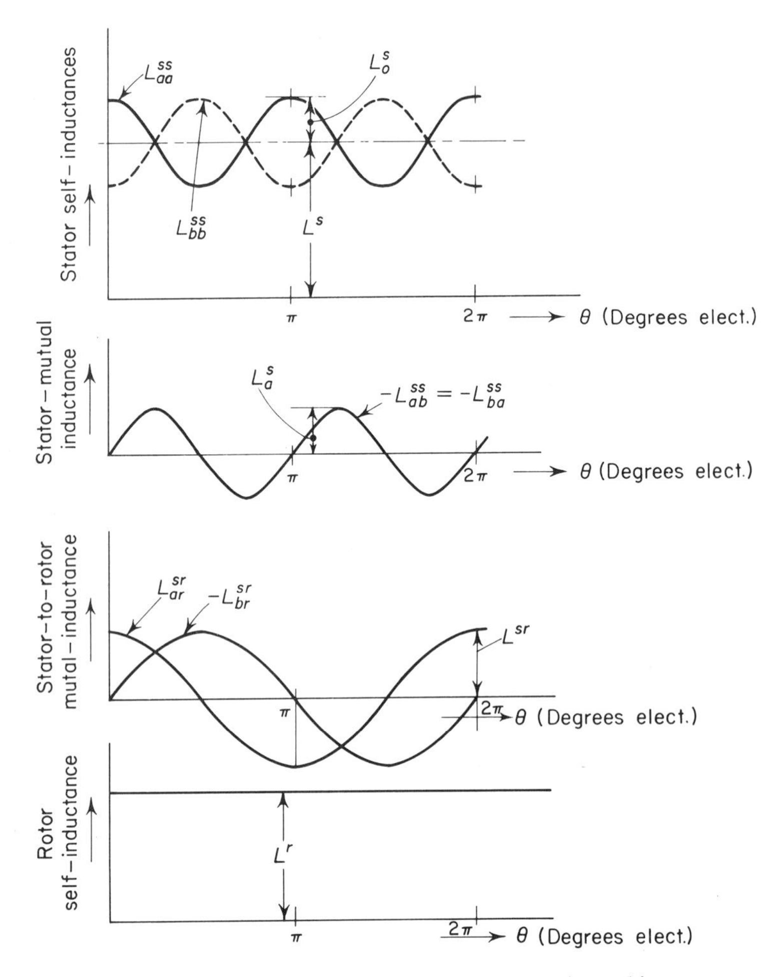

Fig. 8-2 Inductance variations for a salient-pole machine.

In these equations each term denotes a matrix. In order to expand these equations, consider the second term in Eq. (8.4a):

$$\begin{aligned} S_{dq}^{-1}(pL)S_{dq}i_{dq} &= S_{dq}^{-1}\left[\left(\frac{\partial L}{\partial \theta}\dot{\theta}\right)S_{dq}i_{dq} + L\frac{\partial}{\partial \theta}(S_{dq})\dot{\theta}i_{dq} + LS_{dq}(pi_{dq})\right] \\ &= \left[S_{dq}^{-1}\frac{\partial L}{\partial \theta}S_{dq} + S_{dq}^{-1}L\frac{\partial}{\partial \theta}(S_{dq})\right]\dot{\theta}i_{dq} + (S_{dq}^{-1}LS_{dq})pi_{dq} \\ &= \left\{S_{dq}^{-1}\left[\frac{\partial}{\partial \theta}(LS_{dq})\right]\dot{\theta} + [S_{dq}^{-1}LS_{dq}]p\right\}i_{dq} \end{aligned} \tag{8.4c}$$

It can be verified from Eqs. (8.2b), (8.3a), and (8.3d) that

$$S_{dq}^{-1}\left[\frac{\partial}{\partial\theta}(LS_{dq})\right] = \begin{bmatrix} 0 & \nu L_q & 0 \\ -\nu L_d & 0 & -\nu L^{sr} \\ 0 & 0 & 0 \end{bmatrix} \quad (8.4d)$$

and

$$S_{dq}^{-1}LS_{dq} = \begin{bmatrix} L_d & 0 & L^{sr} \\ 0 & L_q & 0 \\ L^{sr} & 0 & L^r \end{bmatrix} \quad (8.4e)$$

where

$$L_d = L^s + L_o^s \quad (8.4f)$$

$$L_q = L^s - L_o^s \quad (8.4g)$$

Substituting Eqs. (8.4d) and (8.4e) in Eq. (8.4c) yields

$$S_{dq}^{-1}(pL)S_{dq}i_{dq} = \begin{bmatrix} L_d p & \nu\dot{\theta}L_q & L^{sr}p \\ -\nu\dot{\theta}L_d & L_q p & -\nu\dot{\theta}L^{sr} \\ L^{sr}p & 0 & L^r p \end{bmatrix}\begin{bmatrix} i_d^s \\ i_q^s \\ -I^r \end{bmatrix} \quad (8.4h)$$

From Eqs. (8.4a) and (8.4h), therefore,

$$\begin{bmatrix} v_d^s \\ v_q^s \\ \hline V^r \end{bmatrix} = \left[\begin{array}{cc|c} R^s + L_d p & \nu\dot{\theta}L_q & L^{sr}p \\ -\nu\dot{\theta}L_d & R^s + L_q p & -\nu\dot{\theta}L^{sr} \\ \hline L^{sr}p & 0 & R^r + L^r p \end{array}\right]\begin{bmatrix} i_d^s \\ i_q^s \\ \hline -I^r \end{bmatrix} \quad (8.5a)$$

As shown in Example 5-5, the torque developed by the machine is given by

$$T_e = \nu[i_d^s i_q^s(L_d - L_q) - i_q^s I^r L^{sr}] \quad (8.5b)$$

The inductances L_d and L_q are respectively known as the *direct-axis* and *quadrature-axis inductances.*

Notice that the angular dependence has been eliminated from the equations of motion by the *dq* transformation, and for constant-speed operation, Eq. (8.5a) is linear.

8.1.2 Equivalent Circuits

The circuit representation of Eq. (8.5a) is shown in Fig. 8-3. Notice that a speed voltage, identified by the coefficient $\nu\dot{\theta}$, is induced in the d axis by the current in the q axis, and speed voltages in the q axis are induced by the rotor current and the d-axis current. The equivalent circuit (of Fig. 8-3) can be put in a slightly modified, and perhaps more meaningful, form by defining the following flux linkages:

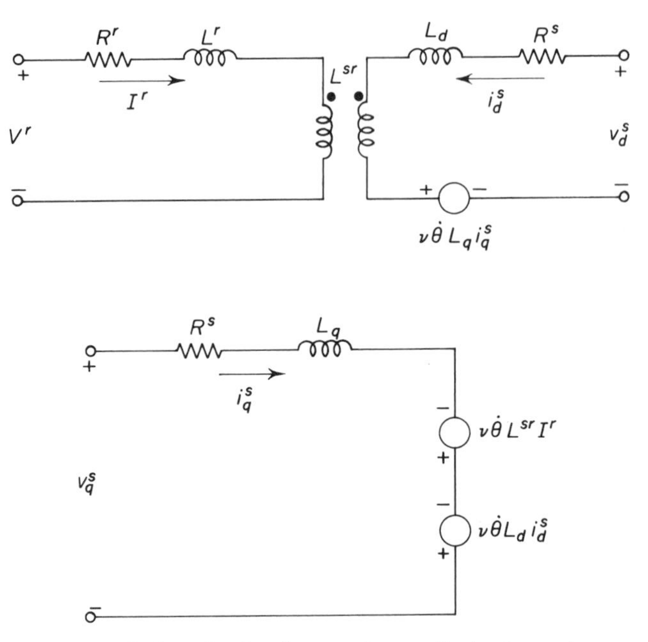

Fig. 8-3 Equivalent circuits of a two-phase, salient-pole synchronous machine.

$$\begin{bmatrix} \lambda_d^s \\ \lambda_q^s \\ \lambda^r \end{bmatrix} = \begin{bmatrix} L_d & 0 & L^{sr} \\ 0 & L_q & 0 \\ L^{sr} & 0 & L^r \end{bmatrix} \begin{bmatrix} i_d^s \\ i_q^s \\ i^r \end{bmatrix} \tag{8.6}$$

By defining these flux linkages we have identified the transformer voltages, accompanied by the coefficient p $(= d/dt)$. Based on Eq. (8.6), the d-axis and q-axis equivalent circuits take the form shown in Fig. 8-4.

The circuits shown in Figs. 8-3 and 8-4 are quite general. These can be modified for various operating conditions. For instance, the synchronous machine operates at constant speed (under steady-state conditions) and the stator voltages are often balanced. In this case, the equivalent circuit can be considerably simplified by using the following constraints:

$$\dot{\theta} = \omega_m = \frac{\omega}{\nu} \tag{8.7a}$$

$$\theta = \omega_m t + \delta \tag{8.7b}$$

Because the voltages are balanced, we assume

$$v_a^s = V^s \sin \omega t \tag{8.8a}$$

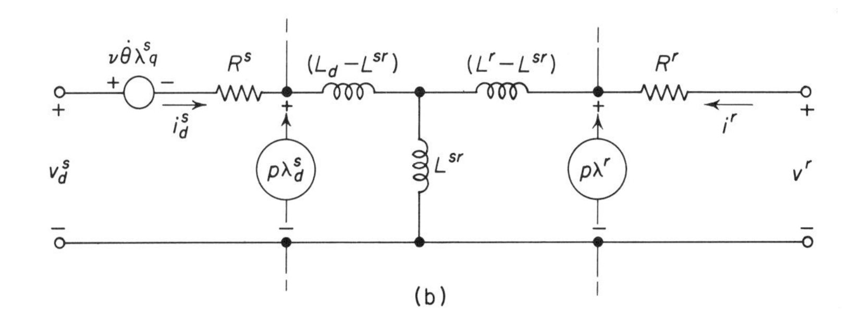

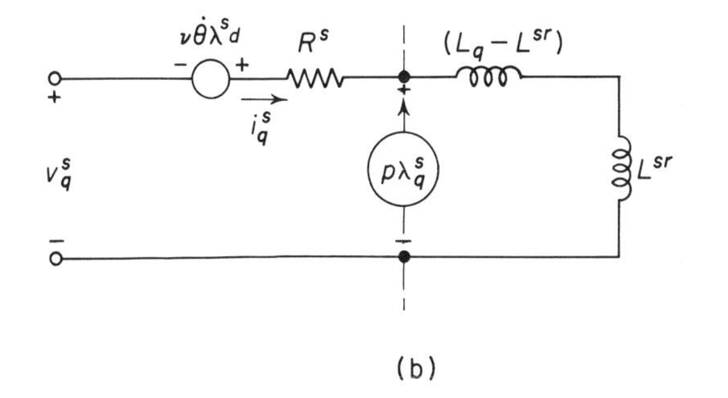

Fig. 8-4 (a) d-axis and (b) q-axis equivalent circuits of a salient-pole synchronous machine.

$$v_b^s = V^s \cos \omega t \tag{8.8b}$$

From Eqs. (8.3c) and (8.8), therefore,

$$\begin{bmatrix} v_d^s \\ v_q^s \end{bmatrix} = V^s \begin{bmatrix} \cos \nu\theta & -\sin \nu\theta \\ \sin \nu\theta & \cos \nu\theta \end{bmatrix} \begin{bmatrix} \sin \omega t \\ \cos \omega t \end{bmatrix} \tag{8.8c}$$

which when expanded and used in conjunction with Eqs. (8.7a,b) yields

$$v_d^s = -V^s \sin \nu\delta \tag{8.9a}$$

$$v_q^s = V^s \cos \nu\delta \tag{8.9b}$$

Notice, from Eqs. (8.9a,b) that, for a balanced 2-phase synchronous machine operating under steady-state conditions, the applied ac voltages are equivalent to dc voltages and the effects of inductances vanish. The equivalent circuits shown in Fig. 8-3 reduce to purely resistive circuits shown in Fig. 8-5. It can readily be verified that the circuits shown in Fig. 8-5 are consistent with the general volt-amp equations, Eq. (8.5a).

Having obtained the equations of motion and the equivalent circuits, we can now proceed to obtain some of the steady-state and transient performance characteristics of the synchronous machine.

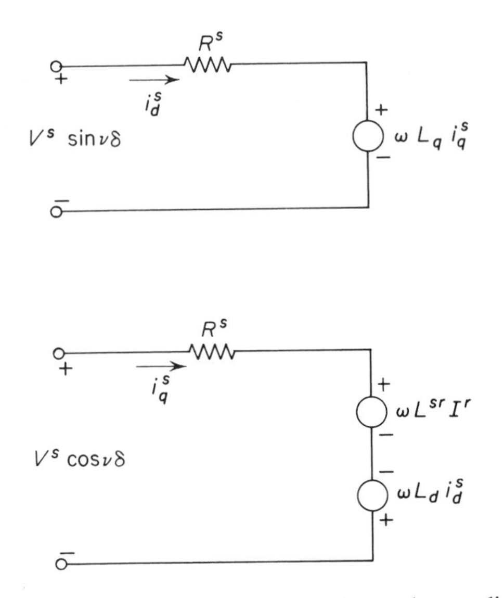

Fig. 8-5 Equivalent circuits of a balanced two-phase, salient-pole synchronous machine under steady state.

8.2 Steady-State Performance Characteristics

In order to obtain the steady-state behavior of the machine, we rewrite Eqs. (8.5a) in conjunction with Eqs. (8.9a,b) to obtain (for $\nu = 1$)

$$-V^s \sin\delta = R^s I_d^s + \omega L_q I_q^s \tag{8.10a}$$

$$V^s \cos\delta = -\omega L_d I_d^s + R^s I_q^s + V_0 \tag{8.10b}$$

where

$$V_0 = \omega L^{sr} I^r \tag{8.10c}$$

In most machines the armature resistance is negligible compared with the inductive reactances ωL_d and ωL_q. Consequently, Eqs. (8.10a,b) can be simplified to

$$-V^s \sin\delta = \omega L_q I_q^s \tag{8.11a}$$

$$V^s \cos\delta = -\omega L_d I_d^s + V_0 \tag{8.11b}$$

Solving for the currents I_d^s and I_q^s we have

$$I_d^s = \frac{1}{\omega L_d}(V_0 - V^s \cos\delta) \tag{8.12a}$$

$$I_q^s = -\frac{V^s \sin\delta}{\omega L_q} \tag{8.12b}$$

Substituting Eqs. (8.12a,b) in Eq. (8.5b), the expression for the electromagnetic torque becomes

$$T_e = \frac{V^s V_0}{\omega^2 L_d} \sin\delta + \frac{(V^s)^2}{2\omega^2}\left(\frac{1}{L_q} - \frac{1}{L_d}\right) \sin 2\delta \tag{8.13}$$

The power developed by the machine is, therefore,

$$P_d = T_e\omega = \frac{V^s V_0}{\omega L_d} \sin\delta + \frac{(V^s)^2}{2}\left(\frac{1}{\omega L_q} - \frac{1}{\omega L_d}\right) \sin 2\delta \tag{8.14a}$$

Or, putting $X_d = \omega L_d$ and $X_q = \omega L_q$,

$$P_d = T_e\omega = \frac{V^s V_0}{X_d} \sin\delta + \frac{(V^s)^2}{2}\left(\frac{1}{X_q} - \frac{1}{X_d}\right) \sin 2\delta \tag{8.14b}$$

The variation of the developed power as a function of δ is shown in Fig. 8-6. Notice that the resultant power is composed of power due to saliency (or reluctance power) and power due to the field excitation (or rotor current). The quantities X_d and X_q in Eq. (8.14b) are respectively known as the *direct-*

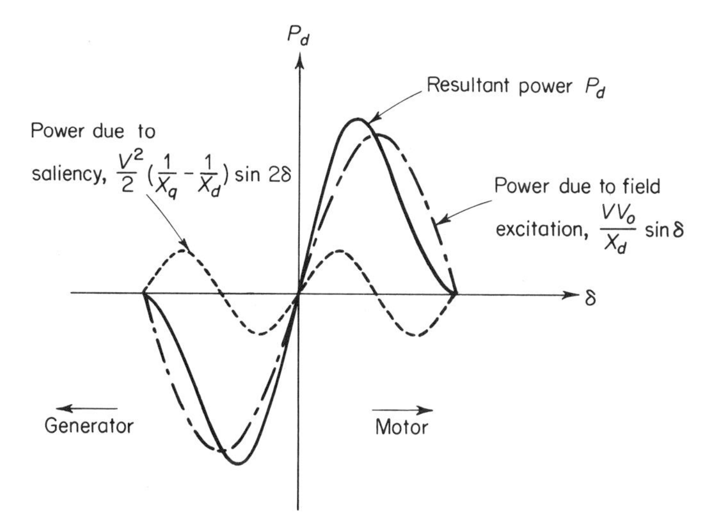

Fig. 8-6 Power-angle characteristic of a salient-pole synchronous machine.

axis and *quadrature-axis reactances.* In this respect, we might notice the similarity between an unexcited salient-pole synchronous machine and the reluctance machine (discussed in Example 2-6). The angle δ is known as the *power angle*, and is a measure of the power developed by the machine. Finally, it should be pointed out that, because the constant-speed and steady-state constraints are used in deriving Eq. (8.14b), this equation gives the steady-state power developed by the machine.

Turning now to the volt-amp equations, Eqs. (8.10a–c), we have, from Eqs. (8.8c) and (7.23a), for sinusoidal excitation,

$$\mathbf{V}^s = V_d^s + jV_q^s = \frac{1}{\sqrt{2}}(v_d^s + jv_q^s) \tag{8.15a}$$

$$\mathbf{I}^s = I_d^s + jI_q^s = \frac{1}{\sqrt{2}}(i_d^s + ji_q^s) \tag{8.15b}$$

so that Eqs. (8.10a–c) under steady-state conditions become (neglecting R^s)

$$-V^s \sin\delta = jX_q I_q^s \tag{8.16a}$$

$$V^s \cos\delta = -jX_d I_d^s + V_0 \tag{8.16b}$$

$$V_0 = \omega L^{sr} I^r \tag{8.16c}$$

For generator operation these equations can be represented by the phasor diagram shown in Fig. 8-7. Here we can see the geometrical significance of

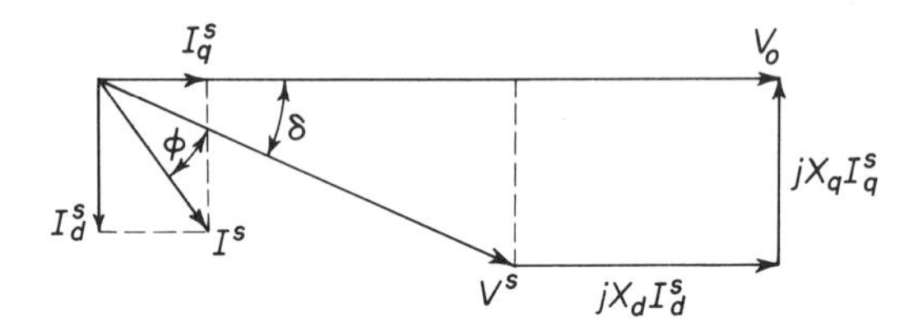

Fig. 8-7 Phasor diagram of a salient-pole synchronous generator, neglecting armature resistance.

the dq components. Clearly, the line current is resolved into the dq components I_d^s and I_q^s, and the direct-axis and quadrature-axis reactances are respectively the maximum and minimum values of the armature reactance for two corresponding rotor positions. These reactances can be measured experimentally, as we shall show in a later section. The preceding equations, phasor diagram, and equivalent circuits for the balanced machine are valid on a per-phase basis. To find the total power, line currents, etc., the values should be multiplied by appropriate constants, as illustrated by the following example.

EXAMPLE 8–1

A 20-kVA, 220-V, 60-cycle, Y-connected, 3-phase salient-pole synchronous generator supplies rated load at 0.707 lagging power factor. The per-phase constants of the machine are armature resistance $R^s = 0.05\ \Omega$; direct-axis reactance $X_d^s = 4.0\ \Omega$; quadrature-axis reactance $X_q^s = 2.0\ \Omega$. Calculate the developed power and per cent voltage regulation at the specified load.

$$V^s = \frac{220}{\sqrt{3}} = 127 \quad \text{V}$$

$$I^s = \frac{20{,}000}{\sqrt{3} \times 220} = 52.5 \quad \text{A}$$

$$\phi = \cos^{-1} 0.707 = 45°$$

From Fig. 8-7 we have

$$I_d^s = I^s \sin(\delta + \phi)$$

$$I_q^s = I^s \cos(\delta + \phi)$$

$$V^s \sin \delta = X_q I_q^s = X_q I^s \cos(\delta + \phi)$$

$$= X_q I^s \cos \delta \cos \phi - X_q I^s \sin \delta \sin \phi$$

Or,

$$\tan \delta = \frac{X_q I^s \cos \phi}{V^s + X_q I^s \sin \phi} \tag{a}$$

Substituting the given values in Eq. (a) yields

$$\tan \delta = \frac{2 \times 52.5 \times 0.707}{127 + (52.5 \times 2 \times 0.707)} = \frac{74.2}{201.2} = 0.37$$

Or,

$$\delta = 20.6°$$

$$I_d^s = 52.5 \sin(20.6 + 45) = 47.5 \quad \text{A}$$

$$X_d I_d^s = 4 \times 47.5 = 190.0 \quad \text{V}$$

$$V_0 = V^s \cos \delta + X_d I_d^s$$

$$= 127 \cos 20.6 + 190 = 308 \quad \text{V}$$

Per cent regulation is defined by

$$\text{regulation} = \frac{V_0 - V^s}{V^s} \times 100\% = \frac{308 - 127}{127} \times 100 = 142\%$$

The power developed per phase is obtained from Eq. (8.14b), which yields

$$P_d = \tfrac{1}{4}(127 \times 308) \sin 20.8 + \tfrac{1}{2}(127)^2(0.25) \sin 41.6$$

$$= (3.48 + 1.34) \quad \text{kW} = 4.82 \quad \text{kW}$$

The total developed power is, therefore,

$$(P_d)_T = 3 \times 4.82 = 14.46 \quad \text{kW}$$

8.3 The Cylindrical-Rotor Machine

We pointed out earlier that the characteristics of the cylindrical-rotor machine can be derived from the equations of motion of the salient-pole machine. The parameter constraint on the cylindrical-rotor machine is that $X_d = X_q = X_s$, known as *synchronous reactance*. Substituting these in Eqs. (8.16a,b) and (8.14a) we obtain the steady-state characteristics of the cylindrical-rotor machine as given by

$$-V^s \sin\delta = jX_s I_q^s \tag{8.17a}$$

$$V^s \cos\delta = -jX_s I_d^s + V_0 \tag{8.17b}$$

$$P_d = \frac{V^s V_0}{X_s} \sin\delta \tag{8.17c}$$

Equations (8.17a,b) are represented by the phasor diagram shown in Fig. 8-8(a); the corresponding equivalent circuit is shown in Fig. 8-8(b). Here again we have neglected armature resistance.

The power-angle characteristic is given by Eq. (8.17c) and, comparing it to Eq. (8.14a), we notice that the power due to saliency is zero. From Fig. 8-8(a) we can express the power delivered to the load as

$$P_d = V^s I^s \cos\phi \tag{8.18a}$$

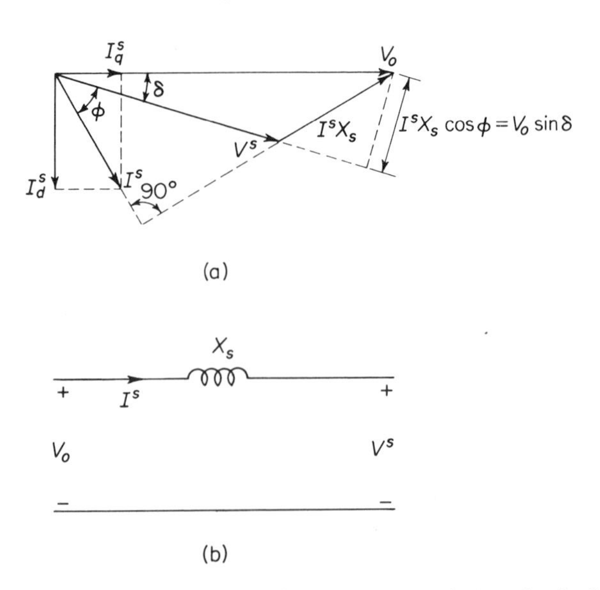

Fig. 8-8 (a) Phasor diagram; (b) approximate equivalent circuit of a cylindrical-rotor synchronous machine.

Also, from Fig. 8-8(a),

$$I^s X_s \cos \phi = V_0 \sin \delta \tag{8.18b}$$

From Eqs. (8.18a,b),

$$P_d = \frac{V_0 V^s}{X_s} \sin \delta \tag{8.18c}$$

which is consistent with Eq. (8.17c). We see, therefore, that the cylindrical-rotor machine is a special case of the salient-pole machine so far as the analysis is concerned. Of course, these two types of synchronous machines differ considerably in physical construction and operation. We shall now digress briefly to consider some of the physical features of synchronous machines and their modes of operation.

8.4 Physical Features of Synchronous Machines

We may now recall that the equations of motion for the synchronous machines derived in Sec. 8.1 are for an idealized machine. We mentioned earlier that the actual machine differs considerably from the idealized machine. Whereas the equations of motion of the idealized machine give quantitative information pertaining to the idealized machine and are approximately valid for the actual machine, there are many physical phenomena in synchronous machines which cannot be taken exactly into account by the equations of motion. One of the most common examples is the effect of saturation. Another example is the armature reaction with varying power factor and saturation. This is not to imply that attempts have not been made to obtain analytical solutions to these problems. However, analytical solutions are either extremely difficult (beyond what we have discussed so far) or only approximate. Moreover, for problems dealing with electromagnetic energy-conversion devices, we should not slavishly follow the mathematics. Rather, analytical methods backed by a thorough understanding of physical phenomena are best for our studies. We shall attempt in this section to outline the constructional features of synchronous machines and then discuss qualitatively some of the pertinent physical phenomena.

A typical salient-pole synchronous motor is shown in Fig. 8-9. The field winding is on the rotor, and is fed by the *exciter*. The exciter is a self-excited dc generator, shown at the extreme left in Fig. 8-9. The rotor also carries the damper windings (mounted on the salient poles) shorted at both ends and is very similar to the cage-type rotor of an induction motor. The stator of the machine carries a 3-phase winding, and is very similar to that of an induction motor. The machine can operate either as a synchronous generator or as a motor. Most commonly it is used as a generator. In fact,

Fig. 8-9 A salient-pole synchronous motor. (Courtesy of General Electric Company.)

the bulk of the power is generated by means of synchronous generators. The salient-pole machine is used for low-speed operation, such as in hydroelectric power stations (where a typical speed might be 120 r/min, and the machine has 60 poles to generate 60 c/s ac). The cylindrical-rotor machine is generally driven by a steam turbine and is operated at high speeds (such as 3600 r/min for a 2-pole, 60-cycle machine).

Normally, the armature windings carry polyphase excitation and the resulting airgap field is a traveling (or rotating) field the speed of which is given by

$$n_s = 120 \frac{f}{p} \tag{8.19}$$

where n_s = synchronous speed, r/min; f = frequency, c/s; and p = number of poles. The field winding carries dc excitation fed by an external source, or exciter (Fig. 8-9). The rotor runs at synchronous speed and, with proper relative polarities of the stator and rotor fields, *pulls into step* with the stator field. Any deviation in the rotor speed from *synchronism* results in the flow of currents in the damper windings, which tends to provide an additional

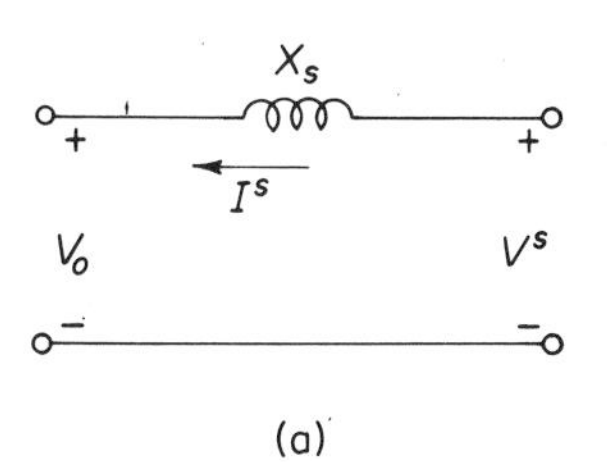

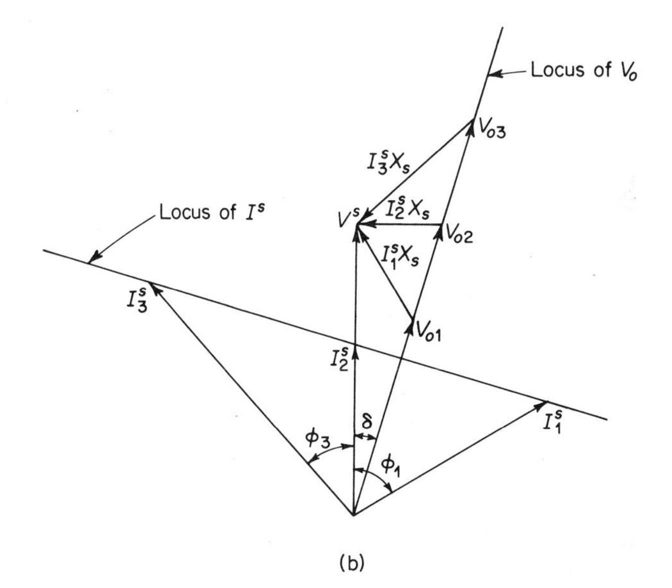

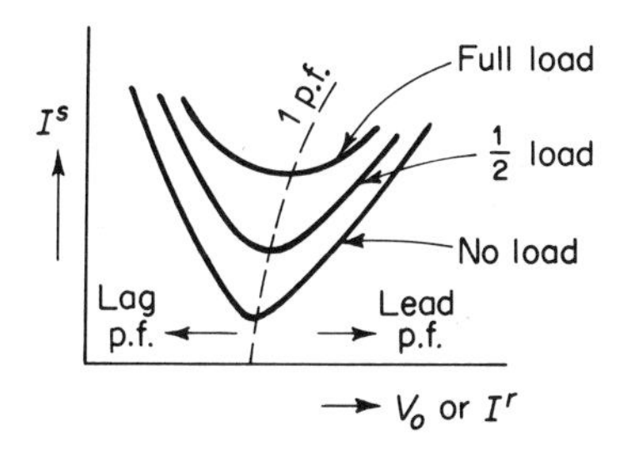

Fig. 8-10 (a) Equivalent circuit of a synchronous motor; (b) phasor diagram for varying V_o; (c) V curves.

torque to restore synchronism. When the machine operates as a generator, it is driven by a prime mover at synchronous speed. Sudden load changes on the machine tend to fluctuate its speed, and the machine may hunt. However, hunting is considerably reduced by the damper windings. When the machine operates as a motor, the damper windings also facilitate its starting. For convenience, in further discussions of the physical features we shall consider only the cylindrical-rotor machine.

We pointed out that synchronous machines are most commonly used as generators. Sometimes the machine is used as a motor running on no-load for power-factor correction. One of the characteristic properties of the synchronous motor is that its power factor depends on the excitation. To make this point clear we may refer to Fig. 8-10(a), which shows an equivalent circuit of a cylindrical-rotor synchronous motor operating at some load corresponding to the power angle δ. Recall from Fig. 8-6 that δ is negative for generator operation; the corresponding phasor diagrams are shown in Figs. 8-7 and 8-8(a). For motor operation, δ being positive, the phasor diagram is shown in Fig. 8-10(b) for various excitation voltages V_{01}, V_{02}, and V_{03}. [*Note*: Recall that V_0 is produced by the field current I^r. See Eq. (8.10c).] If V_0 is less than V^s we call the machine *underexcited*, whereas if V_0 is greater than V^s we term it *overexcited*. From Fig. 8-10(a) we observe that

$$\mathbf{V}^s = \mathbf{V}_0 + j\mathbf{I}^s X_s \tag{8.20}$$

Consequently, from Fig. 8-10(b), we immediately notice that the power-factor angle ϕ changes with $\mathbf{V}_0$. For an underexcited machine, the power factor is lagging (ϕ_1) and for the overexcited machine it is leading (ϕ_3). Of course, the field excitation could be adjusted so that $\mathbf{V}^s$ and $\mathbf{I}^s$ are in the same phase and the machine operates at unity power factor. For different loads on the machine, the variations of the armature current I^s with the field current I^r are shown in Fig. 8-10(c). These curves are known as the *V curves* of the synchronous motor. The fact that the synchronous motor can be used for power-factor improvement is illustrated by the following example.

EXAMPLE 8–2

A 3-phase, Y-connected load takes 50 A current at 0.707 lagging power factor at 220 V between the lines. A 3-phase, Y-connected, cylindrical-rotor synchronous motor having a synchronous reactance of 1.27 Ω per phase is connected in parallel with the load. The power developed by the motor is 33 kW at a power angle of 30°. Calculate (a) the reactive kVA of the motor, and (b) the overall power factor of the load and the motor.

(a) The phasor diagram and the circuit are shown in Fig. 8E-2, where the circuit and the phasor diagram are drawn on a per-phase basis. From Eq. (8.18c) we have

$$P_d = \frac{33{,}000}{3} = \frac{220}{\sqrt{3}} \frac{V_0}{1.27} \frac{1}{2}$$

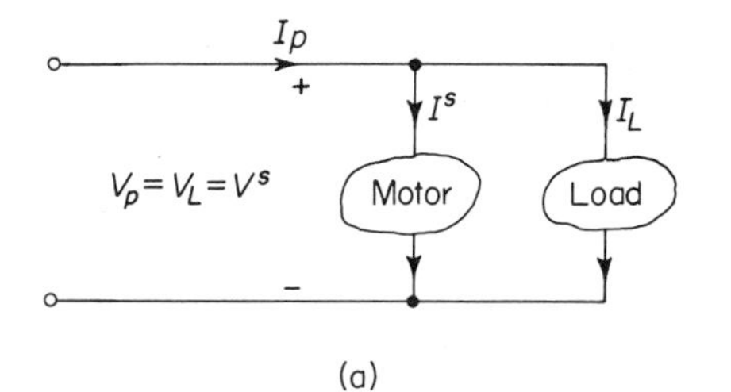

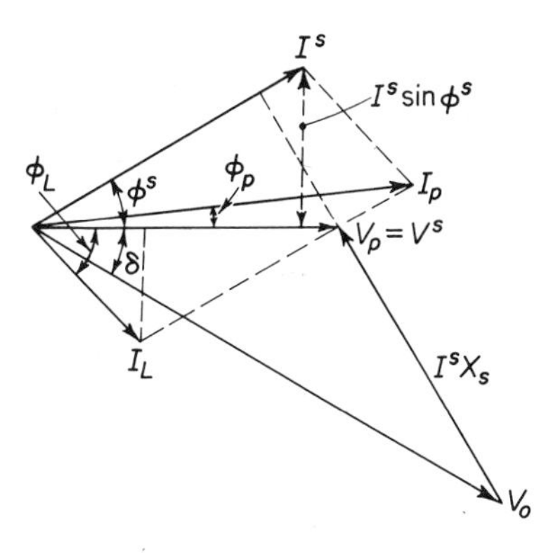

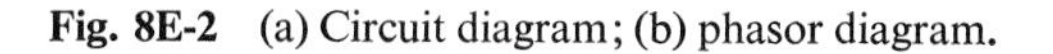

Fig. 8E-2 (a) Circuit diagram; (b) phasor diagram.

Or,

$$V_0 = 220 \quad \text{V}$$

From Fig. 8E-2(b),

$$I^s X_s = 127 \quad \text{V}$$

Or,

$$I^s = \frac{127}{1.27} = 100 \quad \text{A}$$

$$\phi^s = 30^\circ$$

$$\text{reactive kVA of the motor} = \sqrt{3}\,\frac{220}{1000}\,I^s \sin\phi^s \tag{a}$$

$$= 19 \quad \text{kvar}$$

(b) The power-factor angle is ϕ_p, which is obtained from

$$\tan\phi_p = \frac{I^s \sin\phi^s - I_L \sin\phi_L}{I^s \cos\phi^s + I_L \cos\phi_L} = 0.122$$

Or,

$$\phi_p = 7^\circ$$

and

$$\cos\phi_p = 0.992 \quad \text{leading}$$

8.5 Three-Phase Synchronous Machines

Recall, from the last chapter, that the theory developed for the 2-phase induction motor is also applicable to the 3-phase machine. In this respect, the 3-phase synchronous machine is not very much different from the 2-phase machine, except for an additional winding on the armature. Of course, like the induction motor, the 3-phase synchronous machine is much more common than the 2-phase or single-phase machine. In this section, we shall consider the *dq* transformation for the 3-phase machine and elaborate on the concept, introduced previously, of synchronous reactance. For simplicity, we shall consider the cylindrical-rotor machine, but the results are easily extendable to the salient-pole machine.

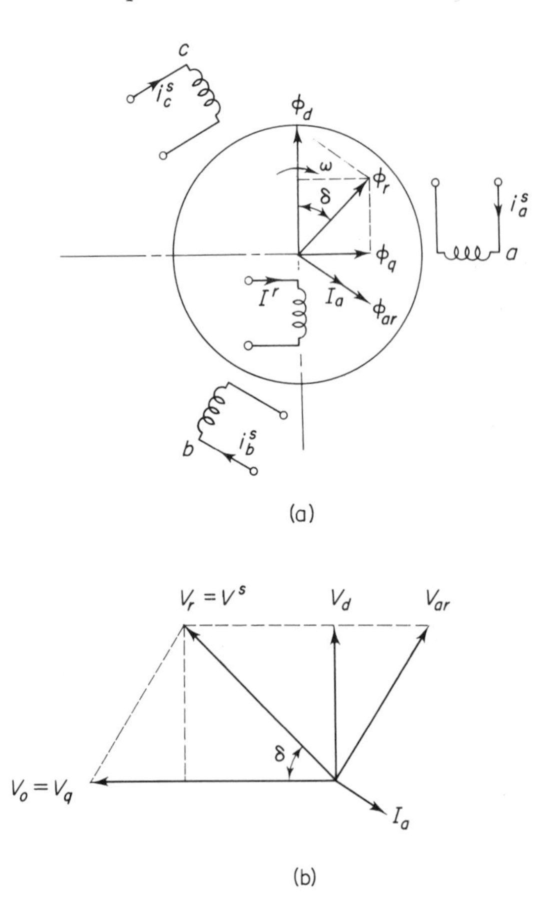

Fig. 8-11 (a) A three-phase synchronous machine, and field and armature fluxes; (b) phasor diagram.

A 3-phase machine is schematically shown in Fig. 8-11(a). We know that the polyphase windings, when excited, produce a rotating magnetic field in the airgap of the machine. Thus, for the machine under consideration, the flux, resulting from the 3-phase excitation, is denoted by ϕ_{ar} and is known as the *armature-reaction flux*. The flux produced by field (or rotor) excitation is ϕ_d; and ϕ_r is the resultant flux obtained by the superposition of the armature and field fluxes. Note that ϕ_d rotates at the synchronous speed of the rotor, as do the fluxes ϕ_q and ϕ_r. The flux diagram [Fig. 8-11(a)] is a space diagram. The displacement between ϕ_d and ϕ_r is δ. Now, each flux produces a corresponding voltage due to the rate of change of the flux linking a particular phase armature-winding. The mutual relationships of the various voltages are shown in Fig. 8-11(b), which is a time diagram. If we consider Fig. 8-11(a) for a particular instant, then the flux diagram can be considered a space diagram and the voltages lead the fluxes (producing the voltages) by 90° in time. To summarize, the voltages $V_q(=V_0)$, V_{ar}, and $V_r(=V^s)$ are respectively produced by the fluxes ϕ_d, ϕ_{ar}, and ϕ_r. We may consider these voltages to pertain to phase "a" of the armature winding. Because the current which produces a flux is in phase with the flux, we denote this current by I_a in Figs. 8-11(a) and (b). Therefore, in Fig. 8-11(b), V_{ar} can be considered an inductive voltage related to I_a by

$$V_{ar} = j\omega L_{ar} I_a \tag{8.21}$$

So far we have assumed that all the fluxes link with both the stator and rotor windings. In fact, however, there is always some flux that links with the stator windings but not with the rotor windings (and vice versa). This flux is called the (stator-)*leakage flux*, and gives rise to the inductance known as *leakage inductance*. The total inductive reactance of a phase of the armature winding is then

$$X_s = X_l + X_{ar} \tag{8.22}$$

and is known as the synchronous reactance, and consists of the *leakage reactance* X_l and *armature-reaction reactance* X_{ar}. We immediately see that the above physical argument leads to the steady-state equivalent circuit for the cylindrical-rotor machine, developed from the equations of motion, as shown in Fig. 8-8. If the armature resistance R^s is also included in the circuit, the total impedance is called the *synchronous impedance* Z_s and is given by

$$Z_s = R^s + jX_s \tag{8.23}$$

So far we have assumed only steady-state operation of the 3-phase synchronous machine. In order to obtain the dynamical equations of motion, we shall now develop the *dqo* transformation.

8.5.1. The dqo Transformation

In Sec. 7.4.2 we developed the 3-phase to *dq*-transformation for the induction machine. Using the same approach, we can derive the *dqo* transformation for the synchronous machine. In this case, however, we consider the resolution of fluxes, rather than MMF's, along the *dq* axis.

The space and time variations of the fluxes are shown in Fig. 8-12. At

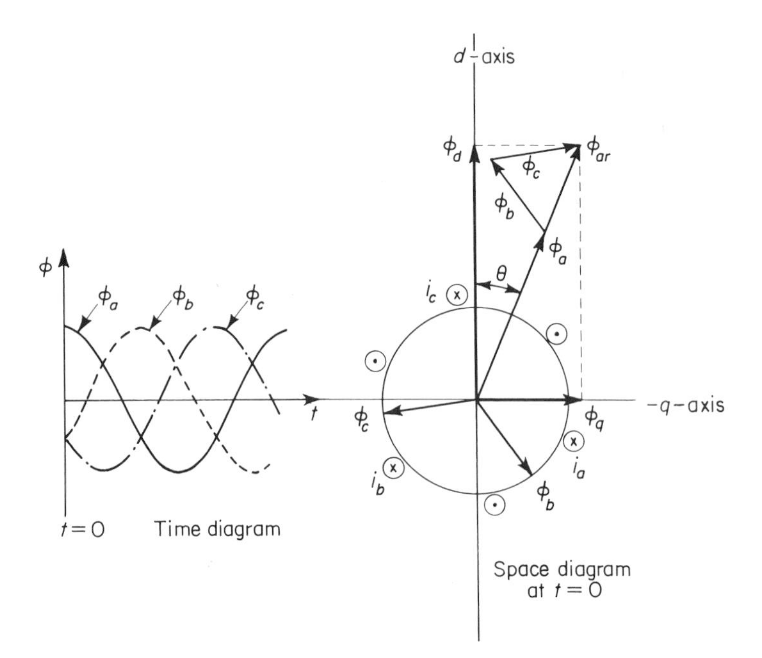

Fig. 8-12 Armature fluxes in a three-phase synchronous machine.

time $t = 0$, the space fluxes are

$$\phi_a = \phi_m \cos\theta \tag{8.24a}$$

$$\phi_b = \phi_m \cos\left(\theta - \frac{2\pi}{3}\right) \tag{8.24b}$$

$$\phi_c = \phi_m \cos\left(\theta + \frac{2\pi}{3}\right) \tag{8.24c}$$

Resolving these fluxes along the d and q axes we have

$$\phi_d = \phi_a \cos\theta + \phi_b \cos\left(\theta - \frac{2\pi}{3}\right) + \phi_c \cos\left(\theta + \frac{2\pi}{3}\right) \tag{8.25a}$$

$$\phi_q = -\phi_a \sin \theta - \phi_b \sin\left(\theta - \frac{2\pi}{3}\right) - \phi_c \sin\left(\theta + \frac{2\pi}{3}\right) \tag{8.25b}$$

In terms of resultant armature-reaction flux, we have

$$\phi_d = \phi_{ar} \cos \theta \tag{8.26a}$$

$$\phi_q = -\phi_{ar} \sin \theta \tag{8.26b}$$

From Fig. 8-12,

$$\phi_{ar} = \tfrac{3}{2}\phi_m \tag{8.26c}$$

Substituting Eqs. (8.24a–c) in Eqs. (8.25a,b) we obtain, in matrix notation,

$$\begin{bmatrix} \phi_d \\ \phi_q \end{bmatrix} = \begin{bmatrix} \cos \theta & \cos\left(\theta - \frac{2\pi}{3}\right) & \cos\left(\theta + \frac{2\pi}{3}\right) \\ -\sin \theta & -\sin\left(\theta - \frac{2\pi}{3}\right) & -\sin\left(\theta + \frac{2\pi}{3}\right) \end{bmatrix} \begin{bmatrix} \phi_a \\ \phi_b \\ \phi_c \end{bmatrix} \tag{8.27a}$$

and

$$\begin{bmatrix} \phi_a \\ \phi_b \\ \phi_c \end{bmatrix} = \frac{2}{3} \begin{bmatrix} \cos \theta & -\sin \theta \\ \cos\left(\theta - \frac{2\pi}{3}\right) & -\sin\left(\theta - \frac{2\pi}{3}\right) \\ \cos\left(\theta + \frac{2\pi}{3}\right) & -\sin\left(\theta + \frac{2\pi}{3}\right) \end{bmatrix} \begin{bmatrix} \phi_d \\ \phi_q \end{bmatrix} \tag{8.27b}$$

Notice that a factor $\frac{2}{3}$ is associated with the second transformation matrix, but not with the first. In order to have the same coefficients associated with the two matrices and at the same time obtain a power-invariant transformation, let us rewrite Eqs. (8.27a,b) as

$$\begin{bmatrix} \phi_d \\ \phi_q \end{bmatrix} = \sqrt{\frac{2}{3}} \begin{bmatrix} \cos \theta & \cos\left(\theta - \frac{2\pi}{3}\right) & \cos\left(\theta + \frac{2\pi}{3}\right) \\ -\sin \theta & -\sin\left(\theta - \frac{2\pi}{3}\right) & -\sin\left(\theta + \frac{2\pi}{3}\right) \end{bmatrix} \begin{bmatrix} \phi_a \\ \phi_b \\ \phi_c \end{bmatrix} \tag{8.28a}$$

and

$$\begin{bmatrix} \phi_a \\ \phi_b \\ \phi_c \end{bmatrix} = \sqrt{\frac{2}{3}} \begin{bmatrix} \cos \theta & -\sin \theta \\ \cos\left(\theta - \frac{2\pi}{3}\right) & -\sin\left(\theta - \frac{2\pi}{3}\right) \\ \cos\left(\theta + \frac{2\pi}{3}\right) & -\sin\left(\theta + \frac{2\pi}{3}\right) \end{bmatrix} \begin{bmatrix} \phi_d \\ \phi_q \end{bmatrix} \tag{8.28b}$$

The currents and voltages also transform according to Eqs. (8.28a,b).

However, allowing for the zero-sequence (noninteracting and nontorque-producing) component, we modify Eqs. (8.28a,b) for currents and voltages as follows:

$$i_{dqo} = S_{dqo} i_{abc} \tag{8.29a}$$

$$v_{dqo} = S_{dqo} v_{abc} \tag{8.29b}$$

where

$$S_{dqo} = \sqrt{\frac{2}{3}} \begin{bmatrix} \cos\theta & \cos\left(\theta - \frac{2\pi}{3}\right) & \cos\left(\theta + \frac{2\pi}{3}\right) \\ -\sin\theta & -\sin\left(\theta - \frac{2\pi}{3}\right) & -\sin\left(\theta + \frac{2\pi}{3}\right) \\ \frac{1}{\sqrt{2}} & \frac{1}{\sqrt{2}} & \frac{1}{\sqrt{2}} \end{bmatrix} \tag{8.30a}$$

and

$$S_{dqo}^{-1} = \tilde{S}_{dqo} \tag{8.30b}$$

Having obtained the transformation matrix, we can derive the dynamical equations of motion of the 3-phase machine in a manner similar to that for the 2-phase machine, discussed in Sec. 8.1.1.

8.6 Transients in Synchronous Machines[3–5]

In Sec. 8.1 we derived the dynamical equations of motion of the idealized synchronous machine having two (armature) windings on the stator and one (field) winding on the rotor. We did not include the effect of damper windings. In Sec. 8.2 and later we concentrated our attention on the steady-state behavior of synchronous machines. We now return to the dynamical equations of motion and study the transient behavior of the machine. We wish to point out, however, that we shall consider only few illustrative cases. The subject of transients in synchronous machines has been studied thoroughly, and extensive literature is available on this topic (for instance References 3–5).

8.6.1 Reactances and Time Constants

In order to obtain quantitative information about the transients in the machine, let us refer to the equivalent circuits shown in Fig. 8-4 and define the *transient reactances* and *time constants*. From Fig. 8-4(a), with $v^r = 0$, the direct-axis *transient operational inductance*, $L_d'p$, at the terminals $p\lambda_d^s$, is given by

$$L'_d p = (L_d - L^{sr})p + \cfrac{1}{\cfrac{1}{L^{sr}p} + \cfrac{1}{(L^r - L^{sr})p + R^r}}$$

which when simplified yields

$$L'_d p = L_d p - \frac{(L^{sr}p)^2}{L^r p + R^r} \tag{8.31}$$

The same result can be obtained by eliminating I^r from Eq. (8.5a) and replacing V^r by v^r. From Eq. (8.5a) we have

$$v_d^s - \frac{L^{sr}p}{L^r p + R^r} v^r = (R^s + L'_d p)i_d + v\dot{\theta} L'_q i_q \tag{8.32a}$$

$$v_q^s + \frac{L^{sr}v\dot{\theta}}{L^r p + R^r} v^r = -L'_d v\dot{\theta} i_d + (R^s + L'_q p)i_q \tag{8.32b}$$

where

$$L'_d p = L_d p - \frac{(L^{sr}p)^2}{L^r p + R^r} \tag{8.33a}$$

$$L'_q p = L_q p \tag{8.33b}$$

Comparing Eqs. (8.31) and (8.33a), we find that they are identical. The quadrature-axis inductance does not change, as shown by Eq. (8.33b). Equations (8.33a,b) are given in the operational form, that is, as functions of p. For sinusoidal time variations, we can substitute $p = j\omega$ in Eqs. (8.33a,b) to obtain

$$L'_d = L_d - \frac{j\omega (L^{sr})^2}{R^r + j\omega L^r} \tag{8.34a}$$

At low frequencies (when $\omega \to 0$), $L'_d \simeq L_d$; but, at higher frequencies (when $\omega \to \infty$),

$$x'_d = \omega L'_d = \omega \left[L_d - \frac{(L^{sr})^2}{L^r} \right] \tag{8.34b}$$

which is known as the *direct-axis transient reactance*. Another way of obtaining Eq. (8.34b) from Eq. (8.34a) is to consider the field resistance to be negligible, that is, $R^r \ll \omega L^r$. The transient reactance, although smaller than the synchronous reactance in magnitude, is important in calculations involving sudden disturbances on the machine, as we shall see later. In fact, in such calculations X_d is replaced by x'_d.

Turning now to the electrical time constants, we note that the two time constants which are most significant in transient calculations are (1) the *open-circuit-field time constant* (with $i_d = 0$), as given by

$$\tau'_{do} = \frac{L^r}{R^r} = \tau^r \tag{8.35a}$$

and (2) the *short-circuit-field time constant*, as given by

$$\tau_d' = \frac{1}{R^r}\left[L^r - \frac{(L^{sr})^2}{L_d}\right] \tag{8.35b}$$

which is also known as the *transient-field time constant*.

So far we have considered only the effects of field and armature windings. We mentioned earlier that most synchronous machines carry damper windings, and that during the beginning of the transient period the behavior of the machine is governed by the damper windings. The location of the damper windings can be seen in Fig. 8-9, which is schematically represented in Fig. 8-13(a); the *dq* representation of this circuit is shown in Fig. 8-13(b).

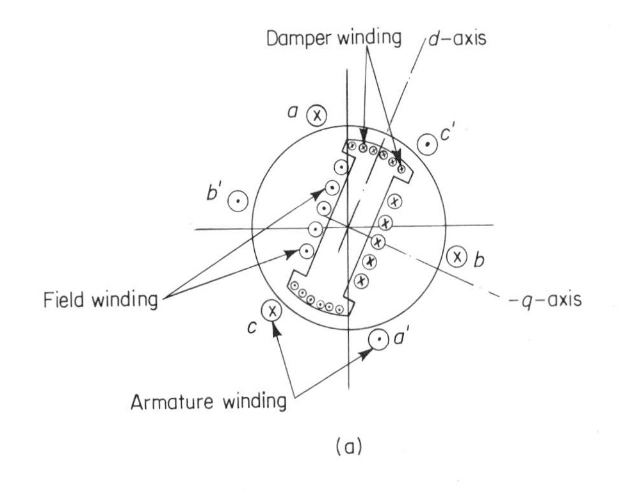

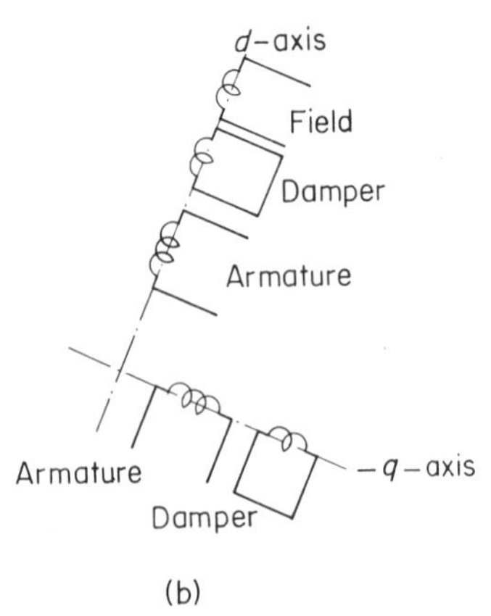

Fig. 8-13 (a) A three-phase, salient-pole machine with damper windings; (b) equivalent *dq*-representation.

An analytical determination of the operational reactances of this circuit is beyond the scope of this book, but using the approximation that the frequencies are very large, we define

$$x_d'' = \lim_{\omega \to \infty} X_d = \textit{direct-axis subtransient reactance}$$

$$x_q'' = \lim_{\omega \to \infty} X_q = \textit{quadrature-axis subtransient reactance}$$

For a typical cylindrical-rotor machine,

$$X_d = X_q = X_s = 1.0$$

$$x_d'' \simeq 0.5 x_d' \simeq 0.1 X_s$$

For a salient-pole machine,

$$X_d = 1.0$$

$$x_d'' \simeq 0.4 x_d' \simeq 0.25 x_d$$

$$x_q'' \simeq 0.4 x_q' \simeq 0.4 X_q$$

8.6.2 *An Approximate Analysis*[6]

Having identified the various transient and subtransient reactances and time constants, we now present an approximate analysis of transients in a synchronous machine. The assumption we make here is that the transients are slow. A typical example of such a case is the oscillation of the rotor of the machine due to sudden load changes. The frequency of these oscillations is very small compared with the line frequency. For slow variations, therefore, we put $p \ll \nu\dot{\theta}$ and $\nu\dot{\theta} = \omega$ in Eq. (8.5a), so that

$$\left.\begin{aligned} L_d p &\ll \omega L_q \\ L^{sr} p &\ll \omega L_q \\ L_q p &\ll \omega L_d \\ L_q p &\ll \omega L^{sr} \end{aligned}\right\} \tag{8.36}$$

Recalling $\omega L_d = X_d$ and $\omega L_q = X_q$, substituting Eq. (8.36) in Eq. (8.5a) and neglecting R^s we obtain

$$\begin{bmatrix} v_d^s \\ v_q^s \\ v^r \end{bmatrix} = \begin{bmatrix} 0 & X_q & 0 \\ -X_d & 0 & -\omega L^{sr} \\ L^{sr} p & 0 & R^r + L^r p \end{bmatrix} \begin{bmatrix} i_d^s \\ i_q^s \\ i^r \end{bmatrix} \tag{8.37}$$

The last equation in Eq. (8.37) is

$$v^r = R^r i^r + p\lambda^r i^r \tag{8.38a}$$

where

$$\lambda^r = L^{sr} i_d^s + L^r i^r \tag{8.38b}$$

Combining Eqs. (8.34a) and (8.38b) yields

$$\frac{\omega L^{sr}}{L^r}\lambda^r = \omega L^{sr} i^r + \omega(L_d - L_d')i_d^s \tag{8.39a}$$

In Eq. (8.39a), the various terms denote voltages along the q axis, so that we may rewrite this equation as

$$v_q' = v_q + (X_d - x_d')i_d^s \tag{8.39b}$$

where

$$v_q' = \frac{\omega L^{sr}}{L^r}\lambda^r \tag{8.39c}$$

The transient voltage v_q' denotes the initial condition after the load disturbance.

EXAMPLE 8–3

We wish to determine the phase current of a 2-phase, salient-pole synchronous generator symmetrically short-circuited. The generator was initially on no-load, and Eq. (8.37) is applicable.

Under short-circuit conditions, for $t > 0$,

$$v_d^s = v_q^s = 0$$

Under open-circuit conditions, for $t < 0$,

$$v_d^s = 0$$

$$v_q^s = v_0 = \omega L^{sr} i_0^r$$

$$v_0^r = R^r i_0^r$$

$$i_d^s = i_q^s = 0$$

Substituting these conditions in Eq. (8.37) we have, for $t > 0$,

$$0 = X_q i_q^s$$

$$0 = -X_d i_d^s - \omega L^{sr} i^r$$

$$v_0^r = L^{sr} p i_d^s + (R^r + L^r p)i^r$$

From the above last two equations we have

$$v_0^r = \left[L^{sr} p - \frac{X_d}{\omega L^{sr}}(R^r + L^r p)\right] i_d^s$$

Or, multiplying both sides by $\omega L^{sr}/(R^r + L^r p)$ yields

$$-\frac{\omega L^{sr}}{R^r + L^r p}\, v_0^r = \left[X_d - \frac{\omega (L^{sr})^2 p}{R^r + L^r p}\right] i_d^s \tag{a}$$

Equation (a) can be expressed in terms of various time constants such that

$$\frac{\omega L^{sr} i_0^r}{1 + \tau_f p} = X_d \frac{(1 + \tau_d' p)}{(1 + \tau_f p)}\, i_d^s = x_d' p i_d^s \tag{b}$$

where

$$\tau_d' = \frac{1}{R^r}\left[L^r - \frac{(L^{sr})^2}{L_d}\right] = \text{transient time constant}$$

But $\omega L^{sr} i_0^r = v_0$, so that from Eq. (b) we have

$$(1 + \tau_d' p) i_d^s = \frac{v_0^r}{X_d} \tag{c}$$

The solution to Eq. (c) is

$$i_d^s = \frac{v_0}{X_d} + C \exp\left(-\frac{t}{\tau_d'}\right) \tag{d}$$

The constant C is found by substituting the initial condition $i_d^s(0+) = v_0/x_d'$ in Eq. (d) to yield

$$C = \frac{v_0}{x_d'} - \frac{v_0}{X_d}$$

Thus the *dq* components of the transient current are

$$i_d^s = \frac{v_0}{X_d} + \left(\frac{v_0}{x_d'} - \frac{v_0}{X_d}\right) \exp\left(-\frac{t}{\tau_d'}\right) \tag{e}$$

$$i_q^s = 0 \tag{f}$$

The phase current is, therefore,

$$\begin{aligned} i_a^s &= i_d^s \cos\theta - i_q^s \sin\theta \\ &= v_0 \cos\theta \left[\frac{1}{X_d} + \left(\frac{1}{x_d'} - \frac{1}{X_d}\right) \exp\left(-\frac{t}{\tau_d'}\right)\right] \end{aligned} \tag{g}$$

The current i_a^s, as a function of time t, is sketched in Fig. 8E-3. As we shall see later, the current oscillogram, such as the one shown in Fig. 8E-3, is used to determine experimentally the transient reactance x_d'.

Observe that in the approximate analysis we have not taken into account the effect of the subtransient reactance x_d''. Actually, the initial short-circuit current during the first few cycles is limited only by x_d''. In a later section, we shall indicate the discrepancy between the result obtained by the approximate analysis and the experimentally obtained oscillogram of the short-circuit current.

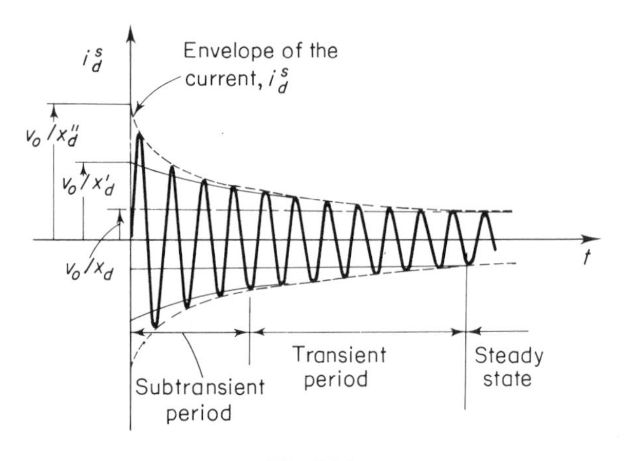

Fig. 8E-3

8.6.3 Mechanical Transients

The mechanical equation of motion of the synchronous machine is

$$J\ddot{\theta}_m + b\dot{\theta}_m = T_e + T_m \tag{8.40}$$

where T_e = torque developed by the machine, T_m = externally applied torque, J = moment of inertia of the rotating system (including the load, or prime mover), and b = friction coefficient, including electrical damping.

For the sake of illustration, let us consider a 2-pole cylindrical-rotor machine, and assume that the frequency of mechanical oscillations is small, so that the steady-state power-angle characteristics could be used. Note that this analysis is only approximate. The per-phase power developed by the machine, given by

$$P_d = \frac{V_0 V^s}{X_s} \sin \delta \tag{8.41}$$

which is the same as Eq. (8.18c), is plotted in Fig. 8-6—which also represents the electrical torque—to a different scale. Let the changes in θ_m, T_e, and T_m —due to a sudden load change—be represented by $\Delta\theta_m$, ΔT_e, and ΔT_m respectively, so that Eq. (8.40) modifies to

$$(Jp^2 + bp)\Delta\theta_m = \Delta(T_e + T_m) \tag{8.42}$$

The change in the electrical torque is, from Eq. (8.41),

$$\Delta T_e = \Delta\left(\frac{V_0 V^s \sin \delta}{\omega_m X_s}\right) \tag{8.43a}$$

where ω_m = mechanical velocity of the rotor and is the same as synchronous speed under steady-state conditions. In Eq. (8.43), for constant voltages, only sin δ changes for load changes, and for small variations $\Delta(\sin\delta)\simeq\Delta\delta$. Also $\Delta\theta_m = \Delta\delta$, the number of poles being two. Therefore, Eq. (8.43a) becomes

$$\Delta T_e = -K_e\Delta\delta \tag{8.43b}$$

We see from Fig. 8-6 that ΔT_e is negative for generator operation, so that in Eq. (8.43b), $K_e = -(V_0V^s/\omega_mX_s)$. From Eqs. (8.42) and (8.43b) we have

$$(Jp^2+bp+K_e)\Delta\delta = \Delta T_m \tag{8.44}$$

which is a linear second-order differential equation in terms of the power angle δ. Comparing this with the second-order differential equation of Example 2-12, the natural frequency of oscillation and the damping ratio are respectively

$$\omega_n = \sqrt{\frac{K_e}{J}} \tag{8.45a}$$

$$\zeta = \frac{b}{2\sqrt{K_eJ}} \tag{8.45b}$$

EXAMPLE 8-4

A 30-hp, 220-V, 3-phase, Y-connected, 60-cycle, 3600 r/min, cylindrical-rotor machine, on no-load, is brought up to the rated speed by an auxiliary motor, and is then suddenly connected to a 220-V, 3-phase source (with the proper phase sequence). Study the mechanical transient from the following data:

synchronous reactance/phase = 2.0 Ω
excitation voltage V_0 of Eq. (8.41) = 150 V/phase
moment of inertia of rotating parts = 1.5 MKS units
damping torque b of Eq. (8.40) = 12 N·m/rad/s

Denoting $\Delta\delta$ by δ', the equation of motion is, from Eq. (8.44),

$$J\frac{d^2\delta'}{dt^2}+b\frac{d\delta'}{dt}+K_e\delta' = 0 \tag{a}$$

In Eq. (a), K_e is known as the *synchronizing torque*. For motor operation, K_e (for the 3-phase machine) is given by

$$K_e = \frac{V_0V^s}{\omega_mX_s}\times 3 \tag{b}$$

For the given machine, $V_0 = 150$ V, $V^s = 220/\sqrt{3} = 127$ V, $X_s = 2.0\ \Omega$, and $\omega_m = 120\pi$ rad/s. Substituting these in (b), we obtain

$$K_e = \frac{150\times127\times3}{120\pi\times2} = 756 \quad \text{N·m/rad}$$

Equation (a) therefore becomes

$$(1.5p^2+12p+756)\delta' = 0 \tag{c}$$

From Eqs. (8.45a,b) and Eq. (c) we have

$$\omega_n = \sqrt{\frac{756}{1.5}} = 22.5 \quad \text{rad/s}$$

$$\zeta = \frac{12}{2\sqrt{756\times 1.5}} = 0.178$$

In terms of cycles per second, the natural frequency of oscillation is obtained from

$$2\pi f_n = \omega_n$$

Or,

$$f_n = \frac{\omega_n}{2\pi} = \frac{22.5}{2\pi} = 3.6 \quad \text{c/s}$$

In most machines,

$$0.2 < f_n < 2$$

$$\zeta \simeq 0.2$$

Knowing ζ and ω_n, we can obtain the mechanical behavior from the equation

$$\frac{\delta'}{\delta'_{ss}} = 1 - \frac{1}{\sqrt{1-\zeta^2}} e^{-\zeta\omega_n t} \sin\left(\sqrt{1-\zeta^2}\,\omega_n t + \phi\right) \tag{d}$$

where

$$\phi = \tan^{-1}\frac{\sqrt{1-\zeta^2}}{\zeta} \tag{e}$$

and δ'_{ss} = steady-state power angle. Note that Eq. (d) is the solution to Eq. (a) for $\zeta<1$. (See Problem 8-4.)

8.7 Determination of Machine Reactances

The synchronous reactance X_s of a cylindrical-rotor machine can be obtained from the open-circuit and short-circuit tests on the machine. The open-circuit saturation curve and the steady-state, short-circuit armature current are shown in Fig. 8-14 on a per-phase basis. For a 2A field current, the short-circuit current is 25A, whereas the open-circuit voltage is 57 V. Consequently, the synchronous impedance is 57/25 = 2.48 Ω. Neglecting armature resistance, $Z_s \simeq X_s = AC/BC = 2.48$. As shown in Fig. 8-14, X_s varies with saturation.

For the salient-pole machine, it is necessary that we know the values of both X_d and X_q. The physical significance of these reactances was discussed earlier, and these are respectively the maximum and minimum values of the armature reactance (for different rotor positions). These reactances are

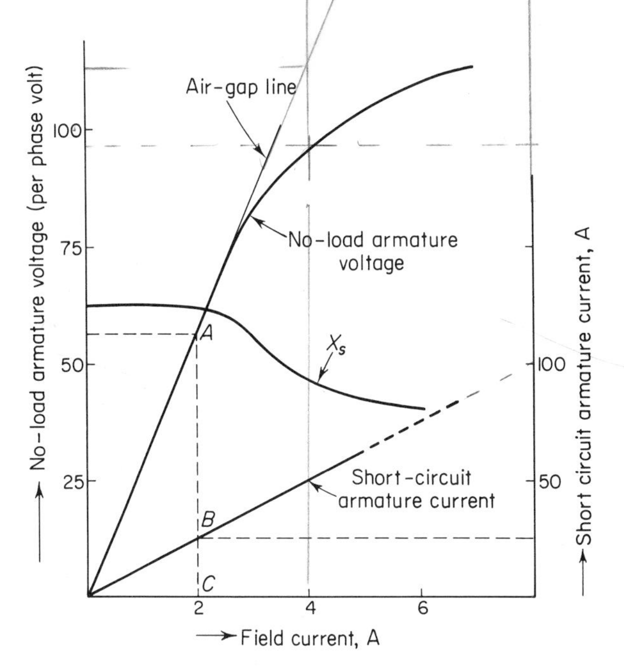

Fig. 8-14 No-load and short-circuit test data on a synchronous machine, and variation of X_s with field current.

determined by the *slip test*. In this test, the machine is excited by a 3-phase source (for a 3-phase machine) and driven mechanically at a speed slightly different from the synchronous speed, with the field winding unexcited and open-circuited. Oscillograms are taken of the armature current, armature voltage, and the (induced) field voltage. These are shown in Fig. 8-15. The ratio of maximum to minimum armature current yields the ratio X_d/X_q.

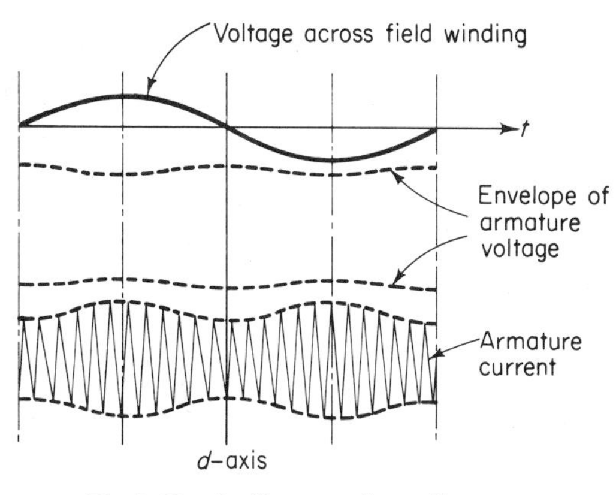

Fig. 8-15 Oscillograms from slip test.

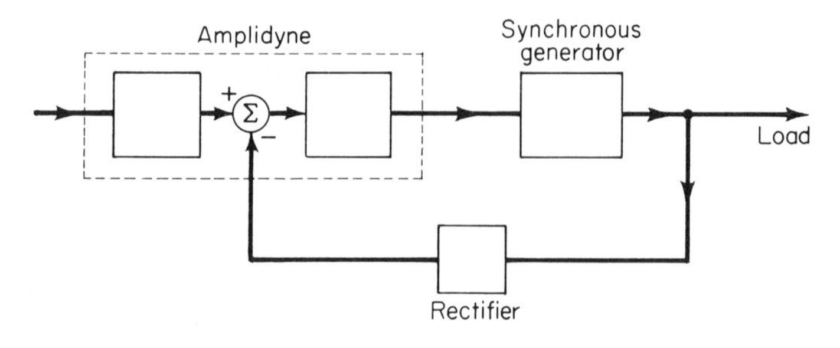

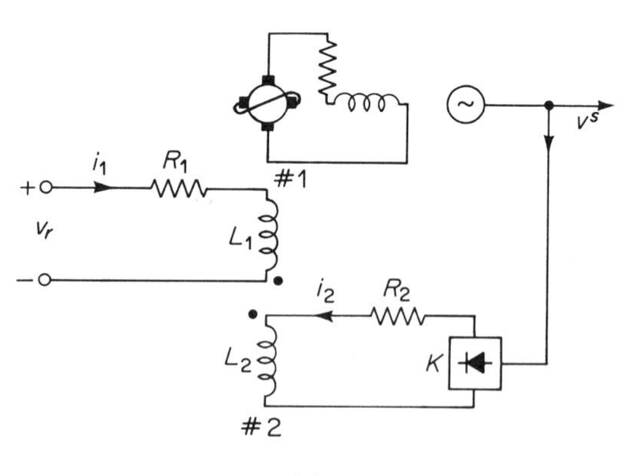

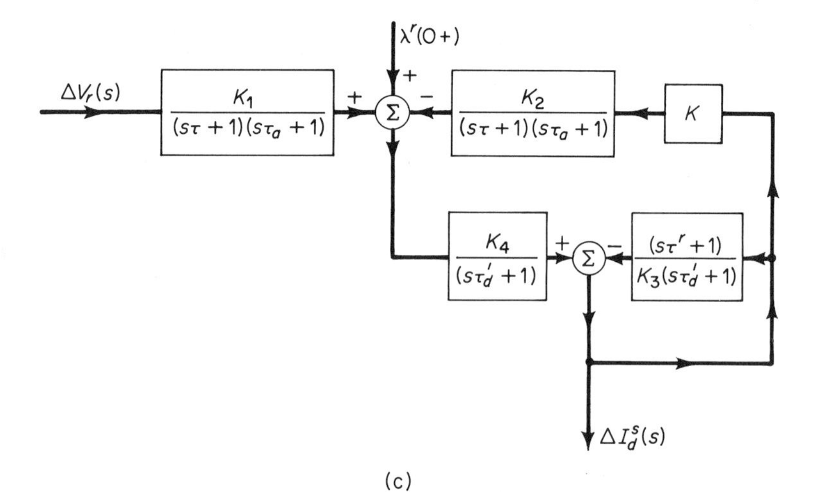

Fig. 8E-5 (a) Functional block diagram; (b) block diagram showing components; (c) overall block diagram showing component transfer functions.

For instance, from the diagram we find that $X_d/X_q = 1.6$. Knowing X_d from the open-circuit and short-circuit tests (described above for the cylindrical-rotor machine), we can calculate X_q. There are other methods available for the determination of these reactances; the interested reader should consult Reference 7.

The transient and subtransient reactances x'_d and x''_d are determined by recording the 3-phase currents when a sudden short circuit is applied to the machine running on no-load and at rated speed. Generally, the currents of the various phases are not symmetrical about the time axis. However, x'_d and x''_d can be determined either (1) by eliminating the dc component, or (2) from an oscillogram such as the one shown in Fig. 8E-3.[7]

EXAMPLE 8–5[8]

In Chapter 1 (Fig. 1-3) we mentioned the example of the synchronous machine as an element in a voltage-regulating system. Let us now consider this example in some detail. A functional block-diagram representation of this system is given in Fig. 8E-5(a). The dynamic behavior of the system is to be obtained by deriving the component transfer function and the overall transfer function. Incremental variations of the armature current and power angle, and constant rotor speed, are assumed. Armature resistance is negligible.

From Eq. (8.5a) we have, in terms of the flux linkages λ's,

$$\left.\begin{aligned} v_d^s &= p\lambda_d^s + \omega_m\lambda_q^s \\ v_q^s &= -\omega_m\lambda_d^s + p\lambda_q^s \\ v^r &= p\lambda^r + R^r i^r \end{aligned}\right\} \tag{a}$$

where

$$\left.\begin{aligned} \lambda_d^s &= L_d i_d^s + L^{sr} i^r \\ \lambda_q^s &= L_q i_q^s \\ \lambda^r &= L^{sr} i_d^s + L^r i^r \end{aligned}\right\} \tag{b}$$

If we now assume constant flux linkages in the armature during a disturbance in the field circuit, and take Laplace transforms of Eqs. (a) and (b) we obtain

$$\left.\begin{aligned} V_d^s(s) &= \omega_m \Lambda_q^s(s) \\ V_q^s(s) &= -\omega_m \Lambda_q^s(s) \\ V^r(s) &= s\Lambda^r(s) - \lambda^r(0+) + R^r I^r(s) \end{aligned}\right\} \tag{c}$$

and

$$\left.\begin{aligned} \Lambda_d^s(s) &= L_d I_d^s(s) + L^{sr} I^r(s) \\ \Lambda_q^s(s) &= L_q I_q^s(s) \\ \Lambda^r(s) &= L^{sr} I_d^s(s) + L^r I^r(s) \end{aligned}\right\} \tag{d}$$

Eliminating $I^r(s)$ from Eqs. (c) and (d) yields

$$\left[L_d - \frac{s(L^{sr})^2}{R^r + sL^r}\right] I_d^s(s) = -\frac{V_q^s(s)}{\omega_m} - \frac{L^{sr}}{R^r + sL^r}[V^r(s) + \lambda^r(0+)] \quad \text{(e)}$$

Using the time constants defined by Eqs. (8.35a,b), we can now reduce Eq. (e) to

$$I_d^s(s) = -\frac{V_q^s(s)}{\omega_m L_d}\frac{s\tau^r + 1}{s\tau_d' + 1} - [V^r(s) + \lambda^r(0+)]\frac{L^{sr}}{R^r L_d(s\tau_d' + 1)} \quad \text{(f)}$$

Or, in terms of individual transfer functions, Eq. (f) can be expressed as

$$I_d^s(s) = -V_q^s(s)K_1G_1(s) - V^r(s)K_2G_2(s) - \lambda^r(0+)K_2G_2(s) \quad \text{(g)}$$

where the component transfer functions for the synchronous machine are

$$K_1G_1(s) = \frac{s\tau^r + 1}{\omega_m L_d(s\tau_d' + 1)} \quad \text{(h)}$$

$$K_2G_2(s) = \frac{L^{sr}}{R^r L_d(s\tau_d' + 1)} \quad \text{(i)}$$

The incremental change in the direct-axis current is obtained from Eq. (g) as

$$\Delta I_d^s(s) = -\Delta V_q^s(s)\ K_1G_1(s) - \Delta V^r(s)\ K_2G_2(s) \quad \text{(j)}$$

And the change in the quadrature-axis current is, from Eqs. (c) and (d),

$$\Delta I_q^s(s) = \frac{\Delta V_d^s(s)}{\omega_m L_q} \quad \text{(k)}$$

Recall that the transfer function of the amplidyne was found in Chapter 6, and is given by Eq. (6.35); and when an amplidyne is used as an element in a feedback system, the transfer function is given by Eq. (a), Example 6-5. For convenience, let us redraw a simplified form of Fig. 1-3, as shown in Fig. 8E-5(b), from which we can write the following equations for incremental variations:

For reference field no. 1

$$\Delta V_r(s) = (L_1 s + R_1)\Delta I_1(s) - L_{12}s\Delta I_2(s) \quad \text{(l)}$$

For control field no. 2

$$K\Delta V^s(s) = (L_2 s + R_2)\Delta I_2(s) - L_{12}s\Delta I_1(s) \quad \text{(m)}$$

Assuming $L_1 = L_2 = L$ and defining a time constant

$$\tau_a = \frac{L}{R_1 R_2}(R_1 + R_2) \quad \text{(n)}$$

we have, from Eqs. (l)–(n),

$$\Delta[I_1(s) - I_2(s)] = \Delta V_r(s)\frac{1}{R_1(s\tau_a + 1)} - K\Delta V^s(s)\frac{1}{R_2(s\tau_a + 1)} \quad \text{(o)}$$

But the change in amplidyne-generated voltage in the q axis is related to the change in amplidyne field currents by

$$\Delta V_a^s(s) = K_a\Delta[I_1(s) - I_2(s)] \quad \text{(p)}$$

where K_a is the constant of proportionality. The change in the current through the amplidyne q axis is

$$\Delta I_a^s(s) = \frac{\Delta V_a^s(s)}{R_a(s\tau+1)} \tag{q}$$

where R_a is the amplidyne q-axis resistance and τ is its time constant.

Now, the change in the generator field voltage is given by

$$\Delta V^r(s) = K_{DQ}\Delta I_a(s) = \frac{\Delta V_r(s)K_3G_3(s)}{R_1} - \frac{K\Delta V^s(s)K_3G_3(s)}{R_2} \tag{r}$$

where K_{DQ} = quadrature-to-direct-axis transfer ratio of the amplidyne and

$$K_3G_3(s) = \frac{K_aK_{DQ}}{R_a(s\tau+1)(s\tau_a+1)} \tag{s}$$

The expression for $\Delta V^r(s)$ from Eq. (r) can now be substituted in Eq. (j) to obtain the overall transfer function of the system.

The transfer function is represented in the form of a block diagram in Fig. 8E-5(c), in which $K_1 = K_aK_{DQ}/R_1R_a$, $K_2 = R_1K_1/R_2$, $K_3 = \omega_m L_d$, and $K_4 = L^{sr}/R^rL_d$.

For a purely inductive load, $\Delta I_d^s = \Delta I^s$ and the change in the armature current of the synchronous generator, due to changes in the field excitation, can be directly obtained from the block diagram.

Combining Eqs. (r), (s), and (j), the overall transfer function of the system takes the form

$$\frac{\Delta I_d^s(s)}{\Delta V_r(s)} = \frac{K_5}{(s\tau+1)(s\tau_a+1)(s\tau_d'+1)+K_6} \tag{t}$$

where K_5 and K_6 are constants involving K_1, K_2, K_3, and K_4. In an actual system, the incremental change in the armature current, for a purely inductive load, has been found to be of the following form:[8]

$$\Delta i = \Delta V_r[4.88 - 1.52e^{-30t} - 6.9e^{-7.65t}\sin(11.8t+29°)]$$

8.8 Summary

In this chapter we studied synchronous machines, emphasizing their dynamical characteristics. We obtained the dynamical equations of motion using the dq transformation. This transformation was modified to the dqo transformation for the 3-phase machine. Physical meanings were given to the direct-axis and quadrature-axis reactances, and transient and subtransient reactances were introduced to determine the transient performance of the machine. Procedures were given in outline to determine experimentally the machine reactances.

Steady-state power and voltage characteristics were derived for the cylindrical-rotor and salient-pole machines, and excitation characteristics of the synchronous motor were studied. A brief introduction was given to some of the physical features of synchronous machines.

Finally, an example of a synchronous-generator voltage-regulating system was considered and the dynamic behavior of the system was obtained by linearized analysis making use of transfer functions.

PROBLEMS

8–1. Choose a suitable 2-pole, 2-phase model of a salient-pole synchronous machine and derive the inductance matrix as given by Eq. (8.2b).

8–2. The inductance matrix of a 2-phase salient-pole machine is given by Eq. (8.2b). By resolving the MMF's along the d and q axes, obtain the transformation matrix S_{dq}. Obtain the transformed flux-linkage matrix and hence show that the phase voltages are given by

$$v_a = p\lambda_d - \omega\lambda_q$$
$$v_b = p\lambda_q + \omega\lambda_d$$
$$v_f = p\lambda_f + R_f i_f$$

where λ_d, λ_q, and λ_f are the d axis, the q axis, and the field flux linkages, respectively, R_f is the field resistance, and the armature resistance is negligible. Verify that under this transformation $(v_a i_a + v_b i_b) = (v_d i_d + v_q i_q)$.

8–3. The $\alpha\beta$ and dq axes of a 3-phase, salient-pole synchronous machine are shown in Fig. 8P-3. By resolving MMF's along the $\alpha\beta$ axes show that

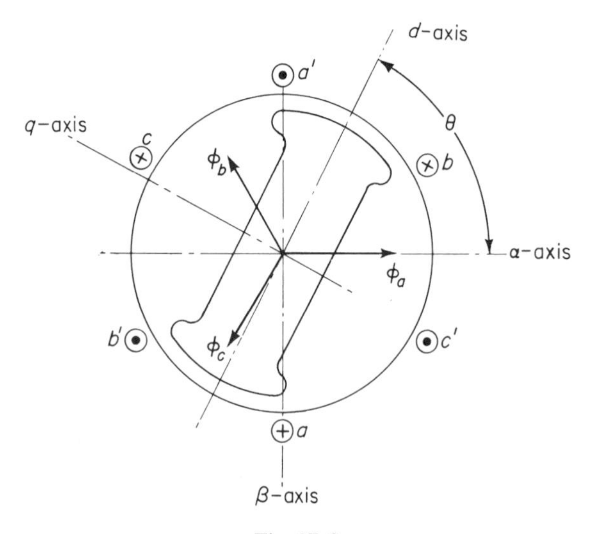

Fig. 8P-3

$$\begin{bmatrix} i_\alpha \\ i_\beta \\ i_0 \end{bmatrix} = \begin{bmatrix} \frac{2}{3} & -\frac{1}{3} & -\frac{1}{3} \\ 0 & -\frac{1}{\sqrt{3}} & \frac{1}{\sqrt{3}} \\ \frac{1}{3} & \frac{1}{3} & \frac{1}{3} \end{bmatrix} \begin{bmatrix} i_a \\ i_b \\ i_e \end{bmatrix}$$

Find $S_{\alpha\beta 0}$ such that $i_{abc} = S_{\alpha\beta 0}\, i_{\alpha\beta 0}$.

8–4. Solve Eq. (a) of Example 8-4 to obtain Eq. (d).

8–5. The power developed by a cylindrical-rotor synchronous machine is given by Eq. (8.18c). Derive an expression for the reactive power developed by the machine.

8–6. The approximate equivalent circuit for a cylindrical-rotor machine is shown in Fig. 8-10(a). Draw the phasor diagrams for (a) generator operation and (b) motor operation and verify that the power developed by the machine in both cases is correctly given by Eq. (8.18c).

8–7. Calculate the percentage regulation of a 20kVA, 220-V, Y-connected synchronous generator operating at full load and (a) 0.8pf lagging, and (b) 0.8pf leading. The per-phase direct-axis and quadrature-axis reactances are 5.0 Ω and 3.0 Ω respectively. Armature resistance is negligible. Calculate the developed power in both cases.

8–8. A 2-phase, 2-pole cylindrical-rotor machine has the following constants: $L_a^s = L_b^s = 0.5$ H; $R_a^s = R_b^s = 0.5$ Ω; $L^{sr} = 0.35$ H. The field current is constant at 4.0 A. The machine is run as a generator at 3600 r/min. Calculate the following:

a. the open-circuit induced voltage per phase (from Fig. 8-14);
b. the power developed by the machine if it supplies a 10 Ω purely resistive load;
c. the power angle; and
d. the maximum power that can possibly be developed by the machine, neglecting losses.

8–9. The problem in Example 8-5 has the following data: $L_1 = L_2 = 1.0$ H; $R_1 = 15.0$ Ω; $R_2 = 500$ Ω; $R_a = 5.0$ Ω; $\tau = 0.1$ s; $R^r = 25.0$ Ω; $\tau_f = 0.1$ s; $\tau_d' = 0.05$ s; $L^{sr} = 0.2$ H; $\omega_m = 314$ rad/s; $L_d = 0.2$ H. With $K = 1$, $K_a K_{DQ} = 2000$, and $\Delta v_r = 5.0$ V, determine Δi^s if a short circuit is suddenly applied at the machine terminals, which were initially connected to an *infinite bus*. (*Note*: The voltage and frequency at an infinite bus are always constant, regardless of the load.)

8–10. Find the transfer function defined by

$$KG(s) = \frac{\Delta\delta(s)}{\Delta V_r(s)}$$

for the system discussed in Example 8-5, where $\delta(s)$ = Laplace transform of the power angle δ. Assume an appropriate conversion constant, not given. Draw the block diagram showing the transfer function of the various components.

8–11. We define the *per-unit value* by

$$\text{pu value} = \frac{\text{actual value (in any unit)}}{\text{base value (in the same unit)}}$$

For a 3-phase, Y-connected, 3000 kVA, 13.2 kV synchronous generator, convert the following pu values to their ohmic values: $R^s = 0.005$; $X_d = 1.1$; and $X_q = 0.8$.

8–12. Calculate the per cent voltage regulation for the machine in Problem 8-11 by (a) using ohmic values, and (b) using pu values. Compare the results and note if either method has any advantage over the other.

8–13. Draw the phasor diagram of an overexcited salient-pole synchronous motor and include the effect of the armature resistance R^s. From the diagram show that

$$\tan\delta = \frac{I^sR^s \sin\theta + I^sX_q \cos\theta}{V^s + I^sX_q \sin\theta - I^sR^s \cos\theta}$$

and

$$X_q = \frac{V^s \sin\delta - I^sR^s \sin(\theta+\delta)}{I^s \cos(\theta+\delta)}$$

REFERENCES

1. White, D. C., and H. H. Woodson, *Electromechanical Energy Conversion.* New York: John Wiley & Sons, Inc., 1959.

2. Seely, S., *Electromechanical Energy Conversion.* New York: McGraw-Hill Book Company, 1961.

3. Vowels, R. E., "Transient Analysis of Synchronous Machines," *Proc. IEE* (London), Vol. 99, Part IV, 1952, pp. 204–216.

4. Ku, Y. H., *Electric Energy Conversion.* New York: The Ronald Press Company, 1959.

5. Crary, S. B., and M. L. Waring, "The Operational Impedances of a Synchronous Machine," *Gen. Elec. Review*, Vol. 35, 1932, pp. 578–582.

6. Messerle, H. K., *Dynamic Circuit Theory.* New York: Pergamon Press, Inc., 1965.

7. *Test Procedures for Synchronous Machines*, IEEE Publication No. 115, March 1965.

8. Hamdi-Sepen, D., "Transfer Functions of Loaded Synchronous Machine," *AIEE Trans.*, Vol. 78, Part II, 1959, pp. 19–24.

9

Diverse Topics and Realistic Considerations

In the last eight chapters we have attempted to present a coherent theory of electromagnetic devices with special reference to their energy-conversion properties. Naturally, we make no claim to have presented an intensive study of every device; rather, this work should be considered the beginning of a very fruitful, lively, and extensive field—the study of electromagnetic energy-conversion devices. Throughout the preceding chapters we made simplifying —at times somewhat unrealistic—assumptions so that devices would be amenable to mathematical analysis. In some cases our end results, obtained from the analysis of the idealized model, might not be compatible with reality. Now, with the availability of computers (especially digital computers), it is possible to include secondary effects and obtain solutions more refined than those presented so far. In fact, the current trend in research in the area of electromagnetic energy-conversion devices is to obtain solutions which include as many realistic considerations as possible. It is not practically possible to present these various approaches in detail in this chapter. We shall rather illustrate, by means of a number of examples, the applications of analog and digital computers in solving some pertinent problems. Because graphical field mapping has found extensive use in many problems related to electromagnetic devices, we shall briefly review this topic also and illustrate its application by one or two examples.

Briefly, then, in this chapter we shall study the applications of analog and digital computers and of graphical field mapping to energy-conversion problems. As pointed out in Chapter 5, "realistic considerations" includes the effects of saturation, harmonics, leakages, etc. However, it should be pointed out that we shall consider these topics in very general terms; for specifics the reader should consult the pertinent references.

9.1 Applications of Analog Computers [1–9]

We can use analog computers to solve mathematical equations. Such computers are particularly valuable in the study of the effects of changes in the parameters of equations and in obtaining the solutions to nonlinear differential equations. Recall that in the study of induction machines in Chapter 7, for example, the equations of motion are nonlinear differential equations, for which explicit analytical solutions are quite difficult to obtain. While in our analyses we used certain constraints to simplify matters— such as constant speed or slow transients—it is possible to obtain more realistic solutions on analog computers. Certain examples can be found in References 1–7. Before we consider any example here, however, let us briefly review the fundamentals of analog computers.

The analog computer consists of a number of building blocks for simulating the mathematical model of a system to be studied. The building blocks, or computer elements, perform the following mathematical operations: addition (or subtraction); multiplication by a constant or a variable; differentiation; integration; and function generation. With these "computing elements" available, the problem to be solved on the analog computer can be set up in the following manner.

First, the problem at hand is formulated in terms of mathematical equations. These equations may be differential equations (linear or nonlinear), difference equations, a set of linear algebraic equations, partial-differential equations, and so forth. After the equations have been formulated, they are appropriately rearranged for computer solution. For example, in solving a differential equation, generally the rearrangement consists of solving for the highest-order derivative in the equation. The next step is to develop a block-diagram representation of the equations. Here the computer elements are represented symbolically and their interconnections shown. On the block diagram, inputs and outputs are identified. The final step, before setting up the problem on the computer, is scaling—substituting computer variables of voltage and time for problem variables; that is, adapting the magnitude and range of variation of a given problem variable, to the total excursion of (say) 0 to 100 volts and a few seconds, which are the computer variables. Scaling can be done (a) by transforming the problem equations to computer equations; or (b) by including a scale factor with each problem variable.

The steps outlined above will become clear when we consider some examples later. For the present we shall briefly describe the basic computing operations of a general-purpose analog computer.

9.1.1 Basic Computing Operations

The basic component of the electronic analog computer is a high-gain dc amplifier, known as the *operational amplifier*. Using operational amplifiers we can perform a number of mathematical operations. The gain of an operational amplifier is constant up to a frequency of several thousand cycles per second, and its transfer function can be represented by $-A$ in this region. The magnitude of A of a typical operational amplifier is of the order of a few millions. Its input impedance is several hundred megohms, and the maximum output current is of the order of several milliamps. An amplifier is represented symbolically in Fig. 9-1(a), and a typical use of this amplifier as an operational amplifier is shown in Fig. 9-1(b). By Kirchhoff's law, we have

$$i_1 + i_2 + i_3 = 0 \tag{9.1a}$$

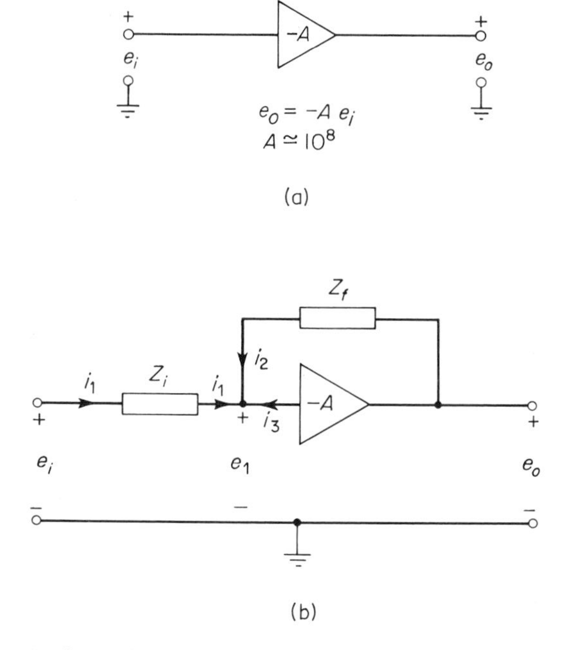

Fig. 9-1 (a) A high-gain dc amplifier; (b) an operational amplifier circuit.

But $i_3 \simeq 0$, since the amplifier has a very high input impedance. Equation (9.1a) then becomes

$$i_1 + i_2 = 0 \tag{9.1b}$$

Now, by Ohm's law,

$$i_1 = \frac{e_i - e_1}{Z_i} \tag{9.2a}$$

$$i_2 = \frac{e_o - e_1}{Z_f} \tag{9.2b}$$

From Fig. 9-1(a),

$$-e_o = Ae_1 \tag{9.2c}$$

From Eqs. (9.1b–9.2c), therefore,

$$\frac{e_o}{e_i} = -\frac{Z_f}{Z_i}\left[\frac{1}{1+\dfrac{1}{A}\left(\dfrac{Z_f}{Z_i}+1\right)}\right] \tag{9.3a}$$

For $A \simeq 10^8$ and $Z_i \simeq Z_f$, Eq. (9.3a) reduces to

$$e_o = -\frac{Z_f}{Z_i} e_i \tag{9.3b}$$

where Z_i and Z_f are respectively the input impedance and feedback impedance in operational form. Equation (9.3b) is the basic equation from which various computing operations can be derived. For most applications, Z_f is a resistor, a capacitor, or an RC circuit.

The most common mathematical operations that can be performed by means of operational amplifiers in conjunction with passive elements are summarized in Fig. 9-2. The following remarks relate to some of the nomenclature and equations pertinent to various operations represented in Fig. 9-2. In Figs. 9-2(a) and (b), β denotes the fraction of the voltages e_i and e_o respectively.

Fig. 9-2(a): $$e_o = -\beta \frac{R_f}{R_i} e_i \tag{9.4a}$$

Fig. 9-2(b): $$e_o = -\frac{1}{\beta}\frac{R_f}{R_i} e_i \tag{9.4b}$$

Fig. 9-2(c): $$e_o = -\left(\frac{R_f}{R_1} e_1 + \frac{R_f}{R_2} e_2 + \frac{R_f}{R_3} e_3\right) \tag{9.4c}$$

Fig. 9-2(d): $$e_o = -\frac{R_f}{R_1}(e_1 - e_2) \tag{9.4d}$$

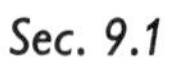

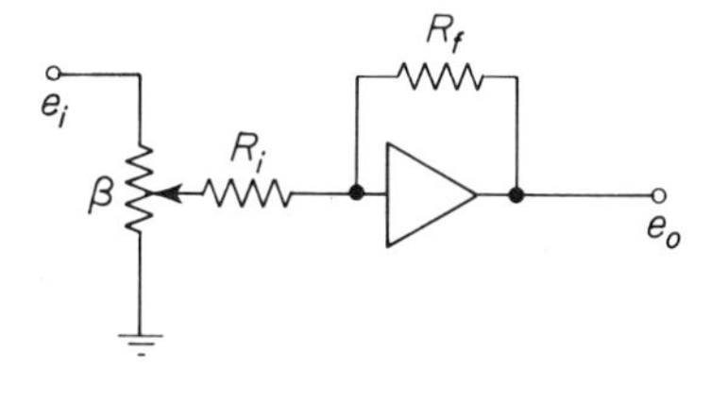

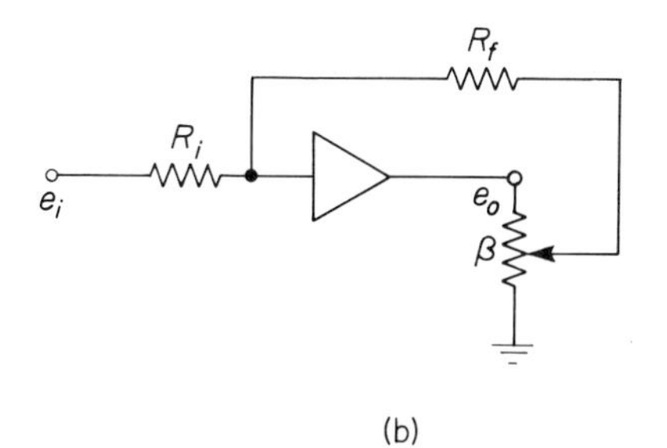

(b)

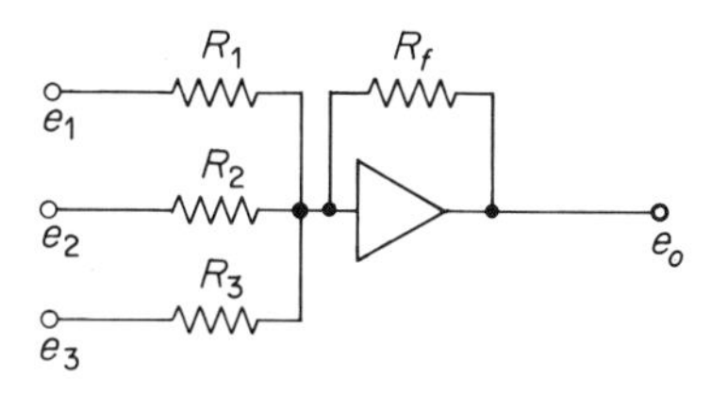

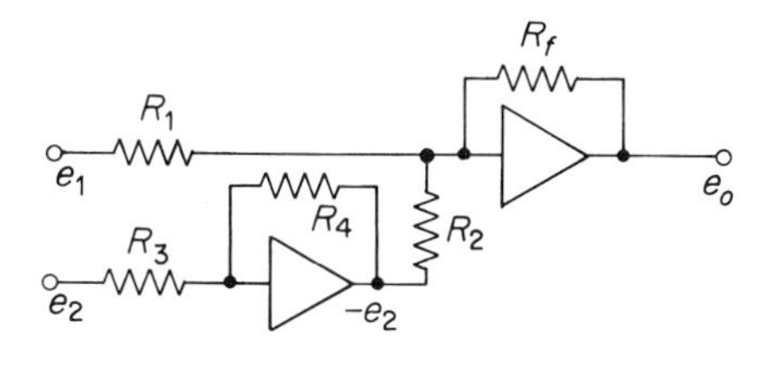

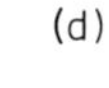

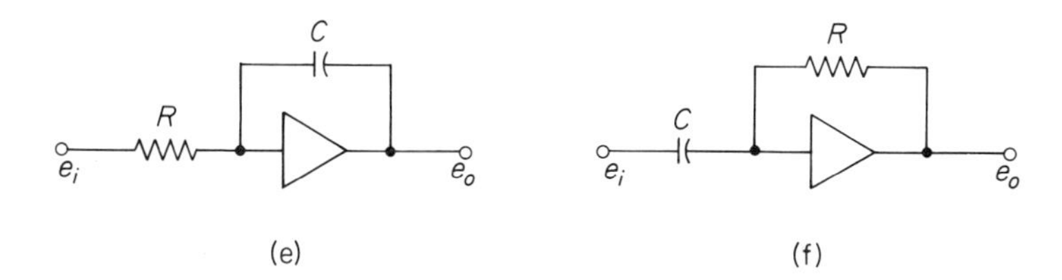

(e) (f)

Fig. 9-2 Basic operations by an operational amplifier: (a) multiplication by a constant; (b) division by a constant; (c) summation; (d) subtraction; (e) integration; (f) differentiation.

$$R_1 = R_2 = R_3 = R_4$$

$$\text{Fig. 9-2(e):} \quad e_o = -\frac{1}{RC}\int_0^t e_i\,dt + e_o(0+) \tag{9.4e}$$

$$\text{Fig. 9-2(f):} \quad e_o = -RC\frac{de_i}{dt} \tag{9.4f}$$

Using these operational blocks, we shall now illustrate the various steps described previously by means of the following example.

EXAMPLE 9–1

Solve for the current i in an RLC series circuit when a 100 V battery is suddenly connected across it. Given: $R = 1\ \Omega$; $L = 100$ mH (millihenrys); $C = 100\ \mu$F (microfarads); $i(0+) = 0$; and $q(0+) = 1000\ \mu$C (microcoulombs).

The mathematical equation for the circuit is

$$L\frac{di}{dt}+Ri+\frac{1}{C}\int i\,dt = v$$

Or, substituting numerical values in this equation yields

$$\frac{di}{dt}+10i+10^5\int i\,dt = 1000$$

Solving for the highest-order derivative, we have

$$\frac{di}{dt} = 1000-10i-10^5\int i\,dt \tag{a}$$

Equation (a) is represented by the block diagram in Fig. 9E-1(a). The initial condition $q(0+)$ is represented as an initial charge on the capacitor of the last integrator. The input voltage is the machine reference voltage. So far, we do not know how many volts at a point in the computer correspond to (say) one ampere of the problem current. The next step is, then, to choose the time and amplitude scales.

In order to choose the time scale, we transform Eq. (a) to machine time by substituting $\tau = \alpha t$, where τ = machine time, t = problem time, and α = a constant. Making this substitution in Eq. (a) gives

$$\frac{di}{d\tau} = \frac{1000}{\alpha}-\frac{10}{\alpha}i-\frac{10^5}{\alpha^2}\int i\,d\tau \tag{b}$$

The computer solution is faster or slower according to whether $\alpha<1$ or $\alpha>1$. Although the choice of α depends on a number of factors—such as permissible response, solution time, response of the recording device at the output—for a second-order system it is recommended that

$$\alpha>1 \quad \text{if} \quad \omega_n \quad \text{is large}$$
$$\alpha<1 \quad \text{if} \quad \omega_n \quad \text{is small}$$

where ω_n = natural frequency of oscillation = $1/\sqrt{LC}$ in the problem under consideration. We note that

$$\omega_n = 10^2\sqrt{10} = \text{large}$$

so that we choose $\alpha = 10$.

Next, to find the scale factor for the dependent variable i, we let $I = \beta i$ in Eq. (b), which then becomes

$$\frac{dI}{d\tau} = \frac{\beta}{\alpha}(1000)-\frac{10}{\alpha}I-\frac{10^5}{\alpha^2}\int I\,d\tau \tag{c}$$

with the initial condition transformed to

$$Q(0+) = \alpha\beta q(0+) = 10^{-3}\alpha\beta \tag{d}$$

The block diagram of Fig. 9E-1(a) then modifies to that shown in Fig. 9E-1(b).

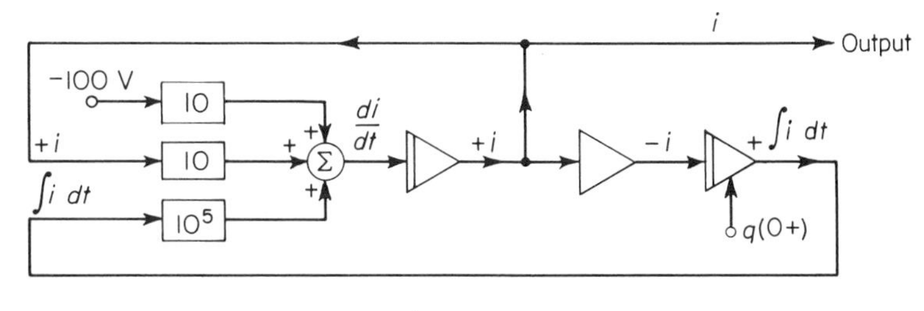

(a)

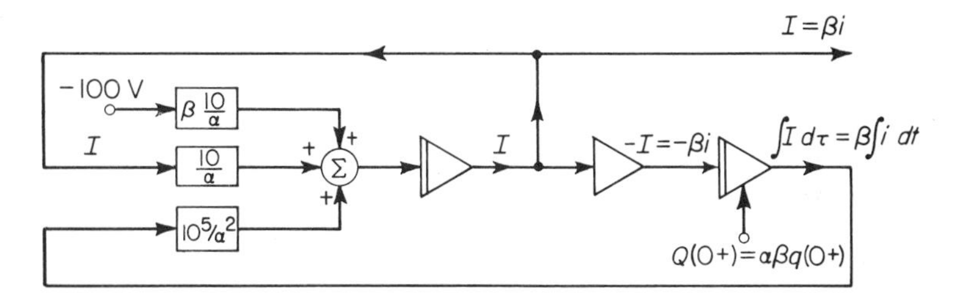

(b)

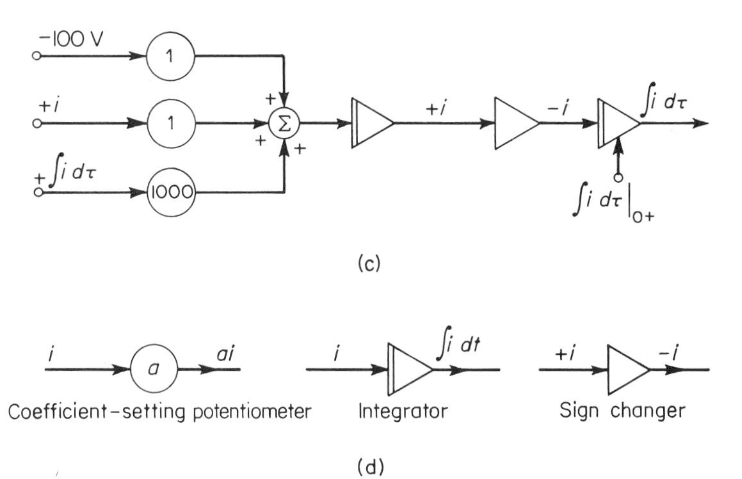

(c)

(d)

Fig. 9E-1

The choice of β depends on the linear region of operation of the amplifiers. If, for example, the linear region of operation is ± 100 volts, then

$$\beta \leqslant \frac{100}{|i_{\max}|} \frac{\text{volt}}{\text{unit}} \tag{e}$$

In our problem, $i_{max} = 100$ (with no charge on the capacitor), so that we can choose $\beta = 1$.

Having chosen the scale factors, we now choose the computing-circuit elements of the computer, keeping in mind that the amplifiers are not overloaded. It is beyond the scope of this book to discuss this matter in detail, but the interested reader should consult a book on analog computers such as Reference 9. With $\alpha = 10$, Eq. (b) becomes

$$\frac{di}{d\tau} = 100 - i + 1000 \int i\, d\tau \tag{f}$$

The final computer representation is given in Fig. 9E-1(c); the nomenclature is explained in Fig. 9E-1(d). Note that the coefficient setting of 1000 in Fig. 9E-1(c) is unrealistic. But this can be achieved by feeding the output $\int i\, d\tau$ into two amplifiers in cascade with gains of 50 and 20 and then feeding the output of the second stage into the coefficient-setting potentiometer set at unity.

The above example illustrates an application of the analog computer and introduces some of its components. We shall now consider some examples of systems involving electromechanical devices.

EXAMPLE 9–2

An antenna-positioning system, involving a number of electromechanical devices studied in preceding chapters, is schematically represented in Fig. 9E-2(a). The equations relating the various inputs and outputs are as follows

$$\text{dc generator:} \quad v_f = (R_f + L_f p) i_f$$

$$v_g = k_g i_f$$

$$\text{dc motor:} \quad v_m = R_a i_a + kn$$

$$T = k_T i_a = \frac{2\pi}{60} J \frac{dn}{dt} + bn$$

$$\text{synchro set:} \quad v_s = k_s(\theta_R - \theta_c)$$

$$\text{magnetic amplifier:} \quad v_f = k_m v_d$$

$$\text{phase-sensitive demodulator:} \quad v_d = k_d v_s$$

The pertinent numerical values are: $k_d = 0.3$ V, dc/V, RMS; $k_m = 1.82$; $k_g = 100$ V/A; $L_f = 20$ H; $R_f = 100\ \Omega$; $k_T = 0.772$ lb-ft/A; $k = 0.11$ V/r/min; $R_a = 0.588\ \Omega$; $b = 0.0025$ lb-ft/r/min; $J_{motor} = 0.5$ slug-ft^2; $J_{antenna} = 10$ slug-ft^2; gear ratio = 50/500; $k_s = 1$ V/degree.

We wish to study the dynamic behavior of this system in response to step input on an analog computer and obtain an explicit mathematical expression for that response.

The total moment of inertia referred to the motor side of the gear box is

$$J = J_{motor} + a^2 J_{antenna} \tag{a}$$

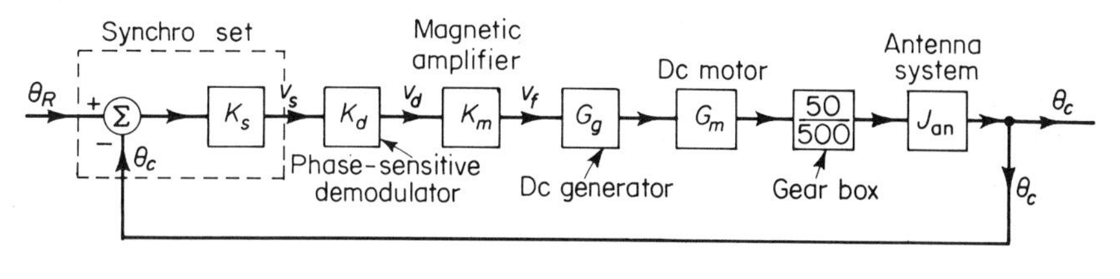

(a)

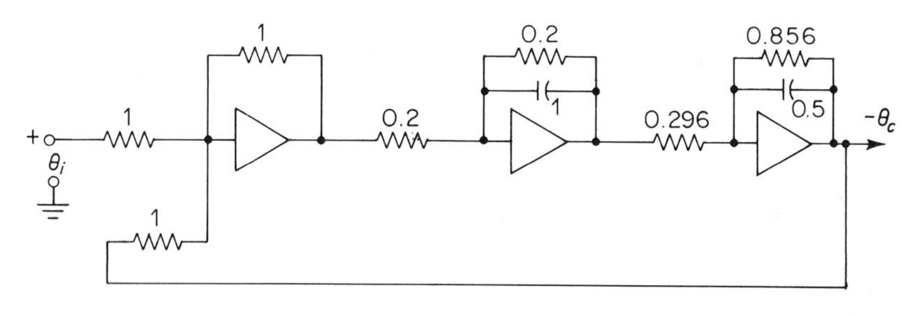

(b)

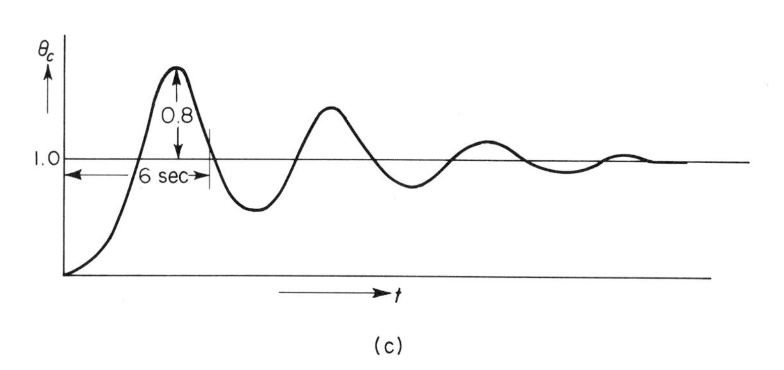

(c)

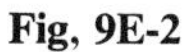

Fig, 9E-2

where a = gear ratio. Substituting numerical values in Eq. (a) yields

$$J = 0.5 + (0.1)^2 \times 10 = 0.6 \quad \text{slug-ft}^2$$

From the given input–output relationships, the overall open-loop transfer function is

$$G(s)H(s) = \frac{34.2}{s(s+5)(s+2.34)} \tag{b}$$

The closed-loop transfer function is

$$\frac{\Theta_c(s)}{\Theta_R(s)} = \frac{C(s)}{R(s)} = \frac{G(s)H(s)}{1+G(s)H(s)} \tag{c}$$

Substituting Eq. (c) in Eq. (b) we find that, for $R(s) = 1/s$,

$$C(s) = \frac{a_1}{s+j(6.35)} + \frac{a_2}{s+0.5+j(2.28)} + \frac{a_3}{s+0.5-j(2.28)} + \frac{a_4}{s} \tag{d}$$

where

$$a_1 = 3.9\times 10^{-3}\angle 180^\circ$$
$$a_2 = 11.8\times 10^{-3}\angle -145^\circ$$
$$a_3 = 11.8\times 10^{-3}\angle 145^\circ$$
$$a_4 = 27.4\times 10^{-3}\angle 0^\circ$$

The output $\theta_c(t)$ is found by taking the inverse transform of Eq. (d), which yields, after some simplification,

$$\theta_c(t) = K[27.4-3.9e^{-6.35t}+23.6e^{-0.5t}\cos(2.28t+2.54)] \tag{e}$$

where K is a scale factor.

Having obtained the analytical solution we now turn to the computer solution. The analog-computer representation is shown in Fig. 9E-2(b). The resistances are in megohms and capacitances are in microfarads. Note that, although θ_c is accompanied by a minus sign, this should not matter if the output is recorded and "inverted." The response to step input is shown in Fig. 9E-2(c). Notice that the settling time is about 6 s and that rise time is 0.8 s. The behavior of the system is thus unsatisfactory; but it can be improved by appropriate compensation. In fact, those familiar with control theory will immediately recognize that a lead compensator should improve performance. For instance, the settling time can be reduced to about 1 s by a suitably designed compensating network. The desired network can easily be found on an analog computer.

EXAMPLE 9–3

We next consider an example of a speed-regulating system of the Ward Leonard type discussed in Chapter 6 (Sec. 6.4). The system is shown in Fig. 9E-3(a). Notice that it is somewhat different from that shown in Fig. 6E-5(a). We wish to obtain an analog-computer representation of this system and to study the effect of gain adjustment on the step response of the system. The time constants of various elements are shown in Fig. 9E-3(a).

The block diagram of the system is shown in Fig. 9E-3(b). In this problem, we do not need to obtain the transfer function of each element because the time constants are specified. From Fig. 9E-3(b), we directly obtain the analog-computer representation shown in Fig. 9E-3(c), in which the resistances are in megohms and the capacitances in microfarads. For each of two values of the gain, $k = 1$ and $k = 0.1$, the step response of the system is shown in Figs. 9E-3(d) and (e). Observe that for larger gains the system tends to be more oscillatory. The advantage of simulating the problem on the computer is clearly seen in the fact that the effect of gain change can readily be observed for all gains.

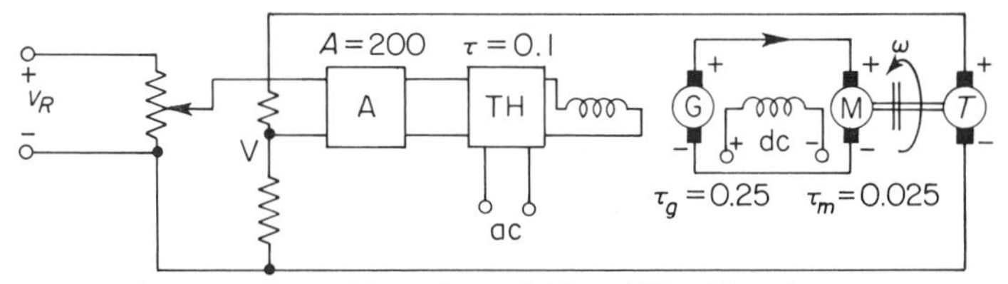

v_R = reference voltage; V = voltage divider; TH = Thyratron;
G = dc generator; M = dc motor; T = dc Tachometer

(a)

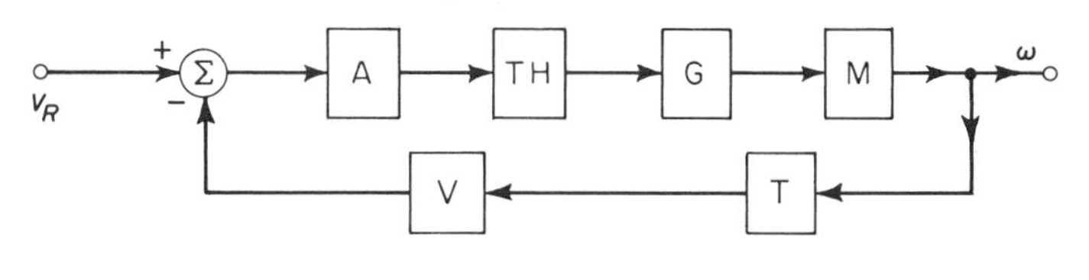

(b)

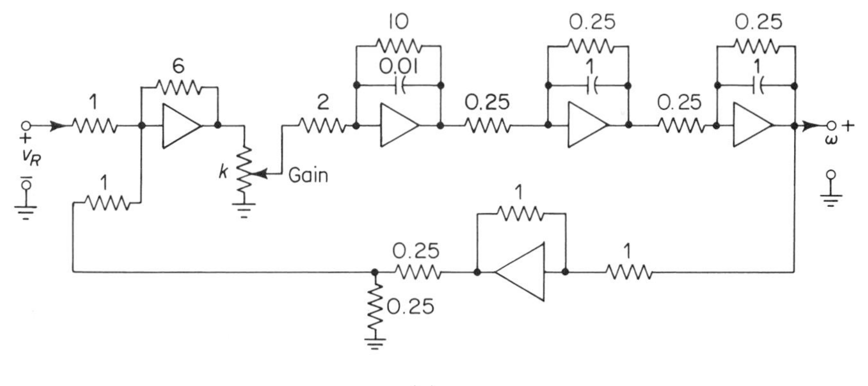

(c)

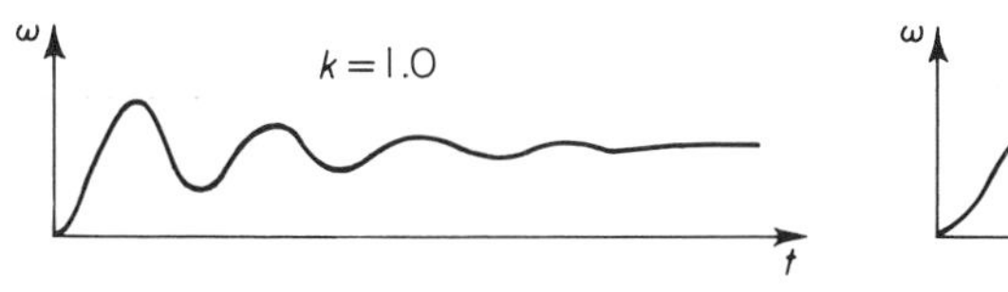

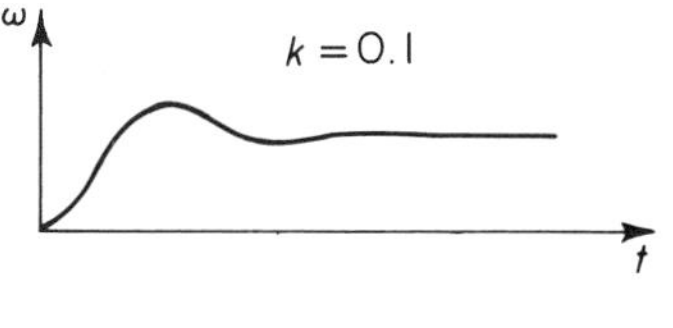

(d)

(e)

Fig. 9E-3

The preceding three examples merely illustrate some of the uses of the analog computer in solving problems involving the dynamics of energy-conversion devices. In this connection, we might recall the analog-computer representations of the induction machine (Fig. 7-22) and of the synchronous machine in a voltage-regulating system study (Fig. 8E-5). The applications of analog computers to such problems are numerous; References 1–8 give some examples.

Before discussing some of the applications of digital computers, we shall review the problems adaptable to digital-computer solutions, keeping in mind that for this purpose the differential equations are to be expressed as difference equations, integrations as summations, and so on.

9.2 Certain Applications of Digital Computers

Digital computers play a key role in practically every profession which involves numerical-data handling and processing. They are revolutionizing engineering practices and will assume more and more importance to the profession in years to come. One competent observer, echoing the opinions of many, has said that exploitation of the digital computer is producing a "second industrial revolution," and that it is having as great an influence on society as the invention of the steam engine did some 200 years ago. It is, therefore, extremely important for us to be conversant with the techniques of using digital computers. Fortunately, programs for a wide range of problems relating to electromagnetic energy-conversion devices and systems can be written in FORTRAN language, which can be self-taught.[10]

The computer finds such a wide range of applications because of its incredible speed, accuracy, memory, and because it works automatically step-by-step after a set of instructions have been fed into the computer. The essential steps performed by the computer are shown in Fig. 9-3. The data

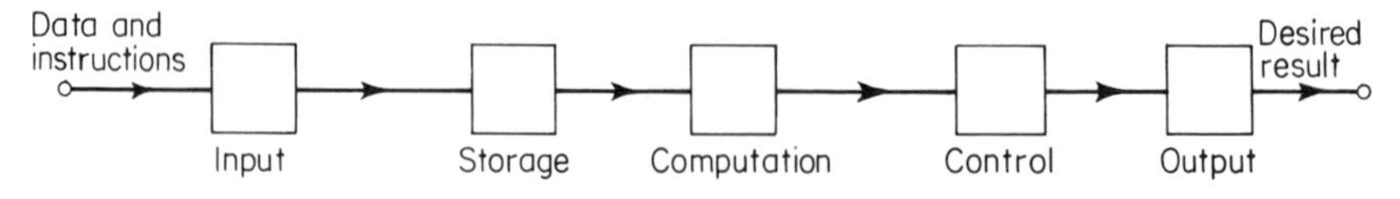

Fig. 9-3 Essential steps performed by a digital computer.

and a set of instructions are fed as input and are stored in the memory of the computer; the computer performs the calculations according to the instructions fed into it. By *control* we mean that the various steps are carried out in an appropriate order. Of course, after the computations have been completed the desired end results constitute the *output*. The various steps involved in solving a problem on a digital computer are, in summary form, as follows:

1. Obtain a mathematical description of the problem.
2. Analyze the problem and prepare a flow chart.
3. Write a program either in
   (a) symbolic (or computer) language, or
   (b) a special language such as FORTRAN or ALGOL.
4. Punch the program onto cards.
5. Punch the data onto cards or paper tape; or record the data on magnetic tape.
6. Feed the program and data into the computer.
7. Obtain the end results.

Because the range of application of digital computers to energy-conversion problems is enormous, it is possible to consider only a few examples here; in later sections we shall consider some more. References 11–20 give samples of other applications. The digital computer has been used in analysis and design (as well as design optimization) of electromagnetic devices; the interested reader should consult References 11–20.

EXAMPLE 9–4

In this example we consider the determination of the speed–torque characteristics of an inertial-guidance gyro-spin motor by conducting acceleration–deceleration tests and then processing the data on a digital computer.

The measurement of shaft torque available to drive a load and the friction and windage power are quantities that cannot easily be measured in certain motors, such as vertical pump motors, very large machines, and very small machines. About the top limit for prony brakes is 300 hp for horizontal motors, and for dynamometers about 2000 hp. When very small motors are being tested, the weight of a prony brake, the side thrust of a dynamometer, or the friction and windage of a coupling often impose loads that constitute a sizable portion of the total shaft output. In such cases, acceleration–deceleration methods may be used to determine the electrical torque being developed, the friction and windage torque, and the net shaft torque available.

The motor to be tested in this experiment is a small gyroscopic-spin motor. In aircraft gyros, the attempt is made to obtain a high spin speed so as to have a maximum ratio of stored energy to weight. For this reason, "inside-out" construction is employed wherein the squirrel-cage rotor is outside the stator winding. This particular motor is designed for operation on 3-phase, 40–80 V, 400-cycle alternating current and has a no-load speed of about 24,000 r/min.

The accuracy of acceleration–deceleration methods is directly dependent upon the accuracy with which the moment of inertia is determined. In the laboratory a substitution method is used. The period of the unknown rotor is measured as a torsion pendulum, and then a known moment of inertia is added to the system to obtain a new rotating system having a different period. The accuracy of the results is also dependent upon the speed measurements. To insure as high accuracy as possible an electric counter which adds up the number of pulses in a prescribed time interval is used. A digital readout indicates the number of times per second that a series of six alternate black and white segments on the rotor pass a given point (a window in the frame), as viewed by a photocell transducer. The counter then reads directly the average speed in tens of r/min.

Having outlined the experimental procedure above, let us now consider the computational details. First, the expression for the moment of inertia by the substitution method, using a torsion pendulum, is obtained from the following equations. If J_1 is a known moment of inertia, attached to a torsion spring of stiffness K to make a torsion pendulum, of time-period τ_1, then

$$\tau_1 = 2\pi \sqrt{\frac{J_1}{K}} \tag{a}$$

If an unknown inertia J is added to the known inertia and the time period changes to τ, then

$$\tau = 2\pi \sqrt{\frac{J+J_1}{K}} \tag{b}$$

From Eqs. (a) and (b), therefore,

$$J = J_1\left[\left(\frac{\tau}{\tau_1}\right)^2 - 1\right] \tag{c}$$

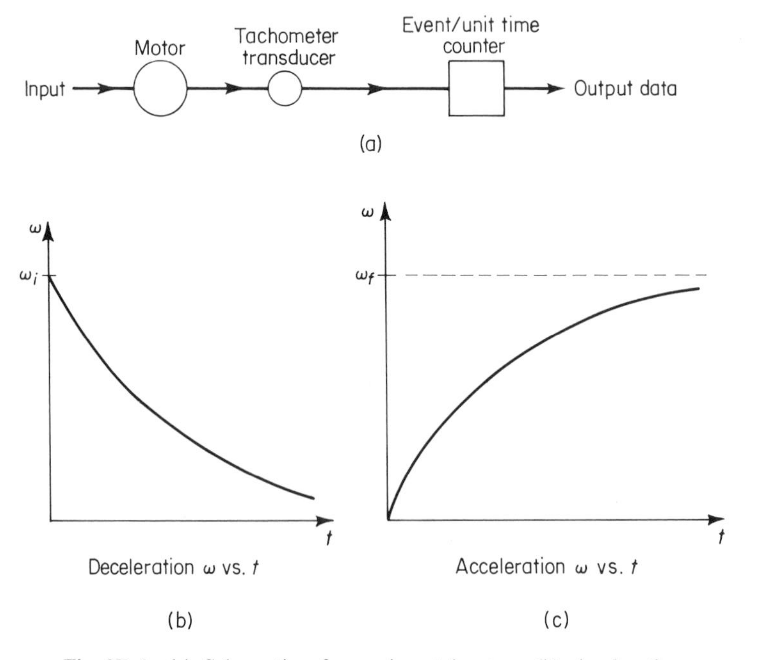

Fig. 9E-4 (a) Schematic of experimental setup; (b) deceleration ω vs. t; (c) acceleration ω vs. t.

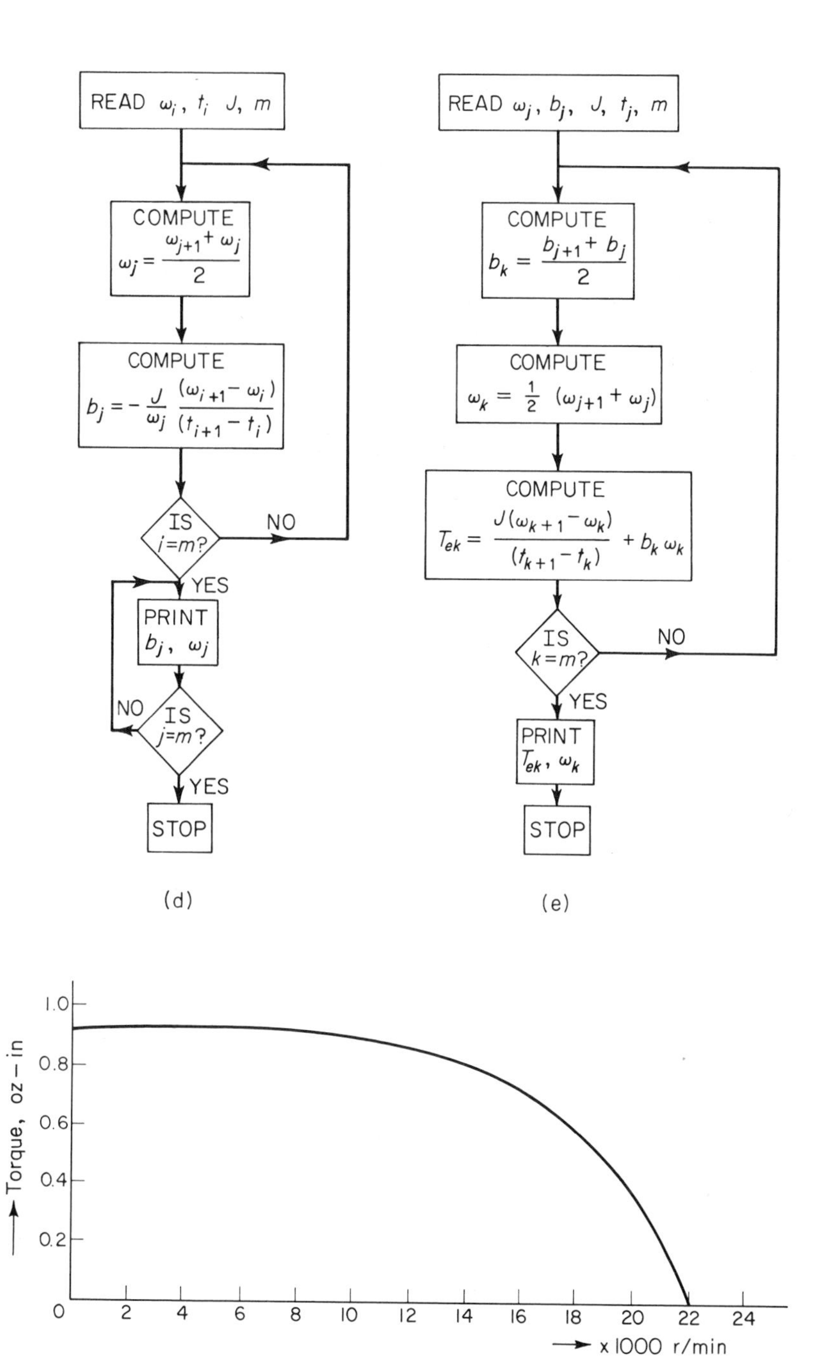

Fig. 9E-4—*Cont.* (d) Flow chart to compute b vs. ω; (e) flow chart to compute T_e vs. ω; (f) speed vs. torque.

From the problem under consideration, we thus find from experiment that the moment of inertia of the motor is

$$J = 3.36 \times 10^{-4} \quad \text{lb-ft-s}^2$$

On no-load, the electrical torque developed by the motor is balanced by the restoring torques due to inertia, friction, and windage. Or,

$$J\dot{\omega} + b\omega = T_e \tag{d}$$

In Eq. (d), b might be a function of ω. To find b, the motor is brought to full speed and then the power input is turned off. Thus, with $T_e = 0$, b as a function of ω can be found by measuring ω (as a function of time) and $\dot{\omega}$. Once b is known, T_e can be found by monitoring ω and $\dot{\omega}$ with the power on. The schematic of the experimental set-up for the collection of data and the pertinent deceleration and acceleration curves are respectively shown in Fig. 9E-4(a), (b), and (c). The equation for acceleration is given as Eq. (d), in which b (which is a function of ω) is found from the deceleration equation

$$J\dot{\omega} + b\omega = 0$$

Or,

$$b = -\frac{J}{\omega}\frac{\Delta\omega}{\Delta t}$$

Knowing b and J, T_e is calculated from Eq. (d). Evidently, for a machine rated at 24,000 r/min an enormous amount of data has to be processed. And, in such cases a digital computer is almost indispensable. We may rewrite the expressions for the quantities to be computed as follows:

$$b = -\frac{J}{\omega}\frac{\Delta\omega}{\Delta t} \tag{e}$$

$$T_e = J\frac{\Delta\omega}{\Delta t} + b\omega \tag{f}$$

We compute b using the deceleration-test data and the flow chart shown in Fig. 9E-4(d). Having obtained b, using the acceleration-test data, Eq. (f), and the flow chart shown in Fig. 9E-4(e), we compute the electrical torque developed by the motor. The torque–speed curve thus obtained is shown in Fig. 9E-4(f). Note that in the flow charts m is the number of ω's read in as data.

In some cases $b \simeq 0$, and T_e is then obtained from acceleration-test data alone using the relationship

$$T_e = J\frac{\Delta\omega}{\Delta t} \tag{g}$$

A FORTRAN program for Eq. (g) is given in Appendix C.

9.2.1 *Numerical Solution of Laplace's Equation*[21–23]

Recall from Chapters 2 and 5 the usefulness of Laplace's equation such as

$$\nabla^2 \mathbf{A} = 0 \tag{9.5a}$$

$$\nabla^2 B_x = \nabla^2 B_y = 0 \tag{9.5b}$$

in solving for electromagnetic fields in source-free regions of energy-conversion devices. Also recall that analytical solutions to Laplace's equation are in the forms of series involving trigonometric functions, Bessel functions, etc. To obtain numerical values from these series solutions is often cumbersome. On the other hand, by using a digital computer, numerical solutions of Laplace's equation of sufficient accuracy for engineering applications can be obtained for specified boundaries and boundary conditions. Furthermore, for irregular boundaries (as illustrated in the next example) analytical solutions are not possible at all.

There are a number of numerical methods available in the literature[21,22] for solving various field equations, including Laplace's equation. We briefly review here the solution of Laplace's equation by the iteration method. Let the equation to be solved be

$$\frac{\partial^2 \psi}{\partial x^2} + \frac{\partial^2 \psi}{\partial y^2} = 0 \tag{9.6}$$

in a region, a portion of which is shown in Fig. 9.4. This region is divided into a square mesh of sides h. If the unknown ψ's at the indicated points are as shown and if h is sufficiently small, then

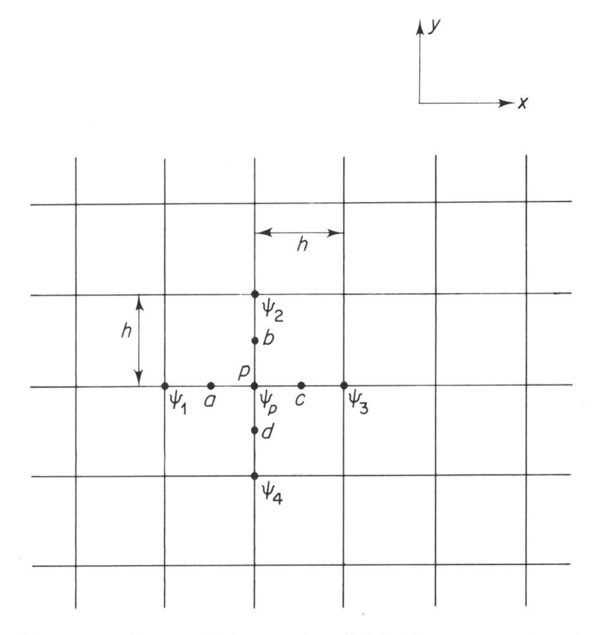

Fig. 9-4 Unknown ψ's in a region divided into a square mesh.

$$\left.\frac{\partial\psi}{\partial x}\right|_a \simeq \frac{1}{h}(\psi_1-\psi_p) \tag{9.7a}$$

$$\left.\frac{\partial\psi}{\partial x}\right|_c \simeq \frac{1}{h}(\psi_p-\psi_3) \tag{9.7b}$$

so that, from Eqs. (9.7a,b), at the point P

$$\left.\frac{\partial^2\psi}{\partial x^2}\right|_p \simeq \frac{1}{h}\left(\left.\frac{\partial\psi}{\partial x}\right|_a - \left.\frac{\partial\psi}{\partial x}\right|_c\right) \simeq \frac{1}{h^2}[(\psi_1-\psi_p)-(\psi_p-\psi_3)] \tag{9.8a}$$

Similarly,

$$\left.\frac{\partial^2\psi}{\partial y^2}\right|_p \simeq \frac{1}{h^2}[(\psi_2-\psi_p)-(\psi_p-\psi_4)] \tag{9.8b}$$

Consequently, from Eqs. (9.6) and (9.8a,b), we have

$$\psi_p \simeq \tfrac{1}{4}(\psi_1+\psi_2+\psi_3+\psi_4) \tag{9.9}$$

In the computations, therefore, ψ_p is found from Eq. (9.9) at the corner of every square in the mesh. The process is repeated over the entire region until the values of ψ's do not change over a complete cycle of computation. It is best to carry out such computations on a digital computer. A FORTRAN program for the iteration method for the following example is given in Appendix C.

EXAMPLE 9–5

A system of two electrodes is shown in Fig. 9E-5. The electrodes are separated

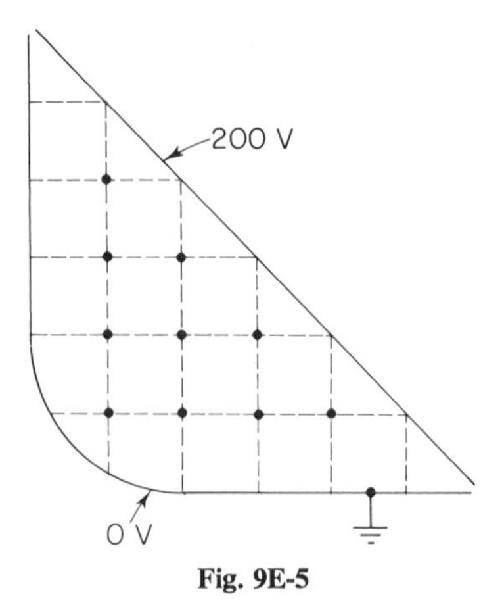

Fig. 9E-5

from each other by infinitesimal gaps, and are respectively maintained at 200 V and 0 V potential. Solve for the potential within the region bounded by the electrodes.

We know that the potential within the region satisfies the equation

$$\frac{\partial^2 V}{\partial x^2} + \frac{\partial^2 V}{\partial y^2} = 0$$

Even for this simple problem, notice that the geometry of the electrodes does not permit an analytical solution. We may, then, divide the region into squares (and triangles) as shown. To find the potentials at the marked points we use the program given in Appendix C. The computer results rounded off to the third decimal place are tabulated below.

0.000						
0.000	200.000					
0.000	119.470	200.000				
0.000	77.888	143.483	200.000			
0.000	48.618	96.061	143.483	200.000		
0.000	20.548	48.618	77.888	119.470	200.000	
0.000	0.000	0.000	0.000	0.000	0.000	0.000

Clearly, the accuracy of the results can be increased by decreasing the size of the mesh and then changing the dimension statement in the program.

9.3 *Graphical Field Mapping*[22–24]

In the last section we discussed the numerical solution of Laplace's equation and illustrated the procedure by an example. In many two-dimensional field problems involving complex geometrical boundaries, approximate field distribution in regions of interest can be estimated by graphical field-mapping techniques. This is a powerful method and finds numerous applications in electromagnetic problems concerning energy-conversion devices.[24–27] Leakage fluxes, flux-density distribution, and permeance functions for electrical machines can be determined by the method of graphical field mapping. We shall briefly review the method of field plotting and then illustrate it by an example. Because most energy-conversion devices contain magnetic materials, we shall focus our attention on magnetic-field (rather than electric-field) plotting. The rules of magnetic-field mapping are derived from the following facts:[28,29]

1. In a source-free region of finite permeability, such as air, the magnetic flux lines and equipotentials intersect orthogonally, as shown in Fig. 9-5, and form curvilinear squares.
2. The magnetic flux passing through a tube of length l, width w and thickness unity is given by Eq. (2.47), or

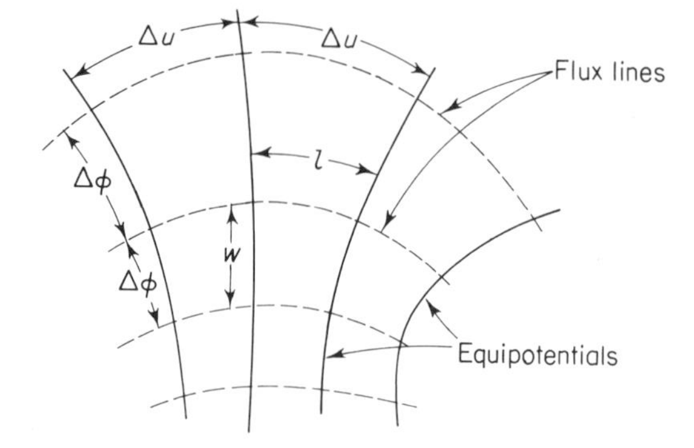

Fig. 9-5 Orthogonal intersections of magnetic flux lines and equipotentials forming curvilinear squares.

$$\Delta\phi = \Delta u\mu\frac{l}{w} \tag{9.10a}$$

where μ is the permeability of the region, and is constant.

3. From Eq. (9.10a) it follows that

$$\frac{\Delta\phi}{\mu\Delta u} = \frac{l}{w} = \text{constant} \tag{9.10b}$$

Consequently, the flux lines and equipotentials form curvilinear squares (Fig. 9-5).

4. If there are n squares between two points on an equipotential, the total flux crossing the equipotential between these points is

$$\phi = n\Delta\phi \tag{9.11}$$

And if there are m squares between any two flux lines, then the permeance $\mathscr{P}$ of a field of a region having n squares along an equipotential and m squares along a flux line is, from Eqs. (9.10b) and (9.11),

$$\mathscr{P} = \frac{\phi}{u} = \mu\frac{n}{m} \tag{9.12}$$

5. The flux lines enter an iron surface at right angles, since an iron surface (without any current sheet on it) is an equipotential.

From the above-mentioned properties of flux lines and equipotentials, the following rules for field mapping are formulated:

1. Note the geometric symmetry of equipotentials in the region in which the field is to be plotted.
2. Sketch the equipotentials and flux lines in those portions (of the region) in which the field is uniform.

3. Draw the equipotentials "parallel" to iron boundaries, and draw the flux lines "perpendicular" to equipotentials and iron boundaries, making curvilinear squares as shown in Fig. 9-5.
4. Note that the flux lines are dense at sharp edges and acute angles, and that the converse is true for obtuse angles.

With the above background material, we shall now consider an example of a salient-pole machine illustrating the field plots.

EXAMPLE 9–6[27]

Draw the equipotentials and flux lines for the salient-pole magnetic structure shown in Fig. 9E-6 and obtain the flux-density distribution in the airgap of the machine.

Using rules 1–4, we sketch the flux lines and equipotentials as shown in Fig.

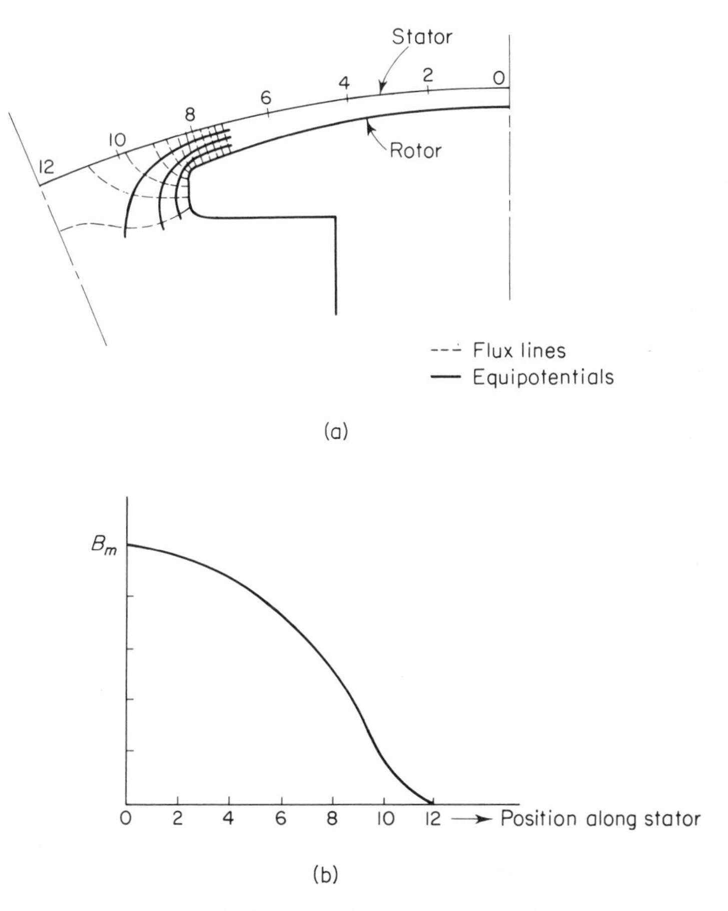

Fig. 9E-6 (a) Field map; (b) flux-density distribution.

9E-6(a), from which we may find and plot the flux-density distribution, as shown in Fig. 9E-6(b).

In Secs. 9.1–9.3 we have briefly presented three tools—analog-computer, digital-computer, and graphical field-mapping techniques—which enable us to extend the scope of the theory developed in earlier chapters. In particular, we shall now attempt to include the effects of space harmonics, saturation, leakage fields, etc. on the performance characteristics of electrical machines.

† 9.4 Space Harmonics[32–36]

Reference was made to space harmonics in Chapter 5, but in our analyses so far we have explicitly assumed that the current sheets are sinusoidally distributed. The airgap fields produced by these current sheets are also sinusoidal. Although it is essential for better machine performance that the airgap fields not contain space harmonics, in many instances space harmonics are inherently present. For a first-order approximation, the effects of harmonics can be neglected, but for a refined solution the harmonics should be taken into account. In some cases, the effects of harmonics must be considered if results compatible with reality are to be obtained.

Recall from Chapter 5 that the various windings on the magnetic structures of a rotating machine produce a magnetic field in the airgap of the machine. Such a magnetic field can be analyzed as a series of harmonic contents. These harmonics are mainly due to nonsinusoidal winding distributions and permeance variations in the airgap due to slot openings. The space harmonics are responsible for effects such as crawling, magnetic noise, vibration, locking, and harmonic voltage ripples. Harmonic fields may travel in either direction—depending upon the number of phases and the order of the harmonic—thereby sometimes giving rise to negative torques and producing cusps and dips in the speed–torque curves of the machine. We shall present here certain approaches to solving the problem of harmonics, a subject that, being very old, has been a topic of extensive research.[32–36]

9.4.1 *Current-Sheet Harmonics in nm-Winding Machines*[34,37]

We introduced the concept of current sheets in Chapter 5 and, assuming sinusoidal current-sheet distributions, derived the inductance formulas for a 2-phase idealized machine (Secs. 5.3 and 5.4). Nonsinusoidal, but periodic, current sheets can be expressed as

$$h(x) = \sum_{k=1}^{m} \sum_{\nu=1}^{\infty} i_k c_{k\nu} \cos(\alpha_\nu \theta - \phi_{k\nu}) \tag{9.13}$$

which leads to the following inductance formulas:[37]

$$L_{km}^{ss} = \frac{1}{2} \zeta \mu_o l \tau_r \sum_\nu c_{k\nu}^s c_{m\nu}^s \cos(\phi_{k\nu}^s - \phi_{m\nu}^s) \frac{\coth \alpha_\nu g}{\alpha_\nu} \tag{9.14a}$$

$$L_{pq}^{rr} = \frac{1}{2} \zeta \mu_o l \tau_r \sum_\nu c_{p\nu}^r c_{q\nu}^r \cos(\phi_{p\nu}^r - \phi_{q\nu}^r) \frac{\coth \alpha_\nu g}{\alpha_\nu} \tag{9.14b}$$

$$L_{kq}^{sr} = \zeta \mu_o l \tau_r \sum_\nu c_{k\nu}^s c_{q\nu}^r \cos(\alpha_\nu x_o + \phi_{q\nu}^r - \phi_{k\nu}^s) \frac{\operatorname{cosech} \alpha_\nu g}{\alpha_\nu} \tag{9.14c}$$

where ζ, μ_o, l, τ_r and g are as defined in Sec. 5.4.1, and the constants c's and ϕ's are obtained from the Fourier analysis of the current sheets. Explicitly, these are given by:

$$\left.\begin{aligned}
c_{k\nu}^s &= a_{k\nu}^2 + b_{k\nu}^2 \\
\phi_{k\nu} &= \tan^{-1} \frac{b_{k\nu}}{a_{k\nu}} \\
\text{where} \quad a_\nu &= \frac{2K_{s\nu}}{\tau_r} \sum_{k=1}^{q} N_k \cos \alpha_\nu (k-1)\tau_s \\
b_\nu &= \frac{2K_{s\nu}}{\tau_r} \sum_{k=1}^{q} N_k \sin \alpha_\nu (k-1)\tau_s \\
\text{and} \quad K_{s\nu} &= \frac{\sin(\alpha_\nu w_s/2)}{\alpha_\nu w_s/2} \qquad \alpha_\nu = \frac{2\pi\nu}{\tau_r}
\end{aligned}\right\} \tag{9.15}$$

In Eqs. (9.14a–c) the effect of space harmonics has been included as identified by the summation over the order of harmonic ν. Clearly, long-hand computation of these equations would be very difficult. However, they can be very easily computed on a digital computer; a computer program for obtaining the inductance coefficients (for a 3-phase machine) is given in Appendix C. Having obtained the machine inductances, we can now formulate the equations of motion.

The equations of motion of the *nm*-winding machine, as given by Eqs. (5.28a,b), can be solved if the symmetry properties of the inductance sub-matrices are used to reduce them to diagonal forms. The various definitions and theorems relating to linear algebra necessary for such matrix manipulations are given in Sec. 5.7.1.

The machine under consideration has n windings on the stator and m windings on the rotor. The inductances of the machine are given by Eqs. (9.14a–c). The windings are symmetrically spaced around the rotor and stator peripheries. Matrices S_n and S_m are now introduced such that $S_n^{-1}L^{ss}S_n$ and $S_m^{-1}L^{rr}S_m$ are reduced to diagonal forms and $S_n^{-1}L^{sr}S_m$ contains only two nonzero terms for every harmonic present (theorem 4, Sec. 5.7.1). Typical elements of S_n and S_m are obtained from theorem 3, Sec. 5.7.1.

The following change of variables is now made in Eqs. (5.28a,b):

$$\left.\begin{aligned} i^s &= S_n v^{s\prime} \\ v^s &= S_n v^{s\prime} \\ i^r &= S_m i^{r\prime} \\ v^r &= S_m v^{r\prime} \end{aligned}\right\} \tag{9.16}$$

where the primed quantities are transformed currents and voltages and the unprimed quantities denote the original currents and voltages. The new equations corresponding to Eqs. (5.28a,b) become

$$\begin{bmatrix} v^{s\prime} \\ \hline v^{r\prime} \end{bmatrix} = \left[\begin{array}{c|c} R^{s\prime}+pL^{ss\prime} & pL^{sr\prime} \\ \hline pL^{rs\prime} & R^{r\prime}+pL^{rr\prime} \end{array}\right] \begin{bmatrix} i^{s\prime} \\ \hline i^{r\prime} \end{bmatrix} \tag{9.17}$$

$$T_e = -\tfrac{1}{2}\,[i^{s\prime *} \quad i^{r\prime *}]\, \frac{\partial}{\partial\theta} \left[\begin{array}{c|c} 0 & L^{sr\prime} \\ \hline L^{rs\prime} & 0 \end{array}\right] \begin{bmatrix} i^{s\prime} \\ \hline i^{r\prime} \end{bmatrix} \tag{9.18}$$

where

$$L^{ss\prime} = S_n^{-1} L^{ss} S_n \tag{9.19a}$$

$$L^{rr\prime} = S_m^{-1} L^{rr} S_m \tag{9.19b}$$

$$L^{sr\prime} = S_n^{-1} L^{sr} S_m \tag{9.20a}$$

$$L^{rs\prime} = S_m^{-1} L^{rs} S_n \tag{9.20b}$$

Equations (9.19a,b) denote the diagonalized submatrices and Eqs. (9.20a,b) each contain two nonzero terms for each harmonic present. Also note that $\tilde{L}^{sr\prime *} = L^{rs\prime}$; that is, the stator-to-rotor transformed inductance submatrix, when transposed and with each element replaced by its complex conjugate, gives the rotor-to-stator transformed inductance submatrix. The elements of the submatrices along the diagonal of Eq. (9.18) are all zero, since $\partial L^{ss\prime}/\partial\theta = \partial L^{rr\prime}/\partial\theta = 0$. It should be noted that the second row in the S matrix is the complex conjugate of the last row. This fact, when used with Eqs. (9.16), yields the complex-conjugate pairs of the transformed currents and voltages for the stator as well as for the rotor.

It may be seen from the above equations that the number of simultaneous equations is reduced from $(n+m)$ to 2 simultaneous equations per space harmonic considered. The total number of equations that may have to be solved depends upon the number of space harmonics present. While space harmonics are included in these equations, it should be remembered that the effects of space harmonics decrease rapidly with their order, because the series in Eqs. (9.14a–c) converge rapidly. Thus, for example, the eleventh harmonic contributes less than 1% to the inductance coefficients.

The second and the last rows of the S matrix form a complex-conjugate pair. Similar relations hold for the third and next-to-last row, and so on. Mathematically, this relationship can be expressed as

$$v_2^{r'} = v_m^{r'*} \tag{9.21a}$$

$$v_3^{r'} = v_{m-1}^{r'*} \tag{9.21b}$$

$$v_4^{r'} = v_{m-2}^{r'*} \tag{9.21c}$$

and so on. Exactly identical relations are true for the transformed currents also.

The volt-ampere equations involving Eq. (9.21b) will be simultaneous and contain rotor-to-stator mutual terms if the second harmonic is present. Similarly, the volt-ampere equations involving Eq. (9.21c) will be simultaneous and include mutual terms if the third harmonic is present, and so on. To show the effects of harmonics and account for the conjugate relationships, a new nomenclature is introduced. Here a voltage, or current, has double subscripts. A subscript ν indicates the order of harmonic and a subscript $+$ is used to distinguish a voltage, or current, from its complex conjugate which has a $-$ subscript. In this notation, for example, $v_{+\nu}^{r'}$ would mean the $(\nu+1)$th transformed rotor voltage, and $v_{-\nu}^{r'}$ implies the corresponding complex conjugate. Similar subscripts are used for the inductance coefficients also, except that in this case a single subscript is necessary, since $L_{+\nu}^{r'} = L_{-\nu}^{r'} = L_\nu^{r'}$. In this new notation the following equation can be written for the νth harmonic:

$$\begin{bmatrix} v_{+\nu}^{s'} \\ v_{-\nu}^{s'} \\ \hline v_{+\nu}^{r'} \\ v_{-\nu}^{r'} \end{bmatrix} = \left[\begin{array}{cc|cc} R^{s'}+L_\nu^{s'}p & & pkL_\nu^{sr'}e^{j\nu\theta} & \\ & R^{s'}+L_\nu^{s'}p & & pkL_\nu^{sr'}e^{-j\nu\theta} \\ \hline pkL_\nu^{sr'}e^{-j\nu\theta} & & R^{r'}+L_\nu^{r'}p & \\ & pkL^{sr'}e^{j\nu\theta} & & R^{r'}+L_\nu^{r'}p \end{array}\right] \begin{bmatrix} i_{+\nu}^{s'} \\ i_{-\nu}^{s'} \\ \hline i_{+\nu}^{r'} \\ i_{-\nu}^{r'} \end{bmatrix} \tag{9.22}$$

where $p = d/dt$ and $k = \frac{1}{2}\sqrt{mn}$, and n and m are large as compared to ν.

In order to eliminate the time dependence of the coefficients in Eq. (9.22) and to refer the rotor quantities to the stator, a further change of variables is introduced in which the transformation matrices are functions of θ. This transformation is the complex rotating transformation, the fb transformation, with which we are already familiar, having used it in Chapters 5 (Example 5-6) and 7 (Sec. 7.3). However, previously the effects of space harmonics were neglected in the analysis. The fb transformation can easily be extended to include the effects of space harmonics by defining the following transformation matrices which operate one by one on each pair of complex-conjugate rotor currents and voltages, Eq. (9.22).

For the νth harmonic, the new fb components are written as

$$v^r_{fb\nu} = \begin{bmatrix} v^r_{f\nu} \\ v^r_{b\nu} \end{bmatrix} \quad \text{and} \quad i^r_{fb\nu} = \begin{bmatrix} i^r_{f\nu} \\ i^r_{b\nu} \end{bmatrix}$$

The fb components are related to the $+\,-$ variables through the transformation matrix $S_{fb\nu}$ as

$$v^{r\prime}_{\pm\nu} = S_{fb\nu} v^r_{fb\nu}$$

where

$$S_{fb\nu} = \begin{bmatrix} e^{-j\nu\theta} & 0 \\ 0 & e^{j\nu\theta} \end{bmatrix} \quad \text{and} \quad S^{-1}_{fb\nu} = S^*_{fb\nu}$$

Consequently, for the νth harmonic,

$$v^{r\prime}_{+\nu} = e^{-j\nu\theta} v^r_{f\nu} \tag{9.23a}$$

$$v^{r\prime}_{-\nu} = e^{j\nu\theta} v^r_{b\nu} \tag{9.23b}$$

Similar relations hold for rotor currents also. Since the angle θ is a function of time, Eqs. (9.23a,b) show that there is a relative motion between the positive–negative and fb components. If the fb components are assumed to be fixed, the $+\,-$ components rotate at an angular velocity $\nu\dot{\theta}$ in opposite directions. If the variations of currents and voltages are sinusoidal, it can be seen that the fb transformation changes the frequency of rotor currents and voltages from ω^r to $\omega^r \pm \nu\omega_m$.

The equations of motion in terms of plus–minus components are now reconsidered and fb transformation is made on rotor currents and voltages. When this operation is performed, Eq. (9.22) becomes

$$\begin{bmatrix} v^{s\prime}_{+\nu} \\ v^{s\prime}_{-\nu} \\ v^r_{f\nu} \\ v^r_{b\nu} \end{bmatrix} = \left[\begin{array}{cc|cc} R^{s\prime} + L^{s\prime}_\nu p & & kL^{sr\prime}_\nu p & \\ & R^{s\prime} + L^{s\prime}_\nu p & & kL^{sr\prime}_\nu p \\ \hline kL^{sr\prime}_\nu (p - j\nu\dot{\theta}) & & R^{r\prime} + L^{r\prime}_\nu (p - j\nu\dot{\theta}) & \\ & kL^{sr\prime}_\nu (p + j\nu\dot{\theta}) & & R^{r\prime} + L^{r\prime}_\nu (p + j\nu\dot{\theta}) \end{array}\right] \begin{bmatrix} i^{s\prime}_{+\nu} \\ i^{s\prime}_{-\nu} \\ i^r_{f\nu} \\ i^r_{b\nu} \end{bmatrix} \tag{9.24}$$

Finally, the torque equation is considered. When the differentiation and part of the matrix multiplication is carried out, Eq. (5.28b) becomes

$$T_e = -\tfrac{1}{2}[i_{+1}^{s\prime *} \quad i_{+3}^{s\prime *} \cdots i_{-1}^{s\prime *} \quad i_{+1}^{r\prime *} \quad i_{+3}^{r\prime *} \cdots i_{-1}^{r\prime *}] \begin{bmatrix} j\dfrac{\sqrt{mn}}{2} L^{sr\prime} e^{j\theta} i_{+1}^{r\prime} \\ j3\dfrac{\sqrt{mn}}{2} L_3^{sr\prime} e^{j3\theta} i_{+3}^{r\prime} \\ \vdots \\ -j\dfrac{\sqrt{mn}}{2} L_1^{sr\prime} e^{-j\theta} i_{-1}^{r\prime} \\ \hline -j\dfrac{\sqrt{mn}}{2} L_1^{sr\prime} e^{-j\theta} i_{+1}^{s\prime} \\ -j3\dfrac{\sqrt{mn}}{2} L_3^{sr\prime} e^{-j3\theta} i_{+3}^{s\prime} \\ \vdots \\ j\dfrac{\sqrt{mn}}{2} L_1^{sr\prime} e^{j\theta} i_{-1}^{s\prime} \end{bmatrix} \qquad (9.25)$$

The total torque, in terms of the $+-$ and fb components, becomes

$$T_e = -j\frac{\sqrt{mn}}{4} \sum_{\nu} \nu L_{\nu}^{sr\prime}[(i_{+\nu}^{s\prime *} i_{f\nu}^{r} - i_{+\nu}^{s\prime} i_{f\nu}^{r*}) + (i_{-\nu}^{s\prime} i_{b\nu}^{r*} - i_{-\nu}^{s\prime *} i_{f\nu}^{r})] \qquad (9.26a)$$

From the properties of the transformation matrices S_{+-} and S_{fb}, Eq. (9.26a) can be written as

$$T_e = -j\,\tfrac{1}{2}\sqrt{mn} \sum_{\nu} \nu L_{\nu}^{sr\prime}(i_{+\nu}^{s\prime *} i_{f\nu}^{r} - i_{+\nu}^{s\prime} i_{f\nu}^{r*}) \qquad (9.26b)$$

To summarize, for a harmonic ν and speed ω_m, the volt-ampere equations are Eq. (9.24) and the torque is given by Eq. (9.26b). Since these equations are far less complicated than the original ones, they can be conveniently solved, and the average torque can be found at different speeds with ν varying as 1, 3, $\cdots$.

The preceding analysis shows that for each harmonic field the motor can be analyzed independently of other harmonics. This is in agreement with the assumptions made heuristically that each particular set of space harmonics exists in a separate machine and that as many separate machines of the same type have to be interconnected as there are space harmonics present. The above discussions provide an analytical basis for this assumption as well as giving the quantitative value of the total torque developed.

For an induction motor, Eq. (9.24) can be represented by an equivalent circuit for steady-state operation using the definition of slip $s_\nu = 1-(\nu\omega_m/\omega^s)$ and the conditions that $p = j\omega^s$ and $\dot{\theta} = \omega_m$. Under these conditions, Eq. (9.24) becomes

$$\left.\begin{aligned}
V^{s'}_{+\nu} &= (R^{s'}+j\omega^s L^{s'}_\nu)I^{s'}_{+\nu}+j\omega^s L^{sr'}_\nu I^r_{f\nu}\\
0 &= j\omega^s L^{sr'}_\nu I^{s'}_{+\nu}+\left(\frac{R^{r'}}{s_\nu}+j\omega^s L^{r'}_\nu\right)I^r_{f\nu}\\
V^{s'}_{-\nu} &= (R^{s'}+j\omega^s L^{s'}_\nu)I^{s'}_{-\nu}+j\omega^s L^{sr'}_\nu I^r_{b\nu}\\
0 &= j\omega^s L^{sr'}_\nu I^{s'}_{-\nu}+\left(\frac{R^{r'}}{2-s_\nu}+j\omega^s L^{r'}_\nu\right)I^r_{b\nu}
\end{aligned}\right\} \tag{9.27}$$

The above equations are represented by the circuit shown in Fig. 9-6. These equations hold for all orders of harmonics independently, and since these are "decoupled," a circuit similar to that shown in Fig. 9-6 can be developed for each harmonic.

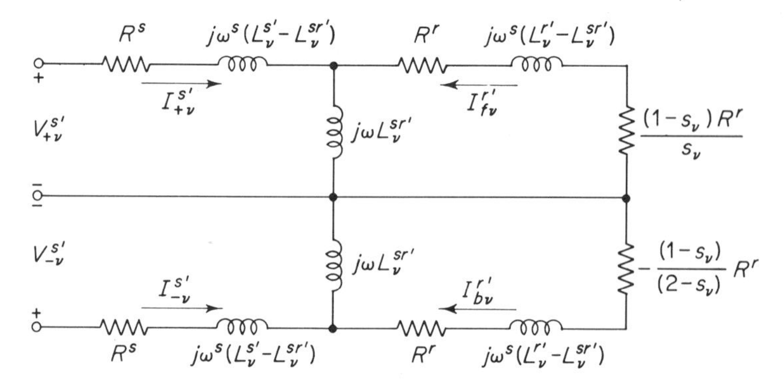

Fig. 9-6 Equivalent circuit for νth harmonic.

The performance calculations for the machine can be made by using either the circuit, or Eqs. (9.24) and (9.27), or all three. The computer program for obtaining the transformed inductances of a machine (with $n = 45$ and $m = 41$) is given in Appendix C.

The preceding analysis is valid if n and m are large and involves the $\pm$ and fb transformations. An analysis based on the dq transformation is available in Reference 38.

9.4.2 *Equivalent Circuits for Harmonic Fields*[32, 39]

Having included the effects of space harmonics in inductance coefficients and having obtained the equations of motion and their solutions by linear transformations, we now turn to a more conventional and simplified approach commonly used to develop equivalent circuits in the presence of space harmonics. We shall confine our remarks to the induction motor alone, although equivalent circuits of all kinds of electrical machines are available in the literature.[39]

The following relations between the number of phases n (on the stator) and the order of ν, the order of harmonic, hold for the forward and backward traveling fields respectively:

$$\nu = (2Kn+1) \tag{9.28a}$$

$$\nu = (2Kn-1) \tag{9.28b}$$

where K is an integer. Accordingly, based on the assumption that each harmonic field is considered to be "acting independently of all others, so far as currents, voltages, and fluxes are concerned,"[32] the circuit shown in Fig. 9-7 is developed for a balanced 3-phase induction motor. Notice that the circuit between points 1 and 2 is identical to that shown in Fig. 7-1, in which no harmonics are present. In the circuits between points 2 and 3, 3 and 4, and so on (Fig. 9-7), the effects of the harmonic fields are reflected as modified circuit parameters. For example, for the νth harmonic the magnetizing reactance to the fundamental, X_{m1}, is modified to $X_{m\nu}$. Similarly, the rotor resistance R_{21} and its leakage reactance X_{21} change to the values shown in Fig. 7-9. In this equivalent circuit, R^s is the per-phase stator resistance and X_l^s is the stator leakage reactance. The method of determining the various constants of this circuit is somewhat involved, but is available in the literature.[32]

The effect of harmonics on the speed–torque characteristics of an induction motor, as derived from Fig. 9-6 or Fig. 9-7, is shown in Fig. 9-8, which shows the resulting cusps and dips in the torque–speed curve which cause asynchronous crawling.

9.5 Saturation[31, 38, 40–48]

In previous chapters we only made references to saturation, otherwise neglecting the quantitative analysis of saturation effects. To recapitulate, in Example 2-4 we included the effect of saturation in magnetic-circuit calculations involving a simple configuration. We also derived the force equation, including the effect of magnetic nonlinearity, and illustrated its application by Examples 2-7 and 3-1. In Chapter 5 we made very general remarks about saturation, and in Chapter 6 we studied qualitatively the effects of saturation on the performance of dc machines. Quantitatively, we introduced the Froelich equation and applied it to Example 6-2. In Chapter 8 we indicated the effect of saturation on the magnitude of the direct-axis synchronous reactance of a synchronous machine. All in all, then, we are somewhat familiar with the nature of the problem at this stage.

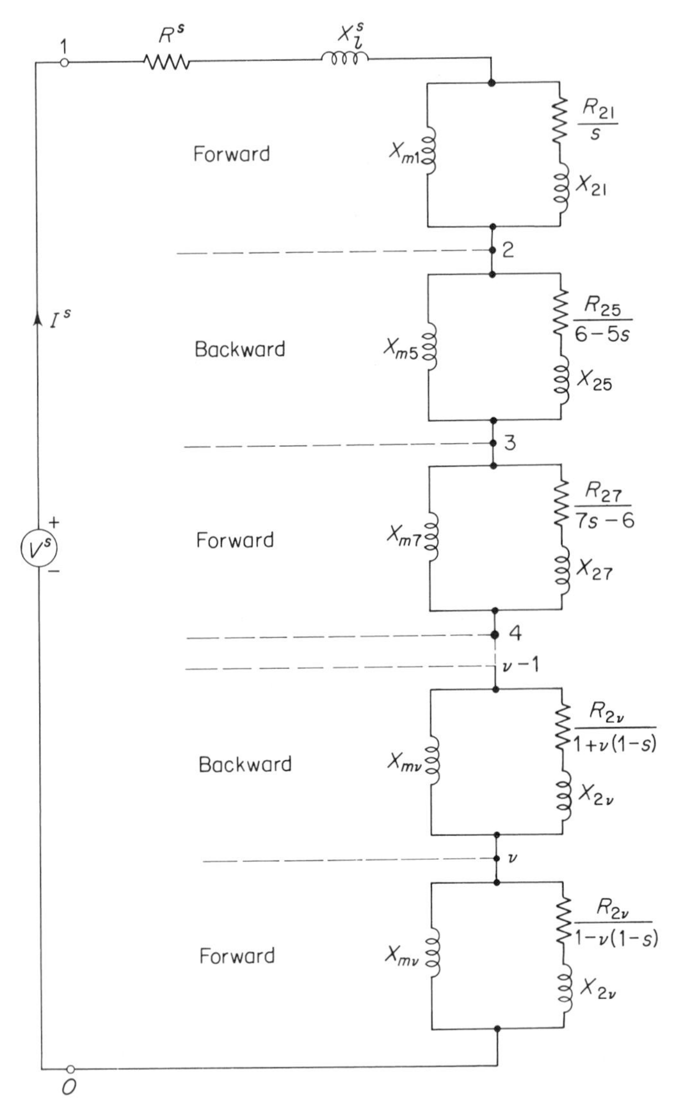

Fig. 9-7 Equivalent circuit of an induction motor, including space harmonics.

We shall now consider the effects of saturation in some detail, because modern electric machines often operate under saturated conditions. The linear theory if applied to such cases does not yield results which are in agreement with tests. In the past, saturation was taken into account using appropriate saturation factors[49] or modified graphical field mapping.[31] Now, with the availability of high-speed digital computers, it is possible to use more precise methods to determine the electromagnetic fields in saturated machines, as

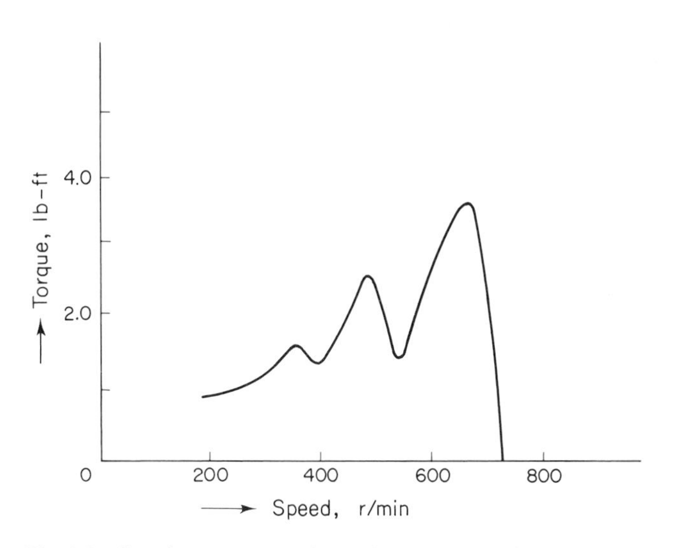

Fig. 9-8 Speed–torque curve for an induction motor, showing the effects of space harmonics.

clearly demonstrated in References 40–45. In the following section we shall present the essence of the method proposed in these papers. Next, in Sec. 9.5.2, we shall discuss the representation of saturation on an analog computer, as proposed in References 38 and 46.

9.5.1 Modified Equation for Magnetic Vector Potential[40–45]

The essence of the method presented in References 40–45 lies in the following:

1. Concept of *reluctivity*, ψ, which is the reciprocal of the permeability μ and a function of the magnetic flux density **B**. Equation (2.22) is therefore expressed as

$$\mathbf{H} = \psi \mathbf{B} \tag{9.29}$$

2. Finding a precise expression for the dependence of ψ on **B**, such as

$$\psi = a(1 - b|B|)^{-1} \tag{9.30}$$

where a and b are constants and depend on the magnetic material. Note that Eq. (9.30) is valid in region for which the reluctivity is nonnegative.

3. Obtaining the partial-differential equation for the magnetic vector potential **A**, for axially (or z-) directed currents alone and for two-dimensional fields. This equation is derived from

$$\mathbf{B} = \nabla \times \mathbf{A}$$
$$\nabla \cdot \mathbf{A} = 0$$

$$\mathbf{J} = \nabla \times \mathbf{H} = \nabla \times \psi\mathbf{B} = \nabla \times (\psi\nabla \times \mathbf{A})$$

which when expanded yields, for $\mathbf{A} = \mathbf{a}_z A$ and $\mathbf{J} = \mathbf{a}_z J$,

$$\frac{\partial}{\partial x}\left(\psi\frac{\partial A}{\partial x}\right)+\frac{\partial}{\partial y}\left(\psi\frac{\partial A}{\partial y}\right) = -J \tag{9.31a}$$

If $J = 0$, that is, in a source-free region, Eq. (9.31a) reduces to

$$\frac{\partial}{\partial x}\left(\psi\frac{\partial A}{\partial x}\right)+\frac{\partial}{\partial y}\left(\psi\frac{\partial A}{\partial y}\right) = 0 \tag{9.31b}$$

4. Defining the boundary conditions.
5. Solving Eq. (9.31b) for the source-free region.
6. Last, obtaining the magnetic flux density at various points within the region from the equations

$$B_x = \frac{\partial A}{\partial y} \tag{9.32a}$$

$$B_y = -\frac{\partial A}{\partial x} \tag{9.32b}$$

and

$$|B| = \sqrt{B_x^2+B_y^2} \tag{9.32c}$$

From steps 1–6, if **H** is known, **B** can be computed. Equation (9.31b), being nonlinear, is solved on a digital computer using a finite difference scheme. A brief discussion of Eq. (9.31b) at this point is in order. To solve this equation, ψ must be known. But to know ψ, from Eq. (9.30), $|B|$ must be known. Again, to know $|B|$, from Eqs. (9.32a–c) we must know A. However, A is unknown and we wish to determine it from Eq. (9.31b). It seems, therefore, that we need one more equation. In order to overcome this difficulty we rely on past knowledge and establish an initial reluctivity to initiate the computations. Typical values of ψ, taken from Reference 40, for a synchronous machine are as follows:

STATOR

B Wb/m²	ψ
0 to 1.5	$44.2(1-0.639B)^{-1}$
1.5 to 3.0	$44{,}883-68{,}862B^{-1}$

ROTOR

B Wb/m²	ψ
0 to 1.47	$228.5(1-0.4948B)^{-1}$
1.47 to 1.697	$152(1-0.5549B)^{-1}$
1.697 to 3.0	$45{,}682-73{,}079B^{-1}$

With the initial reluctivity known, a closed-loop computer program can be prepared and the results of each iteration checked and improved. Excellent results have been obtained by this method, as reported in Reference 40.

9.5.2 *Inverse Saturation Function*[38, 46]

In the last section we discussed the determination of the **B** field in the presence of magnetic saturation. For analog-computer representation of electrical machines, the concept of inverse saturation function has been found to be quite useful.

The saturation function and the inverse saturation function (Fig. 9-9)

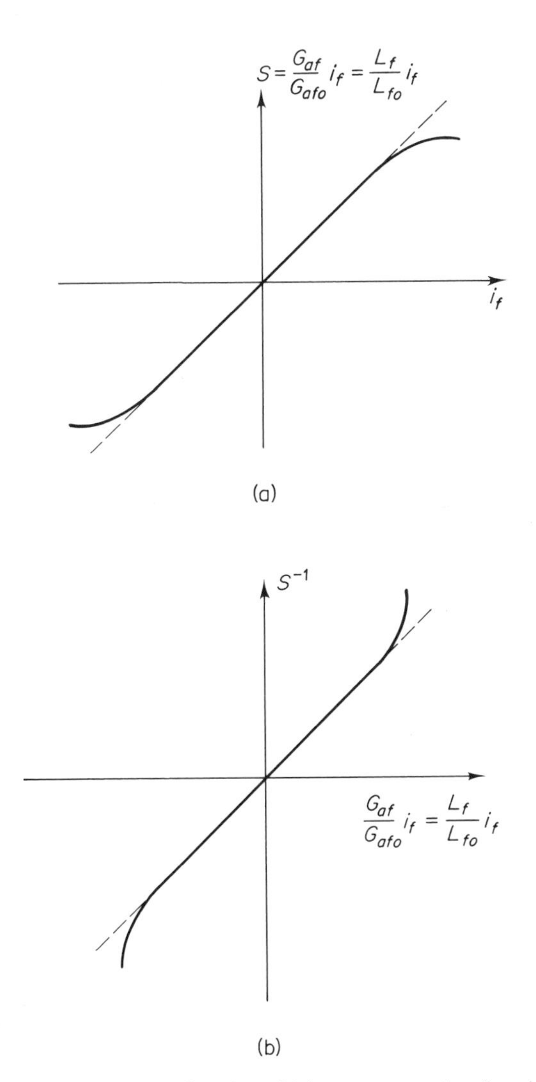

Fig. 9-9 (a) Saturation function; (b) inverse saturation function.

are found from the saturation characteristics of the magnetic material under consideration. For example, for a separately excited dc machine we know (from Chapter 6) that the open-circuit armature voltage as a function of the field current takes the form of Fig. 9-9(a), which also indicates the forms of variation of the mutual flux linkage (between the field and the armature) λ_m. Recall from Eq. (a), Example 6-1, that the open-circuit voltage of the generator is given by

$$v_a = G_{afo}\omega_m i_f \tag{9.33}$$

In Eq. (9.33) we have used the electromechanical-conversion constant G_{afo}, instead of G_{af}, to designate the unsaturated condition. For saturated conditions, however, G_{af} is not a constant. Rather it is a function of the field current i_f. Consequently, we define a *saturation function* S by the equation

$$S = \frac{G_{af}}{G_{afo}} i_f \tag{9.34a}$$

or by the expression

$$S = \frac{L_f}{L_{fo}} i_f \tag{9.34b}$$

where L_{fo} and L_f are respectively the unsaturated and saturated field inductances. From these equations we define the *inverse saturation function* S^{-1} by

$$S^{-1} = \frac{1}{S} \tag{9.35}$$

The usefulness of the inverse saturation function is illustrated by the following example.

EXAMPLE 9–7

A separately excited dc generator is shown in Fig. 9E-7. With the parameters and variables as labeled, a block-diagram representation is to be obtained, including the effect of saturation.

From Fig. 9E-7(a), we have

$$v_f = (R_f + L_f p) i_f \tag{a}$$

$$v_a = G_{af}\omega_m i_f = (R_a + L_a p) i_a + v \tag{b}$$

Or, expressing these equations in terms of saturation functions, defined by Eqs. (9.34a,b), we have

$$v_f \ (R_f + S L_{fo} p) i_f \tag{c}$$

$$v_a = S G_{afo}\omega_m i_f = (R_a + L_a p) i_a + v \tag{d}$$

Equations (c) and (d) can be expressed by the block diagram shown in Fig. 9E-7(b). This representation is quite well suited for simulation on an analog computer.

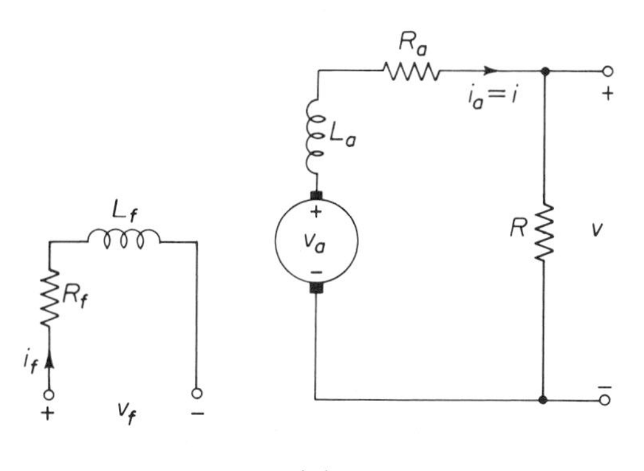

(a)

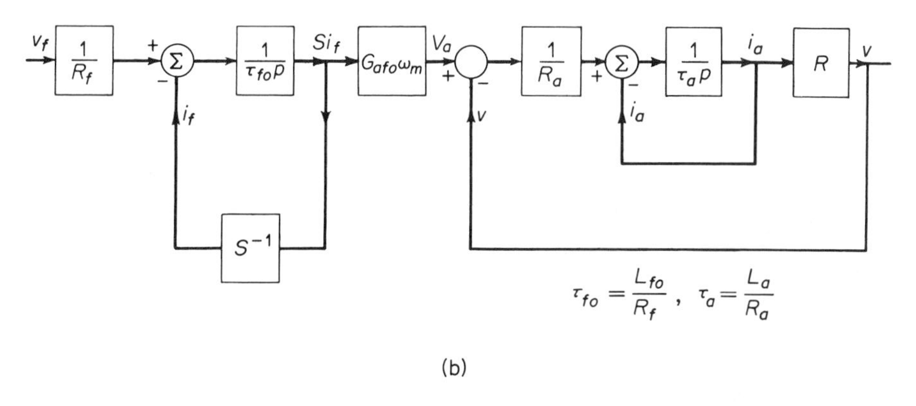

(b)

Fig. 9E-7

The inverse saturation function, of course, has to be obtained from a nonlinear function generator.

Finally, it should be pointed out for the interested reader that magnetic nonlinearity has been taken into account in the equations of motion for dc machines[47] and induction machines.[48] (These are beyond the scope of this book and will not be considered here.)

9.6 Leakage Fluxes and Reactances[26, 32, 50–62]

Leakage fluxes govern the behavior of electromagnetic energy-conversion devices as much as do mutual fluxes. This is true for unconventional devices[50] as well as for conventional rotating machines. Of course, leakage fluxes are considered imperfections and should be reduced to a minimum, if possible. In a perfect (or ideal) machine there are no leakage fluxes and the total flux is the mutual flux. Leakage fields do not contribute to the production of useful torques (or forces) for energy conversion. Nevertheless, they are inherently present in almost every machine and should be taken into account for accurate results. An outstanding example, where ignoring the leakage fields leads to dubious results, is a slow-speed fractional-horsepower induction motor. In such a motor, leakage fluxes are appreciable compared with mutual flux and, if the corresponding *leakage reactance* is not taken into account, it becomes difficult to distinguish between a current-excited and a voltage-excited machine. Leakage fluxes and reactances play an important role in affecting the performance of unconventional devices, such as liquid-metal pumps and linear motors. With a well-designed conventional machine, taking into account the effects of leakage reactances leads to refined solutions.

Whereas it is important to account precisely for the leakage fields, it is also difficult to propose a general method applicable to a wide variety of cases. But this should not be discouraging, for extensive research has been done on various aspects of leakage reactances over the past 70 years and numerous publications are available. (A few are listed at the end of this chapter.) In this section we shall identify the various paths for leakage fluxes which contribute to the overall leakage reactance. Next, we shall refer to some of the methods of determining the various components of leakage reactances.

9.6.1 *Identification*

Let us consider the double-slotted structure shown in Fig. 9-10, which indicates only the flux paths due to stator excitation alone. Similar flux paths can be traced for rotor excitation also. The various leakage fluxes lead to the following components of the leakage reactance:

1. slot-leakage reactance,
2. zigzag-leakage reactance, and
3. belt-leakage reactance.

Reactances (1) and (2) are sometimes combined to constitute the airgap leakage or differential leakage reactance. In addition to the above-mentioned components of leakage reactance, another component, known as (4) the end-connection–leakage reactance, also contributes substantially to the overall leakage reactance of the machine. Whereas the leakage reactance can be subdivided into more components, for our purposes the four components mentioned above will suffice.

In Fig. 9-10, notice also the path of the mutual flux. This is useful flux and leads to the mutual inductance—or magnetizing reactance, as it is commonly called.

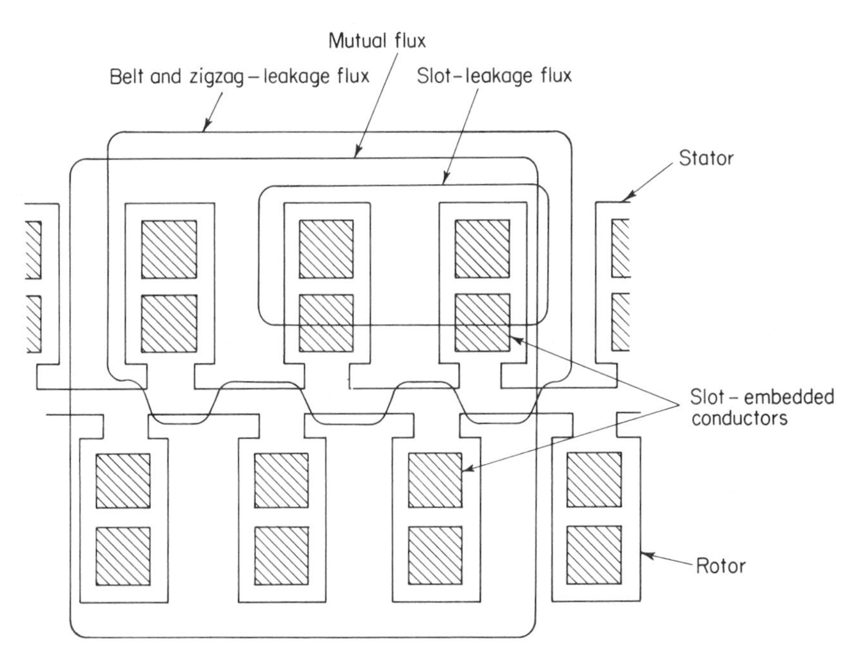

Fig. 9-10 Flux paths in a doubly-slotted structure.

9.6.2 Methods of Calculation

As pointed out earlier, the quantity of literature on the determination of various components of leakage fields and reactances is enormous. In this section we merely wish to point to the references which may be consulted and adapted for ready use.

Reference 32 devotes a full chapter (and a little more) to the reactance calculations. It presents a good summary of the results available until about 1950. References 51–62 give in considerable detail the methods of obtaining the various leakage reactances. Invariably these are fairly involved in analytical details. In the following we consider an approach which takes into account the tooth-tip- and zigzag-leakage reactances of the rotor of an induction motor.[56]

The slotted structure of the rotor and the slot-embedded conductors are shown in Fig. 9-11(a). The dimensions are as shown in Fig. 9-11. The torque–speed characteristic of this machine has been found by using the Lorentz force equation and the concept of wave impedance (Sec. 4.5). Here the tooth-tip- and zigzag-leakage reactances are taken into account by choosing the model shown in Fig. 9-11(b). In the model as well as in the slotted structure, three distinct regions are indicated. The constants for the model are then given by the following expressions:

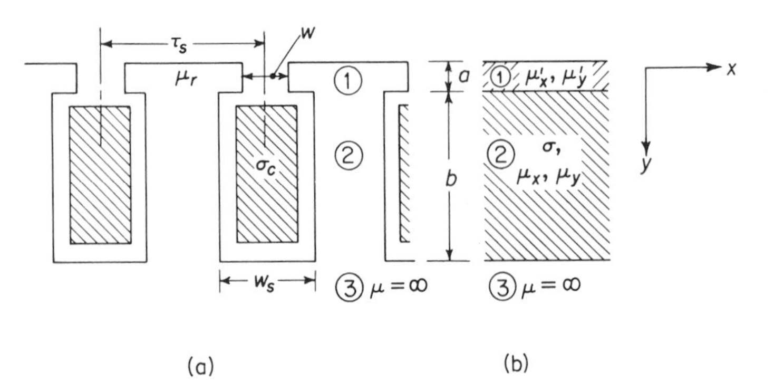

Fig. 9-11 (a) A slotted structure; (b) model to account for tooth-tip and zigzag leakage reactance.

Region 1:

$$\left.\begin{aligned} \mu_x' &= \frac{\mu_o \tau_s}{w} \\ \mu_y' &= \frac{\mu_r \mu_o (\tau_s - w)}{\tau_s} \\ \sigma &= 0 \end{aligned}\right\} \tag{9.36}$$

Region 2:

$$\left.\begin{aligned} \mu_x &= \frac{\mu_o \tau_s}{w_s} \\ \mu_y &= \frac{\mu_r \mu_o (\tau_s - w_s)}{\tau_s} \\ \sigma &= \frac{\sigma_c w_s}{\tau_s} \end{aligned}\right\} \tag{9.37}$$

Region 3:

$$\left.\begin{aligned} \mu_x &= \mu_y = \infty \\ \sigma &= 0 \end{aligned}\right\} \tag{9.38}$$

where μ_o = permeability of free space and μ_r = relative permeability of the tooth tip [region 1, Fig. 3-11(a)] = 4000 (approximately), σ_c = conductivity of the conductor, and other symbols denote the dimensions labeled in Fig. 9-11. Having obtained this model, we can then use the concept of the wave impedance (discussed in Chapter 4) to determine the current distribution and the force density exerted on the rotor for specified airgap fields. It has been found that accounting for the two components of leakage reactances by this method correlates well with test results.

The end-connection–leakage reactances for various machines, especially for turbogenerators, have been calculated by a number of authors,[51, 57–62] and digital computers have been used to compute the cumbersome, formidable expressions to obtain numerical results.

9.7 Summary and Concluding Remarks

In this chapter we illustrated certain applications of analog and digital computers to energy-conversion problems by means of a few examples. We also outlined the techniques of graphical field mapping for electromagnetic fields.

Three important factors which considerably affect the performance of energy-conversion devices—harmonics, saturation, and leakages—were briefly discussed. The references listed at the end of this chapter treat each topic more thoroughly.

Many interesting and useful topics could not be treated here because they are advanced and specialized. These include the numerical computations of fields and their effects on the resistances and reactances of conductors;[63–65, 68] conformal transformations;[27, 52, 66] analogs for fields and field mapping in the presence of magnetic saturation and for other realistic cases.[25, 26, 31, 67] The references cited here constitute a good starting point for fruitful research in this area.

PROBLEMS

9–1. Develop a complete analog-computer representation of the induction motor shown in Fig. 7-22.

9–2. Obtain an analog-computer representation for the voltage-regulating system shown in Fig. 8E-5.

9–3. For the current sheet of Example 5-2, determine the stator inductance matrix including the effect of the first 50 harmonics. Given:

$$g = 0.000416 \text{ m}; \quad \tau_s = 0.00098 \text{ m}; \quad \omega_s = 0.000295 \text{ m}$$
$$\tau_r = 0.881; \quad \text{const } 1 = \tfrac{1}{2}\,\zeta\mu_o l\tau_r = 1.69 \times 10^{-7}$$

Use Eqs. (9.14a) and the computer program given in Appendix C.

9–4. Compare the results of Problem 9-3 with those when the harmonics are neglected and the inductances are calculated for the largest harmonic component only.

9–5. The following design data relate to a single-phase, $\frac{1}{2}$-hp, 110-V, 60-cycle, 1750 r/min induction motor:

	Stator	*Rotor*
Outside diameter	7.25 in	4.35 in
Inside diameter	4.376 in	2.56 in
Length	2.56 in	2.56 in
Number of slots	36	26
Slot opening	0.094 in	0.04 in
Effective airgap	0.0164 in	

The winding layout for the stator is as follows:

	Slot No.													
	1	2	3	4	5	6	7	8	9	10	11	12	13	14
Main		16	30	40	46	46	40	30	16		16	30	etc.	
Start	60	28	23	14			14	23	28	60	28	23	etc.	

The rotor winding is the cage type.

Determine the speed–torque characteristic of this machine under steady-state conditions for the largest harmonic component of the current sheet.

9–6. Write a digital-computer flow chart and a FORTRAN program to obtain the solution of Eq. (9.31b).

9–7. Starting with Maxwell's equations, show that harmonic airgap fields do not interact with each other to produce an average torque.

9–8. Develop a block diagram of a self-excited dc shunt generator, using the inverse saturation function to account for saturation.

9–9. Using the theory developed in Sec. 4.5, derive an expression for the force density of the squirrel-cage–type induction motor shown in Fig. 9-11. Assume

$$B_y = B_m e^{j(\omega t - \beta x)}$$

9–10. Write a computer program for the result derived in Problem 9-9 and, assuming suitable numerical values, compute and plot the force–speed characteristic of the motor.

REFERENCES

1. Riaz, M., "Analog Computer Representations of Synchronous Generators in Voltage-Regulation Studies," *AIEE Trans.*, Vol. 75, Part III, 1956, pp. 1178–1184.

2. Hughes, F. M., and A. S. Aldred, "Transient Characteristics and Simulation of Induction Motors," *Proc. IEE* (London), Vol. 111, 1964, pp. 2041–2050.

3. Krause, P. C., "Simulation of Unsymmetrical 2-Phase Induction Machines," *IEEE Trans.*, Vol. PAS-84 ("Power Apparatus and Systems"), 1965, pp. 1025–1037.

4. Krause, P. C., and C. H. Thomas, "Simulation of Symmetrical Induction Machinery," *IEEE Trans.*, Vol. PAS-84, 1965, pp. 1038–1053.

5. Yu, Yao-Nan, and G. E. Dawson, "Modeling a Four Electric-Machine System on Analog Computer Using Parameters Directly Determined from Tests," *IEEE Trans.*, Vol. PAS–87 ("Power Apparatus and Systems"), 1968, pp. 632–640.

6. Krause, P. C., "Methods of Stabilizing a Reluctance-Synchronous Machine," *IEEE Trans.*, Vol. PAS-87, 1968, pp. 641–649.

7. Jordan, H. E., "Analysis of Induction Machines in Dynamic Systems," *IEEE Trans.*, Vol. PAS-84 ("Power Apparatus and Systems"), 1965, pp. 1080–1088.

8. Lawrenson, P. J., and J. M. Stephenson, "Note on Induction-Machine Performance with a Variable-Frequency Supply," *Proc. IEE* (London), Vol. 113, 1966, pp. 1617–1623.

9. Korn, G. A., and T. M. Korn, *Electronic Analog Computers*. New York: McGraw-Hill Book Company, 1956.

10. McCracken, D. D., *A Guide to FORTRAN Programming*. New York: John Wiley & Sons, Inc., 1961.

11. Novotny, D., and J. J. Grainger, "Digital Computer Analysis of the Instantaneous Reversal Transient in a Single Phase Motor," *IEEE Trans.*, Vol. PAS-83 ("Power Apparatus and Systems"), 1964, pp. 380–386.

12. Jordan, H. E., "Digital Computer Analysis of Induction Machines in Dynamic Systems," *IEEE Trans.*, Vol. PAS-86 ("Power Apparatus and Systems"), 1967, pp. 727–728.

13. Wiederholt, L. F., A. F. Fath, and H. J. Wertz, "Motor Transient Analysis on a Small Digital Computer," *IEEE Trans.*, Vol. PAS-86, 1967, pp. 819–824.

14. Novotny, D. W., and A. F. Fath, "The Analysis of Induction Machines Controlled by Series Connected Semiconductor Switches," *IEEE Trans.*, PAS-87 ("Power Apparatus and Systems"), 1968, pp. 597–603.

15. Slater, R. D., W. S. Wood, F. P. Flynn, and R. Simpson, "Digital Computation of Induction Motor Transient Torque Patterns," *Proc. IEE* (London), Vol. 113, 1966, pp. 819–822.

16. Smith, I. R., and S. Sriharan, "Transient Performance of the Induction Motor," *Proc. IEE*, Vol. 113, 1966, pp. 1173–1181.

17. Saunders, R. M., "Digital Computers as an Aid in Electric Machine Design," *AIEE Trans.*, Vol. 73, Part I, 1954, p. 189.

18. Veinott, C. G., "Synthesis of Induction Motor Design on a Digital Computer," *AIEE Trans.*, Vol. 79, Part III, 1960, p. 12.

19. Godwin, G. L., "Optimum Machine Design by Digital Computer," *AIEE Trans.*, Vol. 78, Part III, 1959, p. 478.

20. Chalmers, B. J., and B. J. Bennington, "Digital-Computer Program for Design Synthesis of Large Squirrel-Cage Induction Motors," *Proc. IEE* (London), Vol. 114, 1967, pp. 261–268.

21. Binns, K. J., and P. J. Lawrenson, *Analysis and Computation of Electric and Magnetic Field Problems*. New York: Pergamon Press, Inc., 1963.

22. Vitkovitch, D., (ed.), *Field Analysis: Experimental and Computational Methods*. Princeton, N.J.: D. Van Nostrand and Co., 1966.

23. Silvester, P., *Modern Electromagnetic Fields*. Englewood Cliffs, N.J.: Prentice-Hall, Inc., 1968.

24. Bewley, L. V., *Two-Dimensional Fields in Electrical Engineering*. New York: Dover Publications, Inc., 1963.

25. Midgley, D., and S. W. Smethurst, "Field Plotting by Fourier Synthesis and Digital Computation," *Proc. IEE* (London), Vol. 112, 1965, pp. 1945–1950.

26. Hawley, R., I. M. Edwards, J. M. Heaton, and R. L. Stoll, "Turbogenerator End-Region Magnetic Fields," *Proc. IEE* (London), Vol. 114, 1967, pp. 1107–1114.

27. Fukushima, K., S. A. Nasar, and R. M. Saunders, "Electromechanical Energy Conversion in Salient Pole Structures," *IEEE Trans.*, Vol. PAS-82 ("Power Apparatus and Systems"), 1963, pp. 760–766.

28. Stevenson, A. R., and R. H. Park, "Graphical Determination of Magnetic Fields—Theoretical Considerations," *AIEE Trans.*, Vol. 46, 1927, p. 112.

29. Wieseman, R. W., "Graphical Determination of Magnetic Fields—Practical Applications," *AIEE Trans.*, Vol. 46, 1927, pp. 141–154.

30. Poritsky, H., "Graphical Field Plotting Methods in Engineering," *AIEE Trans.*, Vol. 57, 1938, p. 727.

31. Poritsky, H., "Calculation of Flux Distribution with Saturation," *AIEE Trans.*, Vol. 70, Part I, 1951, pp. 309–319.

32. Alger, P. L., *The Nature of Polyphase Induction Machines*. New York: John Wiley & Sons, Inc., 1951.

33. Veinott, C. G., "Spatial Harmonic Magnetomotive Forces in Irregular Windings and Special Connections of Polyphase Windings," *IEEE Trans.*, Vol. PAS-83 ("Power Apparatus and Systems"), 1964, pp. 1244–1253.

34. Nasar, S. A., "Electromechanical Energy Conversion in *nm*-Winding Double Cylindrical Structures in Presence of Space Harmonics," *IEEE Trans.*, Vol. PAS-87 ("Power Apparatus and Systems"), 1968, pp. 1099–1106.

35. Smith, I. R., and J. M. Layton, "Harmonic Elimination in Polyphase Machine by Graded Windings," *Proc. IEE* (London), Vol. 110, 1963, pp. 1640–1648.

36. Chalmers, B. J., "AC Machine Windings with Reduced Harmonic Content," *Proc. IEE* (London), Vol. 111, 1964, pp. 1859–1863.

37. Nasar, S. A., "Electromagnetic Theory of Electrical Machines," *Proc. IEE* (London), Vol. 111, 1964, pp. 1123–1131.

38. White, D. C., and H. H. Woodson, *Electromechanical Energy Conversion.* New York: John Wiley & Sons, Inc., 1959.

39. Kron, G., *Equivalent Circuits of Electric Machinery.* New York: John Wiley & Sons, Inc., 1951.

40. Ahamed, S. V., and E. A. Erdelyi, "Nonlinear Theory of Salient Pole Machines," *IEEE Trans.*, Vol. PAS-85 ("Power Apparatus and Systems"), 1966, pp. 61–69.

41. Erdelyi, E. A., S. V. Ahamed, and R. D. Burtness, "Flux Distribution in Saturated DC Machines at No-load," *IEEE Trans.*, Vol. PAS-84 ("Power Apparatus and Systems"), 1965, pp. 375–381.

42. Erdelyi, E. A., S. V. Ahamed, and R. E. Hopkins, "Nonlinear Theory of Synchronous Machines on Load," *IEEE Trans.*, Vol. PAS-85 ("Power Apparatus and Systems"), 1966, pp. 792–801.

43. Ahamed, S. V., and E. A. Erdelyi, "Flux Distributions in DC Machines on Load and Overloads," *IEEE Trans.*, Vol. PAS-85, 1966, pp. 960–967.

44. Jackson, R. F., and E. A. Erdelyi, "Combination and Separation of Coordinates and Modular Programming for DC Machines," *IEEE Trans.*, Vol. PAS-87 ("Power Apparatus and Systems"), 1968, pp. 659–664.

45. Trutt, F. C., E. A. Erdelyi, and R. E. Hopkins, "Representation of the Magnetization Characteristic of DC Machines for Computer Use," *IEEE Trans.*, Vol. PAS-87, 1968, pp. 665–669.

46. Riaz, M., "Dynamics of DC Machine Systems," *AIEE Trans.*, Vol. 74, Part II, 1955, pp. 365–370.

47. Von der Embse, Urban A., "A New Theory of Nonlinear Commutating Machines," *IEEE Trans.*, Vol. PAS-87 ("Power Apparatus and Systems"), 1968, pp. 1804–1809.

48. Silvester, P., "Energy Conversion by Nonlinear Slip-Ring Electric Machines," *IEEE Trans.*, Vol. PAS-84, 1965, pp. 352–356.

49. Fitzgerald, A. E., and C. Kingsley, *Electric Machinery*, 1st ed. New York: McGraw-Hill Book Company, 1951.

50. Laithwaite, E. R., *Induction Machines for Special Purposes.* New York: Chemical Publishing Co., Inc., 1966.

51. Carpenter, C. J., "The Application of the Method of Images to Machine End-Winding Fields," *Proc. IEE* (London), Vol. 107, Part 4, 1960, p. 487.

52. Binns, K. J., "Calculation of Some Basic Flux Quantities in Induction and Other Doubly-Slotted Electrical Machines," *Proc. IEE* (London), Vol. 111, 1964, pp. 1847–1858.

53. Spooner, T., "Analytical Study of the Leakage Field in a Slot," *Gen. Elec. Review*, Vol. 22, 1927, p. 417.

54. Liwschitz-Garik, M. M., and W. H. Formhals, "Some Phases of Calculation of Leakage Reactance of Induction Motors," *AIEE Trans.*, Vol. 66, 1947, pp. 1409–1413.

55. Lloyd, T. C., V. F. Giusti, and S. S. L. Chang, "Reactances of Squirrel-Cage Induction Motors," *AIEE Trans.*, Vol. 66, 1947, pp. 1349–1355.

56. Cullen, A. L., and T. H. Barton, "A Simplified Electromagnetic Theory of the Induction Motor Using the Concept of Wave Impedance," *Proc. IEE* (London), Vol. 105, Part C, 1958, p. 331.

57. Smith, R. T., "End Component of Armature Leakage Reactance of Round-Rotor Generators," *AIEE Trans.*, Vol. 77, Part III, 1958, p. 636.

58. Honsinger, V. B., "Theory of End-winding Leakage Reactance," *AIEE Trans.*, Vol. 78, Part III, 1959, p. 417.

59. Ashworth, D. S., and P. Hammond, "The Calculation of the Magnetic Field of Rotating Machines, Part 2—The Field of Turbogenerator End-Windings," *Proc. IEE* (London), Vol. 108, Part A, 1961, p. 527.

60. Lawrenson, P. J., "The Magnetic Field of the End-Windings of Turbo-generators," *Proc. IEE* (London), Vol. 108, Part A, 1961, p. 538.

61. Reece, A. B. J., and A. Pramanik, "Calculation of the End-Region Field of AC Machines," *Proc. IEE* (London), Vol. 112, 1965, pp. 1355–1368.

62. Tegopoulos, J. A., "Determination of the Magnetic Field in the End Zone of Turbine Generators," *IEEE Trans.*, Vol. PAS-82 ("Power Apparatus and Systems"), 1963, p. 562.

63. Mamak, R. S., and E. R. Laithwaite, "Numerical Evaluation of Inductance and AC Resistance," *Proc. IEE* (London), Vol. 108, Part C, 1961, pp. 252–258.

64. Stoll, R. L., "Numerical Method of Calculating Eddy Currents in Non-magnetic Conductors," *Proc. IEE* (London), Vol. 114, 1967, pp. 775–780.

65. Carpenter, C. J., "Numerical Solution of Magnetic Fields in the Vicinity of Current-Carrying Conductors," *Proc. IEE* (London), Vol. 114, 1967, pp. 1793–1800.

66. Lawrenson, P. J., and S. K. Gupta, "Conformal Transformation Employing Direct-Search Techniques of Minimisation," *Proc. IEE*, Vol. 115, 1968, pp. 427–431.

67. King, E. I., "Equivalent Circuits for Two-Dimensional Magnetic Fields," *IEEE Trans.*, Vol. PAS-85 ("Power Apparatus and Systems"), 1966, p. 927.

68. Silvester, P., "Dynamic Resistance and Inductance of Slot-Embedded Conductors," *IEEE Trans.*, Vol. PAS-87 ("Power Apparatus and Systems"), 1968, pp. 250–256.

Appendix A

Derivation of Lagrange's Equation

Consider an integral I defined by

$$I = \int_{t_1}^{t_2} \bar{L}[q(t), \dot{q}(t), t]\, dt \tag{A.1}$$

where $\bar{L}$ is a function of the unknown variables $q(t)$ and $\dot{q}(t)$ and of the independent variable t. Let $q_0(t)$ be the function which extremizes (maximizes or minimizes—minimizes for all realistic cases) the given integral, and let $q(t)$ be a set of one-parameter "competing" functions defined by

$$q(t) = q_0(t) + \epsilon\eta(t) \tag{A.2}$$

where $\eta(t)$ is an arbitrary twice-differentiable function which is zero at the end points—that is, for which

$$\eta(t_1) = \eta(t_2) = 0 \tag{A.3}$$

and ϵ is the (time-independent) parameter. It is seen from Eqs. (A.2) and (A.3) that no matter which function $\eta(t)$ is chosen, the minimizing function $q_0(t)$ belongs to the family given by Eq. (A.2), for which $\epsilon = 0$. Notice that $\epsilon\eta(t)$ is a variation on the desired function; it should be as small as possible.

Geometrically, a set of competing functions and the minimizing function are shown in Fig. A-1.

From Eq. (A.2),

$$\dot{q}(t) = \dot{q}_0(t) + \epsilon\dot{\eta}(t) \tag{A.4}$$

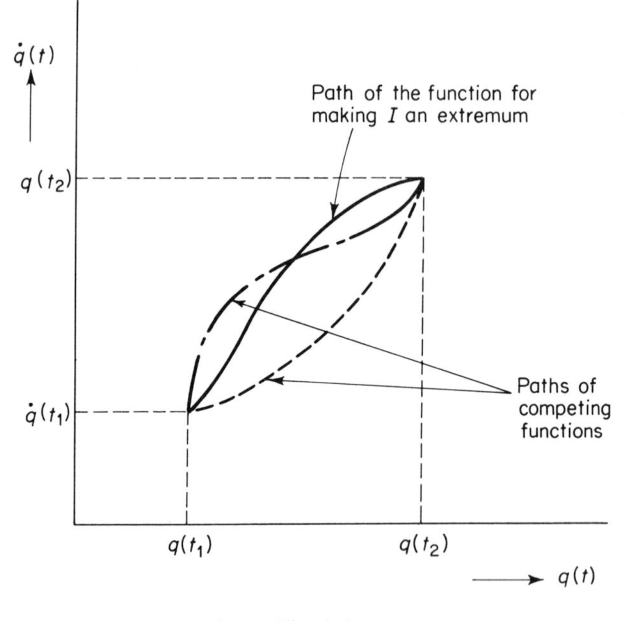

Fig. A-1

and from Eqs. (A.2) and (A.4) an integral

$$I(\epsilon) = \int_{t_1}^{t_2} \bar{L}[q_0(t)+\epsilon\eta(t);\quad \dot{q}_0(t)+\epsilon\dot{\eta}(t);\quad t]\, dt \tag{A.5}$$

is formed. The integral in Eq. (A.5) is a minimum with respect to ϵ, for $\epsilon = 0$, according to the choice of $q_0(t)$ as the minimizing function.

At the outset we assumed that $q_0(t)$ is the minimizing function; that is, $\epsilon = 0$ in Eq. (A.2) implies that the necessary condition for a minimum, the vanishing of the first derivative of I with respect to ϵ, must hold for $\epsilon = 0$. Or,

$$\left.\frac{dI(\epsilon)}{d\epsilon}\right|_{\epsilon=0} = I'(0) = 0 \tag{A.6}$$

Now let us recall the rule for the differentiation of an integral, which says that if

$$I = I(\epsilon) = \int_{x_1(\epsilon)}^{x_2(\epsilon)} f(x,\epsilon)\, dx$$

then

$$\frac{dI}{d\epsilon} = I'(\epsilon) = f(x_2,\epsilon)\frac{dx_2}{d\epsilon} - f(x_1,\epsilon)\frac{dx_1}{d\epsilon} + \int_{x_1(\epsilon)}^{x_2(\epsilon)} \frac{\partial f}{\partial \epsilon}\, dx \tag{A.7}$$

Therefore, Eqs. (A.3–A.7) yield

$$\frac{dI}{d\epsilon} = I'(\epsilon) = \int_{t_1}^{t_2} \left(\frac{\partial \bar{L}}{\partial q} \frac{\partial q}{\partial \epsilon} + \frac{\partial \bar{L}}{\partial \dot{q}} \frac{\partial \dot{q}}{\partial \epsilon} \right) dt = \int_{t_1}^{t_2} \left(\frac{\partial \bar{L}}{\partial q} \eta + \frac{\partial \bar{L}}{\partial \dot{q}} \dot{\eta} \right) dt \qquad \text{(A.8)}$$

When $\varepsilon = 0$, $q = q_0$ and $q = \dot{q}_0$. Therefore, from Eq. (A.6),

$$I'(0) = \int_{t_1}^{t_2} \left(\frac{\partial \bar{L}}{\partial q} \eta + \frac{\partial \bar{L}}{\partial \dot{q}} \dot{\eta} \right) dt = 0$$

Integrating by parts, the second term of this integral gives

$$I'(0) = \frac{\partial \bar{L}}{\partial \dot{q}} \eta \Bigg|_{t_1}^{t_2} + \int_{t_1}^{t_2} \left(\frac{\partial \bar{L}}{\partial q} - \frac{d}{dt} \frac{\partial \bar{L}}{\partial \dot{q}} \right) \eta \, dt = \int_{t_1}^{t_2} \left(\frac{\partial \bar{L}}{\partial q} - \frac{d}{dt} \frac{\partial \bar{L}}{\partial \dot{q}} \right) \eta \, dt = 0 \qquad \text{(A.9)}$$

Since Eq. (A.9) holds for all η, therefore

$$\frac{\partial \bar{L}}{\partial q} - \frac{d}{dt} \frac{\partial \bar{L}}{\partial \dot{q}} = 0 \qquad \text{(A.10)}$$

Equation (A.10) is the Euler–Lagrange equation; when used exclusively in connection with dynamical systems, it is known as *Lagrange's equation.*

The preceding discussion is based on a one-dimensional analysis. However, the same procedure can be carried through for all k's $(1, 2, \ldots, n)$ and Lagrange's equation, which is a necessary condition for the integral of Eq. (A.1) to be a minimum, and hence describes the motion—be it electrical or mechanical—of a system, is written as

$$\frac{\partial \bar{L}}{\partial q_k} - \frac{d}{dt} \frac{\partial \bar{L}}{\partial \dot{q}_k} = 0 \qquad k = 1, 2, \ldots, n$$

Appendix B

Matrices

A *matrix* is an array of numbers, or *elements* (real or complex), arranged in a row, a column, or rows and columns. A typical matrix is

$$S = \begin{bmatrix} s_{11} & s_{12} & \cdots & s_{1n} \\ s_{21} & s_{22} & \cdots & s_{2n} \\ \vdots & \vdots & & \vdots \\ s_{m1} & s_{m2} & \cdots & s_{mn} \end{bmatrix}$$

where s_{ij} are the elements of S. The first subscript i denotes the row and the second subscript j denotes the column of S where s_{ij} occurs.

A matrix having m rows and n columns is known as a $m \times n$ (or "m by n") matrix. If $m = n$, the matrix is a *square matrix*. A $1 \times n$ matrix is a *row matrix* and an $m \times 1$ matrix is a *column matrix*.

A matrix S is a *diagonal matrix* if its elements satisfy the condition $s_{ij} = 0$, $i \neq j$. The diagonal matrix having "one's" along the diagonal (that is $s_{ij} = 1$) is a *unit* or *identity matrix*. Thus, the elements of a unit matrix are $s_{ij} = \delta_{ij} = 1$, $i = j$, and otherwise zero, where δ_{ij} is known as the *Kronecker delta*.

If all elements of a matrix are zero, the matrix is a *null matrix*.

A square matrix S is a *symmetric matrix* if its elements satisfy the condition $s_{ij} = s_{ji}$.

The *determinant of a matrix S* is denoted by det S, or $|S|$, where the elements of det S are the same as the elements of the matrix S. If det $S = 0$, S is a *singular matrix*; if det $S \neq 0$, then S is *nonsingular*.

The transpose of a matrix S is denoted by $\tilde{S}$ and the elements of $\tilde{S}$ are s_{ji}, where s_{ij} are the elements of S.

If $\mathbf{S}$ is a square matrix, λ is a scalar and

$$\mathbf{Sx} = \lambda\mathbf{x}$$

then a value of λ for which this equation has a solution ($\mathbf{x} \neq \mathbf{0}$) is called an *eigenvalue* and the corresponding solutions are called *eigenvectors*.

The eigenvalues are obtained by solving the determinantal equation det $(S-\lambda I) = 0$.

If C is the *sum* of two matrices A and B, such that $C = A+B$, the elements of C are given by

$$c_{ij} = a_{ij}+b_{ij}$$

If C is the *product* of two matrices A and B, such that $C = AB$, the elements of C are given by

$$c_{ij} = \sum_k a_{ik}b_{kj}$$

Note that the number of columns in $A = k =$ the number of rows in B. For multiplication the matrices have to be compatible. It can be readily verified that, in general, $AB \neq BA$.

If S is a matrix such that its product with another matrix T satisfies the equation

$$ST = TS = I$$

where $I =$ identity matrix, then T is called the *inverse* of S and is denoted by S^{-1}.

The inverse of a matrix S exists if and only if det $S \neq 0$. The inverse S^{-1} is found from S by replacing each element s_{ij} by its cofactor S_{ij}, then transposing the resulting matrix, and dividing it by det S.

A real matrix S is an *orthogonal matrix* if $\tilde{S} = S^{-1}$.

Appendix C

Computer Programs

```
C   PROGRAM FOR ACCELERATION-TEST DATA REDUCTION
    DIMENSION SPEED (100), RPM (100), SEC (6), TORQUE (100)
    READ 4, (SEC (M, M = 1, 6))
  4 FORMAT (6F4.0)
    DO 45 J = 1,6
    READ 10, SPEED
 10 FORMAT (20F4.0)
    DO 180 I = 1,99
    IF (SPEED(I)) 28, 28, 29
 29 RPM(I) = (SPEED(I)+SPEED(I+1))/2.
180 TORQUE(I) = (SPEED(I+1)-SPEED(I))*0.0593/SEC(J)
    N =100
    GO TO 150
 28 N = I
150 PRINT 41
 41 FORMAT (1H0, 11X, 4H RPM, 15X, 7H TORQUE 111)
 45 PRINT 50, (RPM(I), TORQUE(I), I = 1,N)
 50 FORMAT (10X, F7.0, 10X, F7.4)
    CALL EXIT
    END
  * DATA
```

```
C  SOLUTION OF LAPLACE'S EQUATION BY ITERATION
   DIMENSION A(7, 7), B(7, 7)
   DO 1 J = 1, 7
   A(J, J) = 200.
   A(1, J) = 0
1  A(J, 7) = 0
   DO 2 K = 2, 5
   LFST = K+1
   DO 2 I = LFST, 6
2  A(K, I) = 0
5  DO 4 J = 1, 7
   DO 4 K = 1, 7
4  B(J, K) = A(J, K)
   DO 3 K = 2, 5
   LFST = K+1
   DO 3 I = LFST, 6
   IF (K-2) 10, 11, 10
11 IF(I-6) 10, 12, 10
12 A(2, 6) = ((A(3, 6)+A(2, 5))/(1.+1./.732)+(A(1, 6)+A(2, 7))/1.732)/2.
   GO TO 3
10 A(K, I) = (A(K-1, I)+A(K+1, I)+A(K, I-1)+A(K, I+1))/4
3  CONTINUE
   DO 6 J = 1, 7
   DO 6 K = 1, 7
   IF(ABS(A(J, K)-B(J, K))-.01) 6, 6, 5
6  CONTINUE
8  WRITE (3, 9) (A(I, J), I = 1, J)
9  FORMAT (7(4X, F12.8))
   CALL EXIT
   END
```

```
C  INDUCTANCE MATRIX
   DIMENSION ALPH(50), A(3, 50), B(3, 50), C(3, 50)
 1 PHI (3, 50), N(3, 90), EL(3, 3), EN(3, 90)
   READ 2, L, ISMAX
   READ 3, TAUR, TAUS, WS, GEE1, CONST1
   READ 4, ((N(K, J), J = 1,90), K = 1, 3)
 2 FORMAT (2I4)
 3 FORMAT (5E12.4)
 4 FORMAT (15I4)
   DO 11 K = 1, ISMAX
   DO 6 J = 1, L
 6 EN(K, J) = FLOATF(N(K, J))
   DO 10 NU = 1, 50, 2
   PNU = FLOATF(NU)
   ALPH(NU) = 6.2832*PNU/TAUR
   ARG1 = 0.5*ALPH(NU)*WS
   FACT1 = 2.0*SINF(ARG1)/(ARG1*TAUR)
   A(K, NU) = 0
   B(K, NU) = 0
   DO 7 J = 1, L
   PJ = FLOATF(J)
   ARG2 = ALPH(NU)*(PJ-1.0)*TAUS
   A(K, NU) = A(K, NU)+EN(K, J)*COSF(ARG2)
 7 B(K, NU) = B(K, NU)+EN(K, J)*SINF(ARG2)
   A(K, NU) = A(K, NU)*FACT1
   B(K, NU) = B(K, NU)*FACT1
   C(K, NU) = SQRTF(A(K, NU)**2+B(K, NU)**2)
   AB = B(K, NU)/A(K, NU)
   IF (A(K, NU)) 12, 5, 13
 5 IF (B(K, NU)) 8, 10, 9
 8 PHI(K, NU) = -1.5705
   GO TO 10
 9 PHI(K, NU) = 1.5705
   GO TO 10
12 PHI(K, NU) = 3.14159+ATANF(AB)
   GO TO 10
13 PHI(K, NU) = ATANF(AB)
10 CONTINUE
11 CONTINUE
   DO 17 K = 1, ISMAX
   DO 17 M = 1, ISMAX
   EL(K, M) = 0
   DO 15 NU = 1, 50, 2
```

```
      ARG2 = ALPH(NU)*GEE1
      FACT2 = 1.0/(TANHF(ARG2)*ALPH(NU))
      DIFF1 = PHI(K, NU)-PHI(M, NU)
   15 EL(K, M) = EL(K, M)+C(K, NU)*C(M, NU)*COSF(DIFF1)*FACT2
      EL(K, M) = CONST1*EL(K, M)
   17 CONTINUE
      DO 19 K = 1, 3
   19 PRINT 20, K, (NU, A(K, NU), B(K, NU), C(K, NU), PHI(K, NU),
    1    NU = 1, 50, 2)
   20 FORMAT(1H1, 28X, 27H FOURIER COEFFICIENTS-PHASE I3///
    1    2HNU, 11X, 1HA, 19X, 1HB, 19X, 1HC, 18X, 4H PHI//(I4, 4E20.5))
      PRINT 21, ((EL(K, M), K = 1, 3), M = 1, 3)
   21 FORMAT (1H1, 20X, 17HINDUCTANCE
    1    MATRIX///(3E20.5))
      CALL EXIT
      END
      DATA
```

```
C  TRANSFORMED INDUCTANCES
   DIMENSION ALPH(50), C(45, 50),PHI(45, 50), PM(45), CL(45)
   DIMENSION EL(1, 45), TL(45), P(1), Q(1)
   READ 3, TAUR, WS, GEE1, CONST1
 3 FORMAT (4E 12.4)
   DO 4 M = 1, 45
 4 PM(M) = FLOATF(M)
   DO 10 NU = 1, 50, 2
   PNU = FLOATF(NU)
   ALPH(NU) = 6.2832*PNU/TAUR
   ARG1 = 0.5*ALPH(NU)*WS
   C(1, NU) = 4.0*SINF(ARG1)/(ARG1*TAUR)
   DO 9 M = 1, 45
 9 PHI(M, NU) = 6.2832*PNU*(PM(M)-1.0)/45.0
10 CONTINUE
   DO 17 M = 1, 45
   EL(1, M) = 0
   DO 15 NU = 1, 50, 2
   ARG2 = ALPH(NU)*GEE1
   FACT2 = 1.0/(TANHF(ARG2)*ALPH(NU))
   DIFF1 = PHI(1, NU)-PHI(M, NU)
15 EL(1, M) = EL(1, M)+(C(1, NU)**2)*COSF(DIFF1)*FACT2
   EL(1, M) = CONST1*EL(1, M)
17 CONTINUE
   DO 19 N = 1, 45
   TL(N) = (0., 0.)
   DO 18 K = 1, 45
   P(1) = 0.
   P(2) = 0.1397*FLOATF((1-N)*(K-1))
18 TL(N) = TL(N)+EL(1, K)*EXPF(P)
19 CONTINUE
   DO 12 N = 1, 45
   CL(N) = 0
   DO 13 K = 1, 45
   ARG3 = 0.1397*FLOATF((1-N)*(K-1))
13 CL(N) = CL(N)+EL(1, K)*COSF(ARG3)
12 CONTINUE
   PRINT 20, (M, EL(1, M), CL(M), TL(M), TL(M+45), M = 1, 45)
20 FORMAT (1H1, 12X, 19H XFORMED)
 1    INDUCTANCES///2X, 1HM, 12X, 6HL(1, M), 12X, 5HCL(M),
 2    16X, 6HRE(TL), 17X, 6HIM(TL)// (I4, 4E20.5)
   CALL EXIT
   END
   DATA
```

```
C  MUTUAL L MATRIX
   DIMENSION ALPH(50), A(3, 50), B(3, 50), C(3, 50), PHI(3, 50),
   1  PSI(45, 50), D(1, 50), N(3, 90), EN (3,90), EL(3, 45), PM(45)
   READ 2, L, ISMAX
   READ 3, TAUR, TAUS, WS, GEE1, CONST1
   READ 4, ((N(K, J), J = 1, 90), K = 1, 3)
 2 FORMAT (2I4)
 3 FORMAT (5E12.4)
 4 FORMAT (15I4)
   DO 11 K = 1, ISMAX
   DO 6 J = 1, L
 6 EN(K, J) = FLOATF(N(K, J))
   DO 10 NU = 1, 50, 2
   PNU = FLOATF(NU)
   ALPH(NU) = 6.2832*PNU/TAUR
   ARG1 = 0.5*ALPH(NU)*WS
   FACT1 = 2.0*SINF(ARG1)/(ARG1*TAUR)
   D(1, NU) = 2.0*FACT1
   A(K, NU) = 0
   B(K, NU) = 0
   DO 7 J = 1, L
   PJ = FLOATF(J)
   ARG2 = ALPH(NU)*(PJ-1.0)*TAUS
   A(K, NU) = A(K, NU)+EN(K, J)*COSF(ARG2)
 7 B(K, NU) = B(K, NU)+EN(K, J)*SINF(ARG2)
   A(K, NU) = A(K, NU)*FACT1
   B(K, NU) = B(K, NU)*FACT1
   C(K, NU) = SQRTF(A(K, NU)**2+B(K, NU)**2)
   AB = B(K, NU)/A(K, NU)
   IF (A(K, NU)) 12, 5, 13
 5 IF (B(K, NU)) 8, 10, 9
 8 PHI(K, NU) = -1.5705
   GO TO 10
 9 PHI(K, NU) = 1.5705
   GO TO 10
12 PHI(K, NU) = 3.14159+ATANF(AB)
   GO TO 10
13 PHI(K, NU) = ATANF(AB)
10 CONTINUE
11 CONTINUE
   DO 30 M = 1, 45
30 PM(M) = FLOATF(M)
   DO 20 M = 1, 45
```

```
      DO 16 NU = 1, 50, 2
16    PSI(M, NU) = 6.2832*PNU*(PM(M)-1.0)/45.0
20    CONTINUE
      DO 17 K = 1, ISMAX
      DO 18 M = 1, 45
      EL(K, M) = 0
      DO 15 NU = 1, 50, 2
      ARG3 = ALPH(NU)*GEE1
      FACT3 = 2.0/((EXPF(ARG3)-EXPF(-ARG3))*ALPH(NU))
      DIFF1 = PHI(K, NU)-PSI(M, NU)
15    EL(K, M) = EL(K,M)+D(1, NU)*C(K, NU)*COSF(DIFF1)*FACT3
      EL(K, M) = CONST1*EL(K, M)
18    CONTINUE
17    CONTINUE
      PRINT 41, (M(EL(K, M), K = 1, 3)M = 1, 45)
41    FORMAT (1H1, 30X, 15HMUTUAL L MATRIX///
     1 2H M, 13X,2HL1, 18X, 2HL2, 18X, 2HL3//(I3, 3E20.5)
      CALL EXIT
      END
      DATA
```

Appendix D

Current Sheets and Laplace's Equation

Fourier Analysis of Current Sheets

An arbitrary current-sheet distribution in the form of pulses (of height Ni/w_s and width w_s) is shown in Fig. D-1. We choose the origin at a distance

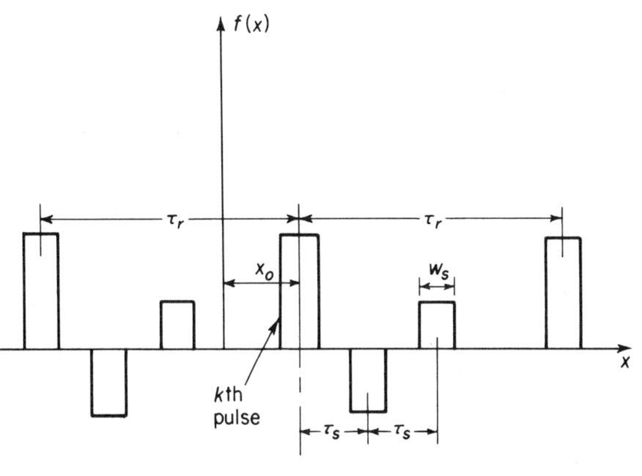

Fig. D-1

x_0 from the kth pulse, which repeats at an interval τ_r. The kth pulse can thus be expressed as a Fourier series as follows:

$$h_k(x) = \frac{N_k i}{w_s} \sum_{\nu=1}^{\infty} (a_{k\nu} \cos \alpha_\nu x + b_{k\nu} \sin \alpha_\nu x) \tag{D.1}$$

where the Fourier coefficients for the νth harmonic of the kth pulse are $a_{k\nu}$ and $b_{k\nu}$. These are evaluated from the orthogonality conditions; that is,

$$a_{k\nu} = \frac{2}{\tau_r}\int_{-\tau_r/2}^{\tau_r/2} f(x)\cos\alpha_\nu x\,dx \qquad \text{(D.2)}$$

Substituting for $f(x)$ and the limits, from Fig. D-1, in Eq. (D.2) we have

$$a_{k\nu} = \frac{2}{\tau_r}\left[\int_{-\tau_r/2}^{x_0}(0)\cos\alpha_\nu x\,dx + \int_{x_0}^{x_0+w_s}(1)\cos\alpha_\nu x\,dx + \int_{x_0+w_s}^{\tau_r/2}(0)\cos\alpha_\nu x\,dx\right]$$

$$= \frac{2}{\alpha_\nu\tau_r}\sin\alpha_\nu x\Big|_{x_0}^{x_0+w_s} = \frac{2}{\alpha_\nu\tau_r}[\sin\alpha_\nu(x_0+w_s)-\sin\alpha_\nu x_0] \qquad \text{(D.3)}$$

Similarly

$$b_{k\nu} = \frac{2}{\alpha_\nu\tau_r}[\cos\alpha_\nu x_0 - \cos\alpha_\nu(x_0+w_s)] \qquad \text{(D.4)}$$

Now, the overall current sheet consists of m pulses separated from each other by τ_s. Then

$$h(x) = \sum_{k=1}^{m} h_k(x) \qquad \text{(D.5)}$$

From Eqs. (D.1) and (D.5) we have

$$h(x) = \sum_{k=1}^{m}\frac{N_k i}{w_s}\sum_{\nu=1}^{\infty}(a_{k\nu}\cos\alpha_\nu x + b_{k\nu}\sin\alpha_\nu x) \qquad \text{(D.6)}$$

In Eq. (D.6) we define

$$a_\nu = \sum_{k=1}^{m} a_{k\nu}N_k \qquad \text{(D.7)}$$

and

$$b_\nu = \sum_{k=1}^{m} b_{k\nu}N_k \qquad \text{(D.8)}$$

We choose the origin such that $x_0 = -w_s/2$, such that Eqs. (D.3) and (D.7) yield

$$a_\nu = \frac{2}{\alpha_\nu\tau_r}\sum_{k=1}^{m}\left\{\sin\alpha_\nu\left[-\frac{w_s}{2}+(k-1)\tau_s+w_s\right]-\sin\alpha_\nu\left[-\frac{w_s}{2}+(k-1)\tau_s\right]\right\}N_k$$

$$= \frac{2}{\alpha_\nu\tau_r}\sum_{k=1}^{m}\left\{\cos[\alpha_\nu(k-1)\tau_s]\sin\frac{\alpha_\nu}{2}w_s\right\}N_k$$

$$= \frac{2K_{s\nu}}{\tau_r}\sum_{k=1}^{m} N_k\cos[\alpha_\nu(k-1)\tau_s] \qquad \text{(D.9)}$$

where

$$K_{sv} = \frac{\sin(\alpha_v w_s/2)}{\alpha_v w_s/2}$$

and

$$\alpha_v = \frac{2\pi v}{\tau_r}$$

Notice that Eq. (D.9) is the same as Eq. (5.10b). A similar procedure leads to the evaluation of the coefficient b_v as expressed by Eq. (5.10d).

Solution of Laplace's Equation

In Sec. 5.4.1 we derived Laplace's equation and included its solution. Following are some of the details of the solution. Let us consider Eq. (5.14a) and rewrite it for a two-dimensional field as

$$\frac{\partial^2 B_x}{\partial x^2} + \frac{\partial^2 B_x}{\partial y^2} = 0 \tag{D.10}$$

Let the solution be of the form

$$B_x(x,y) = X(x)Y(y) \tag{D.11}$$

where $X(x)$ and $Y(y)$ are respectively the functions of x and y alone. Substituting Eq. (D.11) in Eq. (D.10) we have

$$X''Y + Y''X = 0 \tag{D.12}$$

where $X'' = \partial^2 X/\partial x^2$ and $Y'' = \partial^2 Y/\partial y^2$. Now, dividing both sides of Eq. (D.12) by XY we have

$$\frac{X''}{X} + \frac{Y''}{Y} = 0 \tag{D.13}$$

Because Eq. (D.13) must hold for all values of x and y, each term in Eq. (D.13) must be a constant; so we let

$$\frac{X''}{X} = k^2 \tag{D.14}$$

which when substituted in Eq. (D.13) yields

$$\frac{Y''}{Y} = -k^2 \tag{D.15}$$

If k is real, then from Eqs. (D.11), (D.14), and (D.15) the solution takes the form

$$B_x(x,y) = (A_1 \cos kx + A_2 \sin kx)(C_1 \cosh ky + C_2 \sinh ky) \tag{D.16}$$

On the other hand, if k is imaginary, the solution is of the form

$$B_x(x,y) = (A_3 \cosh kx + A_4 \sinh kx)(C_3 \cos ky + C_4 \sin ky) \tag{D.17}$$

Now, let us recall from Eqs. (5.12) and (5.13) that the current sheets are expressed as cosine functions in x. Therefore, it is reasonable to assume that Eq. (D.16) is a solution to Eq. (D.10). Furthermore, the constants A_1 and A_2 can be chosen and combined with the constants C_1 and C_2 such that Eq. (D.16) can be expressed as Eq. (5.15).

Index

A

B

C

D